Constructing Green

Urban and Industrial Environments
Series editor: Robert Gottlieb, Henry R. Luce Professor of Urban and Environmental Policy, Occidental College
For a complete list of books in the series, please see the back of the book.

Constructing Green

The Social Structures of Sustainability

Edited by Rebecca L. Henn and Andrew J. Hoffman

The MIT Press
Cambridge, Massachusetts
London, England

MIT Press books may be purchased at special quantity discounts for business or sales promotional use. For information, please email special_sales@mitpress.mit.edu or write to Special Sales Department, The MIT Press, 55 Hayward Street, Cambridge, MA 02142.

This book was set in Sabon by the MIT Press. Printed and bound in the United States of America.

Library of Congress Cataloging-in-Publication Data

Constructing green : the social structures of sustainability / Rebecca L. Henn and Andrew J. Hoffman ; foreword by Nicole Woolsey Biggart.
pages cm.—(Urban and industrial environments)
Includes bibliographical references and index.
ISBN 978-0-262-01941-5 (hardcover : alk. paper)—ISBN 978-0-262-51962-5 (pbk. : alk. paper)
1. Sustainable development. 2. Social responsibility of business. I. Henn, Rebecca L., 1970–, editor of compilation. II. Hoffman, Andrew J., 1961–, editor of compilation. III. Biggart, Nicole Woolsey, writer of added commentary.
HC79.E5C6183 2013
658.4′083—dc23
2012050544

10 9 8 7 6 5 4 3 2 1

Contents

Foreword: Integrating the Social into the Built Environment
Nicole Woolsey Biggart ix

1 **Introduction 1**
Rebecca L. Henn and Andrew J. Hoffman

I **Emerging Professions and Expertise 33**

2 **Building Expertise: Renovation as Professional Innovation 35**
Kathryn B. Janda and Gavin Killip

3 **LEED, Collaborative Rationality, and Green Building Public Policy 57**
Nicholas B. Rajkovich, Alison G. Kwok, and Larissa Larsen

4 **Beyond Platinum: Making the Case for Titanium Buildings 77**
Jock Herron, Amy C. Edmondson, and Robert G. Eccles

II **Market Structures and Strategies 101**

5 **Why Multinational Corporations Still Need to Keep It Local: Environment, Operations, and Ownership in the Hospitality Industry 103**
Jie J. Zhang, Nitin R. Joglekar, and Rohit Verma

6 **The Evolution of the Green Building Supply Industry: Entrepreneurial Entrants and Diversifying Incumbents 127**
Michael Conger and Jeffrey G. York

7 **Individual Projects as Portals for Mainstreaming Niche Innovations 145**
Ellen van Bueren and Bertien Broekhans

III **Operational, Organizational, and Cultural Change 169**

8 **Empowering the Inhabitant: Communications Technologies, Responsive Interfaces, and Living in Sustainable Buildings 171**
Kathy Velikov, Lyn Bartram, Geoffrey Thün, Lauren Barhydt, Johnny Rodgers, and Robert Woodbury

9 **Building Up to Organizational Sustainability: How the Greening of Places Transforms Organizations 197**
Christine Mondor, David Deal, and Stephen Hockley

10 **Green School Building Success: Innovation through a Flat Team Approach 219**
Michelle A. Meyer, Jennifer E. Cross, Zinta S. Byrne, Bill Franzen, and Stuart Reeve

11 **Generativity: Reconceptualizing the Benefits of Green Buildings 239**
Ronald Fry and Garima Sharma

IV **Perceptions, Frames, and Narratives 261**

12 **Conveying Greenness: Sustainable Ideals and Organizational Narratives about LEED-Certified Buildings 263**
Beth M. Duckles

13 **Challenging the Imperative to Build: The Case of a Controversial Bridge at a World Heritage Site 285**
Olivier Berthod

14 **Incorporating Biophilic Design through Living Walls: The Decision-Making Process 307**
Clayton Bartczak, Brian Dunbar, and Lenora Bohren

Foreword: Integrating the Social into the Built Environment

Nicole Woolsey Biggart

In the 1950s in the United States, air quality in some metropolitan areas was so poor that it was literally difficult to see nearby urban landscapes. At first, air pollution was dismissed as a visible and annoying factor of city life, an accompaniment to postwar economic growth and urbanization. But by the 1960s, air pollution was increasingly a matter of public concern. Smoky fog—"smog"—a chemical reaction of sunlight and the chemical products of the internal combustion engine and industrial production was blamed increasingly for respiratory problems, eye irritation, and other health issues. Although air pollution has a number of sources and causes, automobile emissions were identified as a major contributor. Tailpipe and smokestack emissions of hydrocarbons (HC), carbon monoxide (CO), and nitrogen oxides (NOx)—all precursors to smog—were clearly visible as they spewed into the atmosphere. Automobile-dependent cities such as Los Angeles, which sits in a smog entrapment basin, had persistent periods of noxious air quality.

Los Angeles and the rest of California continue to battle air pollution but substantial, even dramatic improvement has been made to reduce the emissions from automobiles and manufacturing. The Federal Clean Air Act was enacted in 1963, and two years later the Motor Vehicle Air Pollution Control Act was added in recognition of the substantial role cars and trucks have on air quality and health. New regulations on the automotive industry in the form of emissions restrictions and fuel efficiency standards contributed importantly to a reduction in the number of unhealthy air days in the LA Basin from more than two hundred in 1978 to fewer than fifty by 2000. Improving the fuel-burning characteristics of internal combustion engines and improving energy efficiency have had a visible impact on air quality in major cities in the United States. Cars burn cleaner, use less fuel for miles traveled, and thus have fewer negative impacts on health and the wasteful use of a strategic resource—energy.

In this case, a combination of technological innovation, smart regulation, and constrained consumer choices contributed to a major reduction in energy pollution and waste.

Until recently, vehicles have been the largest users of energy, most of it carbon based (although hybrid and electric vehicles promote "fuel switching" they are still less than 3 percent of the nation's total fleet). But as large an environmental impact as vehicles have, they are now the second largest user of energy in the country. The largest is buildings. Today, more than 40 percent of all energy is used to heat, cool, light, and support the communication needs and transport of people in buildings. Not only are buildings energy intensive, they are particularly wasteful users of energy: buildings are like cars in the 1960s—large and inefficient energy consumers.

People tend to be aware of vehicle energy use. They see gasoline stations everywhere and if they own a car routinely fill their tank with gas or diesel as well as motor oil. Like cars, the energy used by buildings may come from petroleum products such as heating oil and may run on coal, natural gas, and other fossil fuels used by the electrical generators that supply energy over electric lines. But people do not typically "fill up" their buildings (although in some regions fuel oil is delivered to homes for winter heating) because much of the energy that workspaces and houses use is produced at a distance and is never seen or smelled and in some cases is not paid for directly but rather is provided as part of a rental arrangement.

Like vehicles, building energy use can contribute to environmental problems and deplete strategic energy resources. However, despite decades of investment in green building technologies, residential and commercial buildings remain stubbornly energy inefficient. According to the Department of Energy, existing buildings could be made 30 percent or more efficient by implementing existing technologies. The problem is not a technological one. It is environmental, economic, and social. It is a problem of technology adoption and it is extraordinarily large, complex, and difficult. Spending more money on developing energy efficient building devices without understanding which get adopted and why will not result in nearly the energy savings that are possible.

Reducing the environmental and economic impact of vehicle use is a continuing challenge, yet far more tractable than greening our building supply for a number of reasons. First, most of the nation's vehicles are retired after ten to fifteen years, allowing for substantial changes in the efficiency of the nation's cars and trucks to be implemented every decade.[1]

In contrast, there are about 5.3 million buildings and 114 million homes in the United States, and only a very small percentage of new ones are added every year. In 2008, for example, new commercial construction contributed only 1.8 percent to the total building area (USDoE 2010). Houses and commercial buildings can last for generations and poorly built buildings can "leak" energy for many years.

Second, the number of automobile and truck manufacturers is relatively small. The capital-intensive nature of vehicle manufacture has concentrated ownership in a few companies that are relatively easy to target for regulations and incentives. The Department of Transportation has effectively mandated CAFE (Corporate Average Fuel Economy) standards nationally requiring manufacturers to meet efficiency targets over their fleet portfolio. Only where states exceed the federal standards can they deviate.

In contrast, the construction industry tends to be regionally and locally based with many small contractors and building trades in the mix. Building regulations are suggested at the national level by organizations such as BOMA (Building Owners and Management Association International) and ASHRAE (American Society of Heating, Refrigerating and Air-Conditioning Engineers) but these model regulations are routinely modified by states and localities, more often to weaken them than to provide more stringent baselines for building inspection.

The economics of vehicles are more tractable in a number of ways. First, many cars and trucks are held by owner-users who benefit financially from fuel savings. Families and businesses that occupy rental space in apartment buildings or in retail malls, industrial parks, and office complexes may not directly benefit from energy efficient space. They may share space with other businesses, for example in strip malls, and have no clear way of separating their energy use from that of other tenants. Ownership patterns in the built environment are often a barrier to the adoption of energy efficient technologies, as building owners—who have the power to make decisions about which technologies are implemented—are not the ones to directly benefit from lower utility costs, in a classic "principal-agent" problem.

Cars are products of mass manufacturing systems, and although a buyer can tweak a design by requesting a different color or optional equipment, efficiency standards are prescribed by model and disclosed to consumers. Some houses are built similarly with limited design options such as "flipping" a floor plan or adding a deck, but commercial buildings are almost always one-of-a-kind projects. Indeed, most are built using a unique blueprint and a unique network of contractors and other building professionals who form constantly changing production networks. The dynamic and uncertain nature of these networks leads developers and contractors to be very conservative, to rely on default solutions to construction problems, and to be disinclined to adopt more energy efficient technologies (Beamish and Biggart 2012). The unique character of many elements in the built environment makes labels and performance disclosures difficult, although LEED certification and Energy Star efficiency ratings for appliances have made some progress.

Buildings, and especially homes, are not just shelters but also status symbols and platforms for social life. They have cultural meanings and the materials they are made of, their design, and especially their size have as much to do with identity and social ambitions as they do with any practical need or economic consideration. White roofs can reflect back the heat of the sun and reduce the cooling energy load and urban heat islands, but "cool roofs" are not "normal" in many regions outside the tropics and may even be banned by local housing codes that protect traditional aesthetics. Treating buildings as a collection of devices will not assure their adoption. Houses have to be seen holistically as a desirable and acceptable place to live.

The list of reasons why buildings are difficult to produce and operate efficiently is very long indeed, but it is not impossible to move the housing and commercial construction markets toward a greener and less wasteful future. Indeed there are examples of buildings that are so efficient that they use only as much energy as they produce: zero net energy buildings. There are widespread efforts to advance more energy efficient design, such as the "passive house," a home that is comfortable in both summer and winter without a conventional heating system. The size, orientation, materials, and window placement of passive houses enable extraordinary energy performance. While these innovative dwellings are growing in number in Germany and Scandinavia, this "solution" may prove less successful in the United States where technologies such as air conditioning are associated with middle-class status.

However, *The Not So Big House* (Susanka 1999) was the first of a popular series of books by Sarah Susanka that advocated design principles that argue for better—not bigger—houses, and lifestyles that focus on quality rather than quantity consumption. The original book inspired a movement and media flurry that have resulted in magazines, DVDs, television shows, community planning seminars and designer "showhouses" demonstrating the desirability of small, well-constructed homes. This design movement, which allies with the New Urbanism community planning school, suggests that at least some Americans are ready to downsize to better built homes. Smaller houses are a critical first step toward residential energy efficiency. Quality construction can be sustainable and efficient if society defines it that way. What values and what demographics appeal to Not So Big House owners? The appeal is not to energy efficiency but to factors of consumption that lead naturally to energy efficient houses. Are anthropologists, marketing researchers, and cultural studies scholars able to analyze the attractive essences of these trends?

The economic obstacles to energy efficient homes and buildings are just beginning to be seen by banks as an opportunity for new products. Wells Fargo Bank, for example, announced in 2012 that it is seeking to lend $30 billion to fund energy efficient buildings and renewable energy projects over the next eight years. The financial conglomerate is also working to increase its own efficiency by 40 percent. Insurance companies such as Swiss Re and Munich Re are considering the impact of sustainability and climate change on the riskiness of their portfolios. How can novel products and policies deal with the principal-agent problems endemic to leased properties? What can economics tell us about the role of ownership and incentives in developing effective energy policies? How can efficiency become a way of hedging risk?

Some cities and states are aggressively pursuing efficiency policies. For example, Washington, D.C., and New York City have mandated the public disclosure of building energy use to create social pressure toward conservation. California has a goal of making all new houses meet zero net energy standards by 2020 and all commercial buildings energy neutral by 2030, and is pursuing building code changes to make this happen. Although exceptional, these cases should be examined to understand the conditions for each location that make these measures possible. What can be learned in states pursuing aggressive energy efficient policies that may be applicable to Texas, Virginia, and Montana? Comparative case analysis can reveal factors that may move policies further toward energy efficiency. Social science is ideally suited to help us understand the politics of sustainability and efficiency.

The list of possible social science and management research topics is broad and deep, and for those who believe that carbon-based energy contributes to climate change and strategic insecurity, it is a moral imperative to understand how energy can be produced and used sustainably. The idea that morality might be involved in social science and business research is a decidedly old-fashioned perspective. It is associated with post

World War II-era scholars such as Philip Selznick, C. Wright Mills, and Chester Barnard. These thinkers were comfortable with the notion that understanding organizations and industries can lead to better social outcomes. This is a stance I share and I am very pleased that the collection of work in this volume offers the kind of understanding upon which better buildings can be built.

This volume appropriately focuses on the energy impact of buildings and their consequences for a sustainable planet as a social problem.[2] Better buildings are those constructed in ways that people will embrace and enjoy and lead to socially and environmentally superior outcomes. They are not devices to be optimized in a social vacuum.

Notes

1. The Department of Transportation's National Highway Safety Administration issues regular policy briefs on the nation's vehicle fleet: http://nhts.ornl.gov (accessed December 27, 2012). National vehicle travel statistics are available at http://nhts.ornl.gov/2009/pub/stt.pdf (accessed December 27, 2012).

2. Solving social problems involves the analysis of institutional arrangements that make problems persist. See Biggart and Lutzehiser 2007.

References

Beamish, Thomas D., and Nicole Woolsey Biggart. 2012. The role of social heuristics in project-centered production networks: Insight from the commercial construction industry. *Engineering Project Organization Journal* 2:57–70.

Biggart, Nicole Woolsey, and Loren Lutzenhiser. 2007. Economic sociology and the social problem of energy inefficiency. *American Behavioral Scientist* 50:1070–1087.

Susanka, Sarah. 1999. *The not so big house*. Newtown, CT: Taunton Press.

U.S. Department of Energy (USDoE). 2010. Annual energy outlook. U.S. Energy Information Administration: 25.

1

Introduction

Rebecca L. Henn and Andrew J. Hoffman

We shape our buildings, and afterwards our buildings shape us.

—Winston Churchill, October 28, 1943

In 1941, German bombs completely destroyed the British House of Commons. In 1943, Prime Minister Winston Churchill implored the members of Parliament to rebuild the House of Commons Chamber, quite specifically to the same shape and dimensions of the original because the "survival of Parliamentary democracy" relied on the room's historic and intimate configuration. Churchill's petition exposes the deep connection between society and its material objects. Though research in technology studies and sociotechnical systems acknowledges this direct relationship (Bourdieu 1990; Molotch 2003; Rip and Kemp 1998), studies of our built environment rarely do. And yet, buildings are critically important cultural artifacts that both exemplify our deeply held values at inception, and then shape our actions and values once made real (Bourdieu 1970, 2005; Cameron 2003; Gieryn 2002; Jones 2009; Lawrence and Low 1990; Low and Lawrence-Zúñiga 2003; Rapoport 1982; Schein 1990). The duality of Winston Churchill's statement—"We shape our buildings, and afterwards our buildings shape us"—reflects the interdisciplinary junction of this book, where changes in both social structures and the built environment recursively shape one another.

In particular, this book examines recent shifts of building processes in reaction to concerns for environmental sustainability; what has been termed "green building." Attention to environmental harms created by our built environment, and more important, investigations into ways to minimize those harms, has led to extensive research aimed at altering building practices, technological systems, and economic parameters. But these foci overly privilege the physical, quantitative, and rational aspects of the building sector. This book expands these investigations and applies a social science lens to consider how green buildings require a shift in our intellectual and cultural loyalties and a reexamination of the ways that buildings alter our sense of ourselves and our relationship to the environment around us—both natural and man-made. As two scholars within the field of organizational theory who also possess previous professional experience in architecture (Henn) and construction management (Hoffman), we recognize the fertile potential at the intersection of our theoretical and empirical selves. To us—and to the contributors of this book—the industry's changing norms for environmental sustainability provide a dynamic and promising frontier for new work in organization studies and empirically related disciplines.

The construction and real estate fields account for 14 percent of U.S. gross domestic product (GDP), yet they are reflected in only 0.3 percent of organizational research (O'Leary and Almond 2009). Why do organizational scholars overlook this important sector of the economy and leave a "glaring gap" in their research stream (Thompson et al. 1957)? A number of organizational scholars note the difficulty in accessing research sites that do not have a direct connection to schools of business management via former MBA and doctoral students working in this sector (Augier, March, and Sullivan 2005; Barley 1990; Hinings and Greenwood 2002; O'Leary and Almond 2009). Most employees in the building design and construction field come from either craft training outside of the university, or from separate professional schools within the university such as architecture, engineering, and construction management. Further, the industry is highly fragmented, with little ongoing coordination among a diverse array of professions, trades, unions, interests, and political domains (Eccles 1981; Hirsch 1972; Stinchcombe 1959). This presents a dilemma in identifying both a "level of analysis" as well as a defined "organization" to study (Henn forthcoming). Finally, the industry is traditionally slow to change. The processes by which we frame our homes, erect our skyscrapers, and examine our aging bridges remained relatively unchanged for nearly one hundred years.

Though some of these reasons for scholarly omission present legitimate barriers to research, the work in this volume shows that the barriers may be overcome so that scholars can add contextual richness and theoretical rigor to their existing work in a number of areas. In this introduction, we highlight opportunities for organizational scholars to explore this rapidly evolving site for empirical study, and provide a new and valuable lens for green building scholars as they examine the greening of the built environment.

Looking within the general field of building design and construction for empirical inquiry, the more specific focus on green building offers an engaging and important domain for a field of practice that is in flux. For example, buildings recently outpaced both the transportation and industrial sectors in their percentage of U.S. energy consumption (40 percent, versus 28 percent and 32 percent, respectively; Kelso 2011, 1–2), highlighting one key driver for change as society seeks to reduce energy consumption. Further, the decision to build with either "green" or conventional techniques and materials has become a requisite choice in new construction or renovation of existing structures since the mid-1990s. This choice, while seemingly technical and economic in nature, encompasses a host of cultural and social dimensions that rarely encounter scholarly study.

In short, green building presents a fundamental challenge to centuries-long traditions and routines relied on by the design and construction industry. At the most basic level, architects, contractors, and clients create the minimum triumvirate of the "temporary organization" (Henn forthcoming; Kenis, Janowicz-Panjaitan, and Cambré 2009) from which our buildings emerge. Concerns for environmental sustainability alter the goals, skill sets, and political arrangements within these relationships, subsequently altering fundamental meaning associated with our built environment (Rapoport 1982). Concurrent development of new digital tools, a quest toward environmentally benign building, and increased time pressure on the design and construction team from highly leveraged building owners all coalesce to produce interactive streams of upheaval in the industry, represented by the new terminology of building information modeling (BIM), integrated design, and fast-track construction. Consequently, traditional role structures that fragmented industries rely on for swift organization are now in flux (Bechky 2006).

The host of cultural and social questions arising from the green building movement include: Who wins and who loses in this dynamic realignment of power, process, and meaning? Are certain professions or organizations poised to profit from the turbulence? If so, which ones? How does this change the role of a building's inhabitant—both as an active participant in the building's design and operation, and as a formative recipient of its cultural influence? The contributors to this book examine these questions, providing a deeper understanding of the social dimensions of both the adoption and implications of the spread of green building practices.

Green Building Progress and Shifting Social Structures

The chapters in this volume all stem from a common knowledge of the existing environmental harm that the built environment creates. Buildings consume 40 percent of the world's materials, 14 percent of all freshwater, 40 percent of U.S. energy, 40 percent of the world's energy, 73 percent of U.S. electricity; they produce 40 percent of U.S. nonindustrial waste, and create 39 percent of the carbon dioxide emissions that cause climate change (Kelso 2011; Roodman and Lenssen 1995; USGBC 2009). Further, the U.S. Environmental Protection Agency (EPA) reports that indoor air often contains pollutant levels two to five times higher than outdoor air (2008). Given that we spend more than 90 percent of our time in buildings—locales that contribute to organizational and employee identity, working arrangements, and quality of workspace—they have a profound impact on human productivity and well-being. And finally, looking at a larger scale, the built environment is rapidly encroaching on available natural spaces—we can document that "urbanized land consumes natural space and agricultural land at a rate 2.6 times the population growth in the United States" (Center for Sustainable Systems 2011).

This (partial) litany of environmental affronts leads scientists, engineers, and designers to innovate ways to reverse course and improve the relationship between our natural and man-made environments. Out of this search, the green building movement was born, comprising specialists, practitioners, researchers, and citizens from myriad professions, organizations, institutions, and communities.

In exploring the social aspects of this movement's call for changes in the industry, we begin by clarifying the terminology used both in the movement and in this book, and discuss how it relates to broader terminology within the building sector. We then illustrate how technical and economic progress undermines arguments that green building is either "too complicated" or "too expensive," documenting that this movement creates a real and permanent shift within the building sector. In so doing, we establish the baseline for research that brings into relief the remaining underexplored areas where our shifting social structures—including our institutions, bureaucracies, norms, values, beliefs, and material practices—shape our buildings and the dynamics by which our buildings in turn shape us.

Terminology and Technology

To shape our buildings, we now use improved technologies that reduce environmental pollution, damage, and waste caused by design, construction, and development (USGBC 2011a). It is impossible to overstate the role that the United States Green Building Council (USGBC, founded in 1993) and its Leadership in Energy and Environmental Design (LEED) Green Building Rating System (launched in 1998) have had in creating the technological changes that emerged in the past decades. LEED, as the dominant green building rating system in the United States, incentivized much of this progress (other programs such as BREEAM in the United Kingdom, Minergie in Switzerland, and CASBEE in Japan drove respective changes around the world). The LEED rating system assigns points to various practices and processes that reduce environmental harm and potentially improve the health and well-being of users and the environment. The points are grouped into six categories: sustainable sites, water efficiency, energy and atmosphere, materials and resources, indoor air quality, and innovation and design process. Depending on the total points accumulated, the project receives certification at increasing levels from LEED certified to Silver, Gold, and Platinum ratings. As of November 2011, there were 32,451 projects registered to pursue some level of LEED certification—from new commercial buildings, retail stores, homes, schools, and renovations, to full neighborhoods (USGBC 2012)—and 11,082 projects that received some level of certification, comprising 2 billion square feet (USGBC 2011b). While their numbers still represent a small portion of the roughly 170,000 commercial buildings and 1.8 million residential homes built each year in the United States (Center for Sustainable Systems 2011), steady growth trends for LEED certifications—both formal and informal (buildings designed to meet the standard but without seeking formal certification)—point to a market in transition.

The green building field is crowded with terms that cover various aspects of its meaning, including "sustainable building," "high-performance building," and "smart building." Each has a slightly separate definition. Though "green building" is now the most commonly used term given the dominance of the USGBC and LEED, distinctions in meaning remain. For example, green building fits under the broader umbrella of sustainable building, which considers both environmental and social equity considerations inherent in the Bruntland Commission's definition of sustainable development to "meet the needs of the present without compromising the ability of future generations to meet their own needs" (WCED 1987, 43); or the very similar definition by the American Institute of Architects

(AIA): "the ability of society to continue functioning into the future without being forced into decline through exhaustion or overloading of the key resources on which that system depends" (Gilman 1992).

While sustainability calls for a "triple bottom line" model of balancing social equity, environment, and economy (i.e., the "three e's," also known as the "three p's" of people, planet, and profits), much existing work in green building narrowly focuses on adequately balancing the last two—environment and economy. But, what the chapters in this book show is that changes in the environmental impact of buildings quite naturally impact the social domain in which people interact, as alluded to by Winston Churchill in the opening quote. While we realign our building goals to benefit the environment, we also realign the cultural and social structures of the people who conceive, design, build, and inhabit these buildings. This realignment alters how we perceive both the built environment and the ways in which it intersects with the natural world.

As a result, the chapters in this book strive to be quite specific in selecting terminology to be clear about their meaning. In so doing, they illustrate the shades of association and meaning that each term entails. For example, "green" often has associations with liberal, granola, hippie, anti-corporation, and so on, despite the corporate nature of the USGBC, while "high-performance" evokes the more broadly acceptable goal of efficiency—even though both aim to reduce energy consumption in a building.

In the aggregate, this book will use the term "green building" to encompass strategies, techniques, and construction products that are less resource intensive or pollution producing than conventional practices. In some cases, this involves merely "doing without" extra space, finishes, or appliances. In others, it substitutes a less polluting product for more polluting ones (e.g., paints with low levels of volatile organic compounds, or low VOC). More integrated strategies reconfigure a space to take advantage of unique site attributes (e.g., facing glass toward the sun path to use natural or "passive" solar heat gain instead of using natural gas or electricity to heat a space) or reconfigure design parameters to take advantage of building system synergies (e.g., downsizing the boiler after extra insulation has been added to the exterior shell) (Hoffman and Henn 2008).

Beyond specific technologies to reduce energy use, green building also reaches into the long chain of responsibility for all types of environmental harms. Green building designers, builders, owners, and operators now understand an entire supply chain to know where their resources come from and under what conditions they were acquired, as well as where the "waste" goes, and what environmental impact it has and on whom. These considerations introduce other technical achievements that include regionally sourced materials; durable and rapidly renewable wood products (such as bamboo and cork flooring); green roofs that provide natural habitat and reduce both energy loss and urban heat-island effects; geothermal heating and cooling; rainwater collection systems and graywater irrigation; motion-sensor corridor lighting; and many other applications of new technology that create individual and collective returns.

Over time, these technologies have improved in performance. Solar panels are now more efficient, new fluorescent lighting has a warmer cast, VOC-free paints are available in almost every color, new windows have a higher insulation value, and more generally anyone looking to integrate green design into their building project has a plethora of legitimate and well-tested options that do not involve some sacrifice in spatial or experiential quality. In short, the technology of the green building sector has matured, being tested and applied in multiple applications for improved operation. This technical progress has been concurrent with improvements in the economic aspects of green building.

Economic Progress

Multiple studies assess the economic costs and benefits of green building. They range from surveys of building owners (i.e., Leonardo Academy 2009) to case studies of a sample of green projects (i.e., Miller and Pogue 2009). While most of these studies use fairly small sample sets of green buildings, and we see a glaring need for large dataset analyses that comprise the entire universe of green buildings (initial forays include Brounen and Kok 2011; Brounen, Kok, and Quigley 2012; Eichholtz, Kok, and Quigley 2010), these analyses create a useful template and baseline for considering the economics of green buildings in four categories (Kibert 2007; USGBC 2005; Wilson 2005).

The first category of green building economic analysis is *first cost savings*, and includes cost differentials from reduced material use, savings in construction waste disposal, savings from downsizing mechanical equipment, tax credits and other incentives, streamlined permitting and approvals, and reduced infrastructure costs. While early studies showed construction cost premiums for LEED buildings that ranged from 0.66 percent to 0.8 percent for a LEED certified building to 6.50 percent to 11.5 percent for a LEED Platinum-rated building (Kats et al. 2003; Turner Green Buildings 2005), more recent studies suggest that these capital cost premiums are coming down if not eliminated altogether. Matthiessen and

Morris (2007), for example, concluded "many project teams are building green buildings with little or no added cost, and with budgets well within the cost range of non-green buildings with similar programs." The Urban Green Council (2009) concluded that "in analyzing high-rise residential buildings [in New York City] . . . there is no statistically significant difference in construction cost between LEED and non-LEED projects."

The second category of green building economic analysis is *reduced operating costs* and includes considerations for lower energy costs, lower water costs, greater durability and fewer repairs, reduced cleaning and maintenance, reduced costs of churn (reconfiguring office spaces and relocating office workers), lower insurance costs, and reduced waste generation within the building. For example, Capital E Analysis Group (Kats et al. 2003) calculated that the twenty-year net present value (NPV) for energy, emissions, water, and operations/maintenance savings exceeded the cost premium to reach a savings range of $11 to $13 per square foot. The U.S. Department of Energy (2003) calculated the average annual energy cost benefit for a high-performance building as compared to a base case building to be 48 percent for lighting, 27 percent for cooling, and 29 percent for heating. Turner Green Buildings (2005) reported that "although executives [655 surveyed] believed that green buildings cost more to construct, virtually all believed that their higher initial costs would be repaid over time through lower operating costs, such as energy savings, increased worker productivity, and other benefits. Executives at organizations with green building experience estimated the payback period to be 8.1 years." McGraw Hill Construction (2008) found a 13.6 percent decrease in operating costs for green buildings, as well as a 10.9 percent increase in building value. CB Richard Ellis and EMEA Research (2009, 3) found "green buildings attract higher rents than conventional ones, and also enjoy higher rates of rental growth." Looking to the future, a report by the Rockefeller Foundation and Deutsche Bank (Fulton et al. 2012) found that a $279 billion investment in the residential, commercial, and institutional market sectors could yield more than $1 trillion of energy savings over ten years in the United States, equivalent to savings of approximately 30 percent of the annual electricity costs in the United States.

The third category of green building economic analysis is *health and productivity benefits* which includes improved health, enhanced comfort, reduced absenteeism, improved worker productivity, improved learning, faster recovery from illness, and increased retail sales. For example, the Heschong Mahone Group (1999a, 1999b) measured improved performance and learning in daylit schools and increased sales in daylit retail stores. Boyce, Hunter, and Howlett (2003) found positive benefits from daylighting on human performance and workplace productivity, human health, and financial return on investment in a range of building types. The Capital E Group (Kats 2006) found financial savings in green schools at roughly $70 per square foot—twenty times as high as the initial cost of building green—with benefits such as lower asthma, cold, and flu rates; increased teacher retention; and energy and water savings. Both Sustainability Victoria and Kador Group (2008) and Wilson (1999) found improved employee productivity in sustainable buildings. Given that office worker salaries comprise upward of 90 percent of total office expenses, this increase in productivity provides significant financial benefits from green building.

The fourth and final category of green building economic analysis is a *general benefits* category and includes increased property value, more rapid lease-out, more rapid sales of homes and condominiums, easier employee recruiting, reduced employee turnover, reduced liability risk, an ability to stay ahead of regulations, and positive public image. Turner Green Buildings (2005), for example, reported that executives involved with green colleges and universities see improved community image (90 percent), ability to attract and retain teachers (71 percent), ability to attract students (70 percent), improved student performance (59 percent), and improved ability to secure research funding (59 percent). These reputational enhancements are often a deciding factor in convincing building owners to commit the additional paperwork costs of LEED certification that can run from $40,000 to $200,000 for a single project (Environmental Leader 2009).

Each of these categories of economic analysis, coupled with technical progress, represents a means for encouraging owners and operators to adopt green building practices. Together, they undermine arguments that green building is too new, risky, or expensive to implement. However, while these data speak to the economic, technical, and quantitative rationales for building green, they do not directly attend to the social aspects of what causes an organization or individual to decide to build green, and what social and cultural changes they encounter once they do. That is the gap that this book intends to fill.

The Missing Social Component

In his analysis of green construction cost premiums, Kats asserts that a strict financial benefit "will largely determine whether green design can make the transition from environmentally motivated niche to cost-conscious mainstream" (Kats, Braman, and James 2010, xv–xvi). But producing financial benefits may not be enough to move the industry. As Brenna Walraven, chair of the Building Owners and Managers Association (BOMA) states, "[Green building] forces the developer to move from a place of 'we've done it this way successfully for 20 years or more' to a place of new players, strategies, and approaches. While most developers like risk, this kind of risk is totally contrary to what they're used to doing" (Lockwood 2007, 119). This statement suggests that financial concerns are not the only barrier to wider green building adoption, and invokes underlying cultural, social, political, and institutional aspects that simultaneously transform in the process.

So despite the elevated attention on the technical and economic aspects of green building, investigators must also focus on the social, cultural, and institutional dimensions of the large-scale change in practices, norms, and processes (Guy and Shove 2000). To examine the deep connection between society and its built environment, we must begin to ask more sociological and organizational questions. First, how does our society structure and construct concepts of green building so that it may implement them in practice? Second, how does the practice of green building restructure relationships, both within the team that designs and constructs buildings, as well as among building users, owners, and broader society? What strategies, relationships, and opportunities change as we change the way that we construct our built environment? Third, how do the changes represent not only a technological and economic shift, but also an institutional and cultural one? And finally, how do buildings transform the user, both in the operational aspects of reducing its environmental impact, and in the cultural aspects of his or her relationship with the external environment? The authors in this volume recognize the interconnectedness of these issues, and use organizational theories to provide tools for understanding the adoption and diffusion of green building practices. In the sections that follow we outline a basic structure for initial forays into this research stream (both existing and emergent), and then present the specific contributions of this volume in building upon it.

Constructing Green: Past and Future Research in Social Structures

Much of the writing on green building begins by illustrating the environmental harm of building and then stating the urgent need to change our market institutions to correct our relationship to the natural world and its resources. These arguments center on the physical sciences in defining the nature of the problem, and on the field of economics in defining the solution. But this framing privileges the material and technical aspects of issues, ignoring the social and cultural aspects of how people interpret the issues and adopt corrective actions. These critical social components of the green building field signal the cultural inertia that resists changes in the institutions of construction. People hold an intimate relationship with their buildings, and research on organizations and social structures is a critical, and understudied, dimension for understanding and examining human dimensions inherent in this field.

Initial Forays

Existing work that connects human activity with green building takes three main approaches. The first approach uses the lens of philosophy—and more specifically ethics—to better understand human conceptions of human-environment relations and the moral reasoning that should lead to a more sustainable built environment (Fisher 2008; Fox 2000; McLennan 2004). Although a main goal of this literature is to understand the structure of these relations, the work also aims to provide framing mechanisms for environmental advocacy. The second approach is more instrumental, based in the field of practice, and is shaped in a technical "how to" format that, again, advocates for policy and practice changes to increase environmental sustainability (Kibert et al. 2011; McLennan 2004; Moore 2010). The third approach is that of the academic scholar. This work incorporates components of the technical and philosophical, but provides disciplinary structure and concepts so as to understand it as a multifaceted phenomenon. It does not strive for advocacy of a particular policy or outcome, but rather seeks to understand the economic, technical, social, and environmental structures that the domain engages.

This volume seeks to engage the practitioners and philosophers of the first and second approaches, but contribute a theoretical analysis and structure that place it within the third approach. Given the powerful empirical and theoretical implications of this work, this is a natural and perhaps necessary hybrid. More specifically, this volume applies a sociological and organizational lens to understand green building as a notion of technological, political, economic, and ultimately, social change. Existing work in this third approach mainly focuses on the sociology of energy efficiency (Biggart and Lutzenhiser 2007; Gann, Wang, and Hawkins 1998; Guy and Shove 2000; Lutzenhiser 1994). Although broader social and organizational work exists on general environmental issues (Bansal and Hoffman 2012), or issues of resource extraction (Zietsma and Lawrence

2010), pollution (Hoffman 2001), and waste (Lounsbury 1997, 2001), it does not focus specifically on the built environment or its processes of design and construction. Within this specific channel of inquiry, there are many more opportunities to contribute to working theories of social structure and change.

Promising Avenues

Through the lens of organizational theory, green building is seen as a new set of practices, a social movement, an innovation, an institutional change, a business opportunity, an emergent industry or market sector, and a symbolic representation of corporate responsibility. Each of these dimensions has a rich tradition of theoretical literatures behind it, and we can better understand the empirical phenomena while extending theoretical inquiry by linking those literatures to the examination of the drivers, process, and outcome that the introduction of environmental issues has wrought within the broader industry of building design and the construction sector. Considering these rich, multifaceted opportunities, we find it even more surprising that so few organizational scholars investigate this field. We now illustrate a few specific possibilities for deeper investigation.

The first domain of organizational and sociological inquiry is that of *social movements*, a field that has linked to issues of the natural environment since the1970s and the advent of environmental sociology (Albrecht and Mauss 1975; Catton and Dunlap 1978). The movement of green building has been driven by powerful social movements actors, who could be defined as institutional entrepreneurs (DiMaggio 1988; Fligstein 1997; Lawrence 1999; Maguire, Hardy, and Lawrence 2004) in shaping the discourse and norms and the structures that guide organizational action (most notably the USGBC). A host of supporting interests enable such entrepreneurs to change the institutions of the construction industry. This movement now includes building material suppliers, construction companies, consultants, environmental groups, professional societies (such as the AIA and the National Association of Home Builders [NAHB]), government agencies (such as the EPA and Department of Energy [DOE]), and many others. The growth and influence of this movement offers a distinctive research site because the agents of change can be new entrants, but are more likely incumbents with previously embedded positions in legitimate organizations in the design and construction fields. In this case, questions of diffusion and change strategies may illuminate how social movements such as corporate social responsibility make their way through dominant organizations in inertial fields. Such inquiry fits within broader research taking place on social movements as they relate to environmental sustainability, and can serve to link management research to investigations of the "social objectives of society" (Walsh, Weber, and Margolis 2003).

A second and distinctive characteristic of the building design and construction industry is its *fragmentation*, which resists traditional hierarchical and managerial line control. Like the movie industry, emergency room surgery, firefighting, and police SWAT teams, building design and construction operates according to a project-based governance structure that relies on swift organization and clear but tacit understanding of roles and task jurisdiction (Abbott 1988; Bechky 2006; Bechky and Okhuysen 2011). As more organizations in other sectors move toward this "disaggregated" or "network" form of organizing (Podolny and Page 1998; Powell 1990; Walsh, Meyer, and Schoonhoven 2006), understandings drawn from the construction industry can inform the field more broadly. But by studying green building in particular, the research agenda quickly moves forward to investigate change within this fragmented industry (Henn forthcoming), which is teeming with temporary organizations (Bakker 2010; Jones and Lichtenstein 2008; Kenis, Janowicz-Panjaitan, and Cambré 2009; Lundin and Söderholm 1995; Packendorff 1995), project management concerns and coordination (Gareis 2010; Lundin 1995, 2011; Modig 2007; Okhuysen and Bechky 2009; Söderlund 2011; van Donk and Molloy 2008), and the diffusion of innovations (Briscoe and Safford 2008; Gann, Wang, and Hawkins 1998; Strang and Meyer 1993; Strang and Soule 1998).

A third domain for organizational inquiry that is closely related to the second involves the complex nature of the role of *professionals and expert work* in society. Not only is professional service a significant sector of the economy, but the study of professional work also helps us to understand the functioning of the new knowledge economy and new forms of identity associated within it (Barley and Kunda 2006; Briscoe 2007; Chreim, Williams, and Hinings 2007; Ferlie et al. 2005; Goodrick and Reay 2010; Owen-Smith 2011; Pratt, Rockmann, and Kaufmann 2006; Van Maanen and Barley 1984; Zald and Lounsbury 2010). In building design and construction, any number of licensed and accredited professionals work together on a project. Scott (2008, 219) suggests that "more so than other types of social actors, the professions in modern society have assumed leading roles in the creation and tending of institutions. They are the preeminent institutional agents of our time." Studying the green building industry pulls this work on professionals away from traditional subjects of law, medicine, accounting, and management to test the boundaries of professional influence (Fincham et al. 2008; Scott et al. 2000; Suddaby and Greenwood 2005; Wallace and Kay 2008). Further, and as illustrated by many chapters in this book, the introduction of green considerations creates new professions and alters roles within existing ones (Henn forthcoming). The traditional project team's triumvirate of architects, contractors, and clients rearranges, with new skills and knowledge creating new relations, power dynamics, and competing interests (Davis et al. 2005; Davis et al. 2008).

This leads to a fourth domain of inquiry regarding *governance structures* within the construction industry. In particular, professional service firms (PSFs) have been called the "firms of the future" because of they are "adaptive, receptive, and generative" (Greenwood, Suddaby, and McDougald 2006, 1). PSFs do not organize according to traditional machine-like efficiency and hierarchy that characterize our declining manufacturing base. Instead, a number of studies look to PSFs in an effort to understand the management of a knowledge-based firm (Hinings and Leblebici 2003; von Nordenflycht 2010), which is quite distinctive from the governance structure of the project itself, as cited earlier. Because of the cultural artifacts created by building industry, the investigation into green building can also provide insights into the operation of creative and cultural industries (Brown et al. 2010; Hargadon and Bechky 2006; Lingo and O'Mahony 2010; De Stobbeleir, Ashford, and Buyens 2011; Townley, Beech, and McKinlay 2009; von Nordenflycht 2007).

Fifth, given the heterogeneous nature of building project teams through both time and expertise, green building provides a richly variegated field of *multiple institutional logics* that are constantly in play and competing for dominance (Dunn and Jones 2010; Henn 2010; Rao, Monin, and Durand 2003; Reay and Hinings 2009; Thornton, Jones, and Kury 2005). The rise of sustainability concerns brings into relief previously settled terrain and re-questions norms, values, and assumptions that recursively relate to new and emergent material practices within and outside of professional jurisdictions (Henn forthcoming).

Finally, the social changes wrought by the green building movement expose rich inquiry into *changing social conceptions of the human and man-made environments*, and their relationship with one another. Within field level dynamics and change, there is ample attention to the ways in which individual organizations can respond strategically to field pressures (Oliver 1991) or may strategically influence the process of field change (Lawrence 1999). But there has been less attention to the interpretation and filtering processes that occur in these circumstances, and even less attention to the final form of altered conceptions of "collective rationality" (DiMaggio and Powell 1983) once change is engaged. This introduces concerns for sense-making, sense-giving, issue interpretation, selective attention, and cognitive framing among field members (Dutton and Dukerich 1991; Gioia and Chittipeddi 1991; Hoffman and Ocasio 2001; Hoffman and Ventresca 2002; Scott and Meyer 1994; Weick 1995). The domain of green building challenges many deeply held beliefs of how our built environment is "supposed to be." "For example, notions of a home in the United States include: a lawn (even if the home is situated in the desert) often grown with imported, non-native species; a garage (even if the home is situated near an urban center or public transportation); within developments (even if urban living is more convenient and communal); and an ever-increasing size" (Hoffman and Henn 2008, 407). Alternatives to single-family homes (like cohousing), alternatives to car ownership (like car sharing), alternatives to a lawn (like xeriscaping) challenge entrenched beliefs on a number of increasingly deep levels for all constituents involved in the built environment: owners, designers, architects, contractors, inspectors, users, and—given that the built environmental touches us all—virtually all of society.

And it is this final point that really drives home the criticality of the building sector as an area of empirical and theoretical importance. Its scope and import for the deepest institutions of our society are paramount. Who we are as humans, how we interact with each other, how we see the natural world around us, and how we interact with it are all mediated by the institutions of our built environment. Attempts to change those institutions by incorporating green considerations involve a concurrent change in fundamental institutions of our society. It is this recognition that exposes a wide range of research options for rich inquiry. The six examples of organizational research domains presented earlier are just a few of the understudied yet promising locations for deeper study. The chapters in this volume tap each of these areas and more, standing as a foundational structure upon which to build future inquiry.

Constructing Green: Contributions of this Volume

Building on this preliminary work, the chapters in this volume investigate how the changes brought about by green building alter our social structures. These include alterations in teams and organizational structures; individual and group identities; the relative power of professional jurisdictions and expertise; decision-making processes and the invocation of multiple ethics; the institutional rules, norms, and beliefs within the mainstream industry; and the alteration of specific roles in the design, construction, and use of buildings. Together, these social domains of inquiry represent the tight linkages among the built environment, new technology, and social structures, showing how changes in one area invoke changes in the others.

The organizational structure of these chapters stems from the organizational research questions they examine. Although individual chapters address many aspects of green building, the section groupings highlight rich generalizing questions that the collection offers. There are five main sections that ask: First, how do new social concerns for green building create new domains of professional expertise? Second, how do these concerns alter industry strategies and structure? Third, how do these concerns alter operational, organizational, and cultural arrangements? Fourth, how do individuals and organizations perceive green buildings, and then what kinds of frames and narratives do they use to describe them? The fifth and final section asks two distinguished thought leaders in the field to consider the question: what does the future hold for this empirical domain?

Emerging Professions and Expertise

The three chapters in part I of this volume examine industry structures and how engagement among firms changes as a result of new societal concerns for environmental sustainability. These chapters acknowledge the loose structure of the building industry that comprises primarily temporary, project-based organizations. Traditional role structures within the industry allow design and construction teams to organize swiftly with little error and high reliability (Bechky 2006; Weick 1987). But changing the industry's product, practice, and focus to include environmentally sensitive design calls these traditional structures into question and therefore creates an upheaval that has yet to be resolved. Chapters 1–3 contribute to theories of professional jurisdiction (Abbott 1988), temporary organization (Kenis, Janowicz-Panjaitan, and Cambré 2009), as well as recent calls to engage process—rather than variance—theories in describing organizational processes (Garud, Dunbar, and Bartel 2011; Tsoukas 2009; Van de Ven and Poole 2005).

Janda and Killip (chapter 2) examine the structure of expertise within the building industry, using a "system of professions" model that assumes an ecosystem-type structure, where each profession lays claim to a "jurisdictional territory" of specific tasks. Each professional jurisdiction borders another one, with the border between them under regular contestation and struggle. As "new" tasks arise—residential energy retrofits in this case—from emergent government incentives or market demand, there may not be an existing claimant to the territory. This presents opportunity, but for whom? The authors investigate existing participants in the UK residential sector, highlighting which professions are best poised to expand their territory through workforce training and development.

Using two contrasting case studies, Rajkovich, Kwok, and Larsen (chapter 3) illustrate the combined power of voluntary regulation and contractual incentives to promote the increased collaboration necessary for environmental building goals to make their way through the design, construction, and operation of a building. This new type of practice—defined less by hierarchy, long-term patterns of routine behavior, and linear decision making—recognizes a diversity of interests, the interdependence of design decisions, and the greater innovation necessary to achieve a high-performance green building. However, Rajkovich and colleagues argue that new integrated project delivery contracts and the voluntary nature of LEED certification are still inadequate to promote the level of collaboration necessary for the highest-performance buildings and communities. Rather, they encourage national institutional leaders such as the USGBC to tap expertise of local policymakers for a more transparent, incremental, and collaborative process to formulate revised building and energy codes in the United States.

Herron, Edmondson, and Eccles (chapter 4) continue this argument for increased measurement and regulation, but extend it further into building operations. They argue that to date, "buildings are seen as nouns, not verbs—things, not processes." To correct this misperception, the authors argue for a new "Titanium" building standard—one that guides not only the building's outcome and projected environmental impact, but also the *actual ongoing* use of materials and energy. Their proposal for a Titanium standard illustrates how tools already exist for increased collaboration and high performance, and, like Janda and Killip (chapter 2), the coauthors of chapter 4 investigate the social institutions that are best poised to structure and scale up this new incentive system.

Market Structures and Strategies

Considering the size of the construction industry, as well as its significance in the U.S. GDP, it is not surprising that major opportunities for profit and market capture exist for firms specializing in green building products and services. The three chapters in part II explore these dynamics, emphasizing relationships between large mainstream organizations and smaller green initiatives and the different strategies employed by different actors within the industry. The chapter authors do this by explaining the role of ownership in green building operation, the role of incumbent supply firms within a market sector, and the role of market niches to produce mainstream innovations.

Zhang, Joglekar, and Verma (chapter 5) demonstrate the classic green tension between behavior incentives and resource efficiency. Using multinational corporations within the hospitality industry, the authors examine the full behavioral supply chain, including consumer behavior, ownership structure, and operating structure, while controlling for locale. They find that the social structures of the supply chain significantly influence the resource efficiency of the property. Though this may be a classic example of the split incentives inherent in landlord/tenant relationships, the authors show how restructuring relationships in this industry can lead to greater financial and environmental performance. Using these data, the authors develop a multi-agent theoretical framework that can assess the internal tensions of a service supply chain, such as those between an owner's investment decisions and the operator's commitment to implementing sustainability practices.

Conger and York (chapter 6) focus their study on a new market sector (green building supply) and contribute to studies on entrepreneurship. They ask how and when firms decide to enter an emerging sector, differentiating between firms incumbent to the building industry that expand their offerings and emergent firms that claim a special expertise in the new market. The authors also outline the timing of entry for each group of firms, suggesting that the green building supply market involved greater uncertainty due to the "spillover of public goods" inherent to the market. This caused incumbent firms to hesitate, but later put full force behind the exploitation of a new market. Contributing to the findings of Zhang, Joglekar, and Verma (chapter 5), this work underscores how specific structures along supply chains can influence a firm's financial and environmental performance.

Van Bueren and Broekhans (chapter 7) look more deeply into the mainstreaming of green innovations through the use of strategic niche management. The authors provide a fine-grained understanding of how niche innovations diffuse into the mainstream or "regime," and how those innovations transmute in the process. Specifically, the authors highlight how an early and integrated focus on environmental goals can result in building performance higher than initial projections. This trend—which emerges again in Mondor, Deal, and Hockley (chapter 9)—changes how mainstream or "regime" actors understand, process, justify, defend, and integrate the "niche" concerns for environmental sustainability. The authors provide three mechanisms for mainstreaming innovations developed within a strategic niche, and suggest that these be more firmly embedded into the project team to provide both increased participant learning and higher performance outcomes.

Operational, Organizational, and Cultural Change

Invariably, innovation in products and processes results in organizational and cultural change, and the four chapters in part III underscore the recursive process by which technical innovations infiltrate our daily engagement with the built environment. These chapters build on the mainstreaming topic of van Bueren and Broekhans (chapter 7) and demonstrate how cultural norms change. By tapping into existing social media tools, interpreting new practices in alignment with organizational mission, authorizing a small management shift, or modifying terminology, all of these chapters engage the power of social norms to facilitate a move toward green building. As a result, however, unexpected positive results provide both incentives and increased internal initiative to push the environmental agenda forward within the social groups and organizations.

Velikov, Bartram, Thün, Barhydt, Rodgers, and Woodbury (chapter 8) use their institutions' entry into the U.S. Department of Energy's Solar Decathlon as a testbed prototype for an energy-positive home. Bringing innovations to live execution, the authors develop an adaptive energy control and communication system to provide users with simple and uncomplicated feedback on their energy-influencing behaviors. Not only does the house kindly and aesthetically "speak" to the users through both ambient and web-based tools, but users may also authorize the house to post energy data to social media sites. Drawing on diverse expertise including environmental psychology, the authors emphasize that for meaningful human engagement, the quality of communication and feedback is paramount.

Mondor, Deal, and Hockley (chapter 9) contend that green building projects can provide the "gravitational assist" to generate momentum toward a broader, organization-wide commitment to sustainability. They define "gravitational assist" as the transformation of organizational culture to include concern for a more holistic definition of sustainability. Using three nonprofit organizations as case studies, chapter 9 describes a systemic approach to sustainability through three types of actions—those concerning people, process, and place. The authors argue that building projects often provide an organization with the initial intimate contact with environmental sustainability, and that this "place-based" initiative provides a pathway for an organization to deepen its knowledge of and commitment to sustainability more broadly defined. Then in conclusion they summarize effective stakeholder strategies to enhance this pathway for a more lasting organizational transformation.

Meyer, Cross, Byrne, Franzen, and Reeve (chapter 10) tackle issues of internal diversity and public organizations, showing how a coherent "high-performance" frame combined with supportive management processes can create systemic change in organizational operations. Moving beyond a single building, the authors investigate a school district that faced negative impressions of fiscal waste associated with green building practices. The district started with a cross-organizational team to increase organizational knowledge but not necessarily change practices, essentially instigating a rich communication incubator. This "Green Team" promoted collaborative development of a green building vision that all individuals could support without changing their personal values or agreeing to a pro-environmental agenda. As team participation increased the environmental literacy of the team and wider staff, buy-in and support for green building happened before resistance to a change in practice could build. As a result, the organization became a leader in producing high-performance buildings, winning awards for energy conservation and sustainability, and becoming a resource for other school districts in the United States.

Fry and Sharma (chapter 11) move the conversation of green buildings far beyond the financial benefits into broader strategic changes that produce a generative, creative, and capacity-enhancing organization. More than increased efficiency and productivity gain, the authors provide illustrations of how green buildings become transformative for multiple stakeholders, similar to the implications shown by Mondor, Deal and Hockley (chapter 9). Specifically, the authors propose generativity—the organization's potential to produce enduring, expansive, and transformative consequences (Carlsen and Dutton 2011)—as a primary benefit of the pursuit of green building. Focusing on positive deviance, the authors produce a multilevel analysis that displays deeper-level shifts in meanings and organizing such that the benefits continually expand upward and outward. As a result, outcomes generated by the practices of green building, such as knowledge and energy at the individual level, are converted into organization-wide outcomes such as enhanced processes of organizing for the growth of the organization and its stakeholders, and expanded meaning of individual and organizational accountability toward the environment such that these outcomes are diffused beyond the organization's boundaries. The authors then propose managerial action that can facilitate desired outcomes.

Perceptions, Frames, and Narratives

As illustrated in chapter 10 by Meyer and colleagues, modifying the framing of a proposal may transform resistance into support. In their case, it was a change from the term "green" to "high performance" building that engaged diverse stakeholders. The three chapters in part IV focus on perceptions, frames, and narratives that consequentially influence an individual or organization's willingness to engage with environmental strategies and practices. These detailed understandings also begin to map specific stakeholders with their associated positions. Knowing one's audience becomes a critical factor in proposing a specific technology or innovation, particularly when debates over public goods spill over into deeply held social and political opinions. These chapters describe how actors understand and interpret environmental goals and practices.

Duckles (chapter 12) analyzes organizational narratives about green buildings to create a five-part justification typology. The first four themes—profit-driven green, practical green, green innovators, and deep green—illustrate how different organizational forms (e.g., corporations, nonprofits, and government education, and religious organizations) integrate their creation of a green building into their organizational narrative. However, Duckles also uncovers a "counter-narrative" of hidden green, which describes rationales for *not* highlighting green achievements in the organizational narrative. This latter finding suggests that there are still societal sectors that associate environmental efforts with risk, unacceptable aesthetics, and fiscal waste. Recognizing a resource dependency on this sector encourages some organizations to hide their green building achievements.

Berthod (chapter 13) presents a case study that questions the imperative to build in the first place. Whereas the LEED program helps us build better buildings, Berthod investigates how organizations—and in this case governments—determine, stabilize, and defend the imperative to build at all. Sociological approaches have long advocated investigating underlying narratives and framing mechanisms when technical rationality and efficiency present a collective assumption of one solution. This case puts the dynamics of collective consensus making to the fore, as the eventual construction of the four-lane bridge in Dresden, Germany, resulted in that city's removal from the UNESCO World Heritage listing. Like Duckles (chapter 12), Berthod begins to map stakeholders with positions and narrative frames, following the path of power and decision making within the city administration. This case also highlights how the large scale and long time frame of building projects—particularly infrastructure projects—creates intransigence in the face of a changed project context. In response, Berthod suggests that for increased sensitivity to environmental and other contextual issues, building projects should retain the ability to modify decisions and trajectories based on new information, as well as allow for more critical evaluation during implementation.

Bartczak, Dunbar, and Bohren (chapter 14) analyze the reasons for innovation adoption. Using the example of interior living-wall installations, the authors illustrate the positive psychological and physical health impacts and examine the myriad justifications for building owners to install such a niche technology. Departing from the energy efficiency arguments made in most green building narratives, the authors show how cost concerns fell lower on the owner's value hierarchy than perceived (but loosely supported) benefits, including aesthetic. In turn, the aesthetic benefit resulted in increased use of the unique indoor space as a popular locale for events. The authors link this attraction to emerging theories on "biophilia"—the innate [human] tendency to focus on life and lifelike processes (Wilson 1984, 1)—that is also addressed by Browning (chapter 16).

Perspectives on the Future

For part V, we asked two thought leaders in the practice of green building design and construction to comment on current trajectories and future potential for research in the industry.

Ponce de Leon (chapter 15) asks designers to consider additional criteria for both the sourcing and disposal of materials when selecting, specifying, and designing their buildings and building products. She also implores relevant government organizations to increase regulation within the building design and construction industry so that environmental commitments expand beyond the voluntary realm. Finally, she suggests that the industry must move more quickly to incorporate existing but underused technology that can significantly decrease the negative environmental impact of the building industry.

Browning (chapter 16) confronts the popular notion that implementing biophilic design is a luxury, unsupported by quantifiable benefits.

Instead, Browning asserts that biophilic design yields concrete financial benefits for businesses accruing from worker productivity. He reviews the components of biophilic design and their scope for implementation, as well as the benefits that can accrue to companies and communities by incorporating nature into construction and master planning.

Conclusion

In conclusion, green building is a new and growing empirical shift that stands to transform an industry that has remained unchanged for decades. The topic is important to society, and relevant to all organizations across the planet. In this capacity, the intersection between green building and social science provides a new and growing field for organizational studies, yielding opportunities to expand existing scholarship in temporary organizations, professions, multiple institutional logics, and of course work on social movements.

We expect this volume to provide a baseline of research and connections that will catalyze this field to yield insights that bridge multiple fields including organizational studies, architecture, construction management, real estate finance, urban planning, and public policy. In so doing, the work that follows will come closer to an accurate representation of the interdisciplinary dynamics within society, and will thus be able to contribute soundly to the application of practice.

Acknowledgments

The editors would like to thank the Erb Institute for Global Sustainable Enterprise, the Graham Environmental Sustainability Institute, the School of Natural Resources and Environment, and the A. Alfred Taubman College of Architecture and Urban Planning, all at the University of Michigan, for funding the initial conference on which this book is based.

References

Abbott, Andrew Delano. 1988. *The system of professions: An essay on the division of expert labor*. Chicago: University of Chicago Press.

Albrecht, Stan L., and Armand L. Mauss. 1975. The environment as a social problem. In *Social problems as social movements*, ed. Armand L. Mauss, 556–605. Philadelphia: Lippincott.

Augier, Mie, James G. March, and Bilian Ni Sullivan. 2005. Notes on the evolution of a research community: Organization studies in Anglophone North America, 1945–2000. *Organization Science* 16 (1): 85–95.

Bakker, René M. 2010. Taking stock of temporary organizational forms: A systematic review and research agenda. *International Journal of Management Reviews* 12 (4): 466–486.

Bansal, Pratima, and Andrew J. Hoffman. 2012. *The Oxford handbook of business and the natural environment*. New York: Oxford University Press.

Barley, Stephen R. 1990. Images of imaging: Notes on doing longitudinal field work. *Organization Science* 1 (3): 220–247.

Barley, Stephen R., and Gideon Kunda. 2006. Contracting: A new form of professional practice. *Academy of Management Perspectives* 20 (1): 45–66.

Bechky, Beth A. 2006. Gaffers, gofers, and grips: Role-based coordination in temporary organizations. *Organization Science* 17 (1): 3–21.

Bechky, Beth A., and Gerardo A. Okhuysen. 2011. Expecting the unexpected? How SWAT officers and film crews handle surprises. *Academy of Management Journal* 54 (2): 239–261.

Biggart, Nicole Woolsey, and Loren Lutzenhiser. 2007. Economic sociology and the social problem of energy inefficiency. *American Behavioral Scientist* 50 (8): 1070–1087.

Bourdieu, Pierre. 1970. The Berber house or the world reversed. *Social Sciences Information* 9 (2): 151–170.

Bourdieu, Pierre. 1990. *The logic of practice*. Stanford, CA: Stanford University Press.

Bourdieu, Pierre. 2005. *The social structures of the economy*. Malden, MA: Polity.

Boyce, Peter, Claudia Hunter, and Owen Howlett. 2003. The benefits of daylight through windows. Lighting Research Center, Rensselaer Polytechnic Institute. http://www.lrc.rpi.edu/programs/daylighting/pdf/DaylightBenefits.pdf (accessed December 31, 2012).

Briscoe, Forrest. 2007. From iron cage to iron shield? How bureaucracy enables temporal flexibility for professional service workers. *Organization Science* 18 (2): 297–314.

Briscoe, Forrest, and Sean Safford. 2008. The Nixon-in-China effect: Activism, imitation, and the institutionalization of contentious practices. *Administrative Science Quarterly* 53 (3): 460–491.

Brounen, Dirk, and Nils Kok. 2011. On the economics of energy labels in the housing market. *Journal of Environmental Economics and Management* 62 (2): 166–179.

Brounen, Dirk, Nils Kok, and John M. Quigley. 2012. Residential energy use and conservation: Economics and demographics. *European Economic Review* 56 (5): 931–945.

Brown, Andrew D., Martin Kornberger, Stewart R. Clegg, and Chris Carter. 2010. "Invisible walls" and "silent hierarchies": A case study of power relations in an architecture firm. *Human Relations* 63 (4): 525.

Cameron, Kim. 2003. Organizational transformation through architecture and design. *Journal of Management Inquiry* 12 (1): 88–92.

Carlsen, Arne, and Jane E. Dutton, eds. 2011. *Research alive: Exploring generative moments in doing qualitative research.* Copenhagen: Copenhagen Business School Press.

Catton, William R., and Riley E. Dunlap. 1978. Environmental sociology: A new paradigm. *American Sociologist* 13 (1): 41–49.

CB Richard Ellis and EMEA Research. 2009. Who pays for green? The economics of sustainable buildings. http://www.nzgbc.org.nz/images/stories/downloads/public/Knowledge/reading/WhoPaysForGreen.pdf (accessed December 31, 2012).

Center for Sustainable Systems. 2011. Residential buildings factsheet. http://css.snre.umich.edu/css_doc/CSS01-08.pdf (accessed December 31, 2012).

Chreim, Samia, B. E. Williams, and C. R. Hinings. 2007. Interlevel influences on the reconstruction of professional role identity. *Academy of Management Journal* 50 (6): 1515–1539.

Davis, Gerald F., Douglas McAdam, W. Richard Scott, and Meyer N. Zald, eds. 2005. *Social movements and organization theory.* New York: Cambridge University Press.

Davis, Gerald F., Calvin Morrill, Hayagreeva Rao, and Sarah A. Soule. 2008. Introduction: Social movements in organizations and markets. *Administrative Science Quarterly* 53 (3): 389–394.

De Stobbeleir, Katleen E. M., Susan J. Ashford, and Dirk Buyens. 2011. Self-regulation of creativity at work: The role of feedback-seeking behavior in creative performance. *Academy of Management Journal* 54 (4): 811–831.

DiMaggio, Paul. 1988. Interest and agency in institutional theory. In *Institutional patterns and organizations: Culture and environment*, ed. Lynne G. Zucker, 3–21. Cambridge, MA: Ballinger Publishing Company.

DiMaggio, Paul J., and Walter W. Powell. 1983. The iron cage revisited: Institutional isomorphism and collective rationality in organizational fields. *American Sociological Review* 48 (2): 147–160.

Dunn, Mary B., and Candace Jones. 2010. Institutional logics and institutional pluralism: The contestation of care and science logics in medical education, 1967–2005. *Administrative Science Quarterly* 55 (1): 114–149.

Dutton, Jane E., and Janet M. Dukerich. 1991. Keeping an eye on the mirror: Image and identity in organizational adaptation. *Academy of Management Journal* 34 (3): 517–554.

Eccles, Robert G. 1981. The quasifirm in the construction industry. *Journal of Economic Behavior & Organization* 2:335–357.

Eichholtz, Piet, Nils Kok, and John M. Quigley. 2010. Doing well by doing good? Green office buildings. *American Economic Review* 100 (5): 2492–2509.

Environmental Leader. 2009. Green costs create roadblock to LEED certification. Environmental & Energy Management News, May 13. http://www.environmentalleader.com/2009/05/13/green-costs-create-roadblock-to-leed-certification/ (accessed December 31, 2012).

Ferlie, Ewan, Louise Fitzgerald, Martin Wood, and Chris Hawkins. 2005. The nonspread of innovations: The mediating role of professionals. *Academy of Management Journal* 48 (1): 117–134.

Fincham, Robin, Timothy Clark, Karen Handley, and Andrew Sturdy. 2008. Configuring expert knowledge: The consultant as sector specialist. *Journal of Organizational Behavior* 29 (8): 1145–1160.

Fisher, Thomas. 2008. *Architectural design and ethics: Tools for survival*. Burlington, MA: Elsevier.

Fligstein, Neil. 1997. Social skill and institutional theory. *American Behavioral Scientist* 40 (4): 397–405.

Fox, Warwick. 2000. *Ethics and the built environment*. New York: Routledge.

Fulton, Mark, Jake Baker, Margot Brandenburg, Ron Herbst, John Cleveland, Joel Rogers, and Chinwe Onyeagoro. 2012. United States building energy efficiency retrofits: Market sizing and financing models. http://www.dbcca.com/dbcca/EN/_media/Building_Retrofit_Paper.pdf. (accessed December 31, 2012).

Gann, David M, Yusi Wang, and Richard Hawkins. 1998. Do regulations encourage innovation?—The case of energy efficiency in housing. *Building Research and Information* 26 (5): 280–296.

Gareis, Roland. 2010. Changes of organizations by projects. *International Journal of Project Management* 28 (4): 314–327.

Garud, Raghu, Roger L. M. Dunbar, and Caroline A. Bartel. 2011. Dealing with unusual experiences: A narrative perspective on organizational learning. *Organization Science* 22 (3): 587–601.

Gieryn, Thomas F. 2002. What buildings do. *Theory and Society* 31 (1): 35–74.

Gilman, Robert. 1992. Sustainability: From the UIA/AIA "Call for Sustainable Community Solutions." http://www.context.org/about/definitions/ (accessed December 31, 2012).

Gioia, Dennis A., and Kumar Chittipeddi. 1991. Sensemaking and sensegiving in strategic change initiation. *Strategic Management Journal* 12 (6): 433–448.

Goodrick, Elizabeth, and Trish Reay. 2010. Florence Nightingale endures: Legitimizing a new professional role identity. *Journal of Management Studies* 47 (1): 55–84.

Greenwood, Royston, Roy Suddaby, and Megan McDougald. 2006. Introduction. In *Research in the sociology of organizations: Professional service firms*, ed. Royston Greenwood and Roy Suddaby, 1–16. San Diego: Elsevier.

Guy, Simon, and Elizabeth Shove. 2000. *A sociology of energy, buildings, and the environment: Constructing knowledge, designing practice*. New York: Routledge.

Hargadon, Andrew B., and Beth A. Bechky. 2006. When collections of creatives become creative collectives: A field study of problem solving at work. *Organization Science* 17 (4): 484–500.

Henn, Rebecca L. 2010. Aftermarkets: The messy yet refined logic of design. *Journal of Corporate Citizenship*, no. 37: 41–54.

Henn, Rebecca L. Forthcoming. Constructing green: Professional jurisdictions, social movements, and temporary organizations. PhD diss., University of Michigan.

Heschong Mahone Group. 1999a. *Daylighting in schools: An investigation into the relationship between daylighting and human performance*. Fair Oaks, CA: Pacific Gas and Electric Company.

Heschong Mahone Group. 1999b. *Skylighting and retail sales: An investigation into the relationship between daylighting and human performance*. Fair Oaks, CA: Pacific Gas and Electric Company.

Hinings, C. R., and Royston Greenwood. 2002. Disconnects and consequences in organization theory? *Administrative Science Quarterly* 47 (3): 411–421.

Hinings, C.R., and Huseyin Leblebici. 2003. Editorial introduction to the special issue: Knowledge and professional organizations. *Organization Studies* 24 (6): 827–830.

Hirsch, Paul M. 1972. Processing fads and fashions: An organization-set analysis of cultural industry systems. *American Journal of Sociology* 77 (4): 639–659.

Hoffman, Andrew J. 2001. *From heresy to dogma: An institutional history of corporate environmentalism*. Stanford, CA: Stanford University Press.

Hoffman, Andrew J., and Rebecca Henn. 2008. Overcoming the social and psychological barriers to green building. *Organization & Environment* 21 (4): 390–419.

Hoffman, Andrew J., and William Ocasio. 2001. Not all events are attended equally: Toward a middle-range theory of industry attention to external events. *Organization Science* 12 (4): 414–434.

Hoffman, Andrew J., and Marc J. Ventresca, eds. 2002. *Organizations, policy and the natural environment: Institutional and strategic perspectives*. Stanford, CA: Stanford University Press.

Jones, Paul. 2009. Putting architecture in its social place: A cultural political economy of architecture. *Urban Studies* 46 (12): 2519–2536.

Jones, Candace, and Benyamin B. Lichtenstein. 2008. Temporary inter-organizational projects. In *The Oxford handbook of inter-organizational relations*, ed. Steve Cropper, Mark Ebers, Chris Huxham, and Peter Smith Ring, 231–255. New York: Oxford University Press.

Kats, Gregory. 2006. Greening America's schools: Costs and benefits. Capital E. http://www.usgbc.org/ShowFile.aspx?DocumentID=2908 (accessed December 31, 2012).

Kats, Greg, Leon Alevantis, Adam Berman, Evan Mills, and Jeff Perlman. 2003. The costs and financial benefits of green buildings. Sacramento: California Sustainable Building Task Force. http://www.usgbc.org/Docs/News/News477.pdf (accessed December 31, 2012).

Kats, Gregory, Jon Braman, and Michael James. 2010. *Greening our built world: Costs, benefits, and strategies*. Washington, DC: Island Press.

Kelso, Jordan D. 2011. 2010 buildings energy data book. Washington, DC: Building Technologies Program, U.S. Department of Energy. http://buildingsdatabook.eren.doe.gov/docs/DataBooks/2010_BEDB.pdf (accessed December 31, 2012).

Kenis, Patrick, Martyna Janowicz-Panjaitan, and Bart Cambré, eds. 2009. *Temporary organizations: Prevalence, logic and effectiveness*. Northampton, MA: Edward Elgar.

Kibert, Charles J. 2007. *Sustainable construction: Green building design and delivery*. Hoboken, NJ: Wiley.

Kibert, Charles J., Martha C. Monroe, Anna L. Peterson, Richard R. Plate, and Leslie Paul Thiele. 2011. *Working toward sustainability: Ethical decision making in a technological world*. Hoboken, NJ: Wiley.

Lawrence, Denise L., and Setha M. Low. 1990. The built environment and spatial form. *Annual Review of Anthropology* 19:453–505.

Lawrence, Thomas B. 1999. Institutional strategy. *Journal of Management* 25 (2): 161–187.

Leonardo Academy. 2009. The economics of LEED for existing buildings for individual buildings. http://www.leonardoacademy.org/download/Economics%20of%20LEED-EB%2020090222.pdf (accessed December 31, 2012).

Lingo, Elizabeth Long, and Siobhán O'Mahony. 2010. Nexus work: Brokerage on creative projects. *Administrative Science Quarterly* 55 (1): 47–81.

Lockwood, Charles. 2007. The green quotient: Q&A with Brenna S. Walraven. *Urban Land* 66 (November/December): 118–119.

Lounsbury, Michael. 1997. Exploring the institutional tool kit: The rise of recycling in the U.S. solid waste field. *American Behavioral Scientist* 40 (4): 465–477.

Lounsbury, Michael. 2001. Institutional sources of practice variation: Staffing college and university recycling programs. *Administrative Science Quarterly* 46 (1): 29–56.

Low, Setha M., and Denise Lawrence-Zúñiga. 2003. *The anthropology of space and place: Locating culture*. Malden, MA: Wiley.

Lundin, Rolf A. 1995. Editorial: Temporary organizations and project management. *Scandinavian Journal of Management* 11 (4): 315–318.

Lundin, Rolf A. 2011. Researchers of projects and temporary organizations—one happy family? *International Journal of Project Management* 29 (4): 357–358.

Lundin, Rolf A., and Anders Söderholm. 1995. A theory of the temporary organization. *Scandinavian Journal of Management* 11 (4): 437–455.

Lutzenhiser, Loren. 1994. Innovation and organizational networks barriers to energy efficiency in the US housing industry. *Energy Policy* 22 (10): 867–876.

Maguire, Steve, Cynthia Hardy, and Thomas B. Lawrence. 2004. Institutional entrepreneurship in emerging fields: HIV/AIDS treatment advocacy in Canada. *Academy of Management Journal* 47 (5): 657–679.

Matthiessen, Lisa Fay, and Peter Morris. 2007. Cost of green revisited: Reexamining the feasibility and cost impact of sustainable design in the light of increased market adoption. Davis Langdon. http://www.davislangdon.com/USA/Research/ResearchFinder/2007-The-Cost-of-Green-Revisited/ (accessed December 31, 2012).

McGraw Hill Construction. 2008. Green outlook 2009: Trends driving change. New York. http://greensource.construction.com/resources/smartMarket.asp (accessed March 18, 2012).

McLennan, Jason F. 2004. *The philosophy of sustainable design: The future of architecture*. Kansas City, MO: Ecotone.

Miller, Norm G., and Dave Pogue. 2009. Do green buildings make dollars and sense? Burnham-Moores Center for Real Estate, University of San Diego, CB Richard Ellis. USD-BMC Working Paper 09-11. http://catcher.sandiego.edu/items/business/Do_Green_Buildings_Make_Dollars_and_Sense_draft_Nov_6_2009.pdf (accessed December 31, 2012).

Modig, Nina. 2007. A continuum of organizations formed to carry out projects: Temporary and stationary organization forms. *International Journal of Project Management* 25 (8): 807–814.

Molotch, Harvey Luskin. 2003. *Where stuff comes from: How toasters, toilets, cars, computers, and many others things come to be as they are*. New York: Routledge.

Moore, Steven A., ed. 2010. *Pragmatic sustainability: Theoretical and practical tools*. New York: Taylor & Francis.

Okhuysen, Gerardo A., and Beth A. Bechky. 2009. Coordination in organizations: An integrative perspective. In *Academy of management annals*, ed. James P. Walsh and Arthur P. Brief, 463–502. London: Routledge.

O'Leary, Michael Boyer, and Bradley A. Almond. 2009. The industry settings of leading organizational research: The role of economic and non-economic factors. *Journal of Organizational Behavior* 30 (4): 497–524.

Oliver, Christine. 1991. Strategic responses to institutional processes. *Academy of Management Review* 16 (1): 145–179.

Owen-Smith, Jason. 2011. The institutionalization of expertise in university licensing. *Theory and Society* 40:63–94.

Packendorff, Johann. 1995. Inquiring into the temporary organization: New directions for project management research. *Scandinavian Journal of Management* 11 (4): 319–333.

Podolny, Joel M., and Karen L. Page. 1998. Network forms of organization. *Annual Review of Sociology* 24: 57–76.

Powell, Walter W. 1990. Neither market nor hierarchy: Network forms of organization. In *Research in organizational behavior*, ed. Barry M. Staw and L. L. Cummings, 295–336. Greenwich, CT: JAI Press.

Pratt, Michael G., Kevin W. Rockmann, and Jeffrey B. Kaufmann. 2006. Constructing professional identity: The role of work and identity learning cycles in the customization of identity among medical residents. *Academy of Management Journal* 49 (2): 235–262.

Rao, Hayagreeva, Philippe Monin, and Rodolphe Durand. 2003. Institutional change in Toque Ville: Nouvelle cuisine as an identity movement in French gastronomy. *American Journal of Sociology* 108 (4): 795–843.

Rapoport, Amos. 1982. *The meaning of the built environment: A nonverbal communication approach*. Beverly Hills, CA: Sage.

Reay, Trish, and C. R. Hinings. 2009. Managing the rivalry of competing institutional logics. *Organization Studies* 30 (6): 629–652.

Rip, Arie, and René Kemp. 1998. Technological change. In *Human choice and climate change: Resources and technology*, ed. Steve Rayner and Elizabeth L. Malone, 328–399. Columbus, OH: Battelle Press.

Roodman, David Malin, and Nicholas K. Lenssen. 1995. *A building revolution: How ecology and health concerns are transforming construction*, ed. Jane A. Peterson. Washington, DC: Worldwatch Institute.

Schein, Edgar H. 1990. Organizational culture. *American Psychologist* 45 (2): 109–119.

Scott, W. Richard. 2008. Lords of the dance: Professionals as institutional agents. *Organization Studies* 29 (2): 219–238.

Scott, W. Richard, and John W. Meyer. 1994. *Institutional environments and organizations: Structural complexity and individualism*. Thousand Oaks, CA: Sage.

Scott, W. Richard, Martin Ruef, Peter J. Mendel, and Carol A. Caronna. 2000. *Institutional change and healthcare organizations: From professional dominance to managed care*. Chicago: University of Chicago Press.

Söderlund, Jonas. 2011. Pluralism in project management: Navigating the crossroads of specialization and fragmentation. *International Journal of Management Reviews* 13 (2): 153–176.

Stinchcombe, Arthur L. 1959. Bureaucratic and craft administration of production: A comparative study. *Administrative Science Quarterly* 4 (2): 168–187.

Strang, David, and John W. Meyer. 1993. Institutional conditions for diffusion. *Theory and Society* 22 (4): 487–511.

Strang, David, and Sarah A. Soule. 1998. Diffusion in organizations and social movements: From hybrid corn to poison pills. *Annual Review of Sociology* 24: 265–290.

Suddaby, Roy, and Royston Greenwood. 2005. Rhetorical strategies of legitimacy. *Administrative Science Quarterly* 50 (1): 35–67.

Sustainability Victoria and Kador Group. 2008. Employee productivity in a sustainable building. http://www.sustainability.vic.gov.au/resources/documents/500_Collins_Productivity_Study.PDF (accessed December 31, 2012).

Thompson, James D., William J. McEwen, R. Vance Presthus, Frederick L. Bates, and Joan S. Dodge. 1957. Editor's critique. *Administrative Science Quarterly* 1 (4): 530–532.

Thornton, Patricia H., Candace Jones, and Kenneth Kury. 2005. Institutional logics and institutional change in organizations: Transformation in accounting, architecture transformation in cultural industries. In *Research in the sociology of organizations: Transformation in cultural industries*, ed. Candace Jones and Patricia H. Thornton, 125–170. San Diego: Elsevier.

Townley, Barbara, Nic Beech, and Alan McKinlay. 2009. Managing in the creative industries: Managing the motley crew. *Human Relations* 62 (7): 939–962.

Tsoukas, Haridimos. 2009. A dialogical approach to the creation of new knowledge in organizations. *Organization Science* 20 (6): 941–957.

Turner Green Buildings. 2005. Market Barometer: Survey of green building plus green building in K-12 and higher education. New York: Turner Construction. http://www.turnerconstruction.com/content/files/Green%20Market%20Barometer%202005.pdf (accessed December 31, 2012).

U.S. Department of Energy. 2003. The business case for sustainable design in federal facilities. Federal Energy Management Program. Washington, DC. http://www1.eere.energy.gov/femp/pdfs/bcsddoc.pdf (accessed December 31, 2012).

U.S. Environmental Protection Agency. 2008. The inside story: A guide to indoor air quality. Office of Radiation and Indoor Air, Consumer Product Safety Commission. EPA 402-K-93–007. http://www.epa.gov/iaq/pubs/insidestory.html (accessed December 31, 2012).

Urban Green Council. 2009. Cost of green in NYC. NYSERDA, Davis Langdon. http://www.davislangdon.com/upload/images/publications/USA/Cost_Study_NYC_10.02.09.pdf (accessed December 31, 2012).

USGBC. 2005. Making the business case for high performance green buildings. https://www.usgbc.org/Docs/Member_Resource_Docs/makingthebusinesscase.pdf (accessed December 31, 2012).

USGBC. 2009. Building impacts. http://www.usgbc.org/DisplayPage.aspx?CMSPageID=1720 (accessed February 12, 2012).

USGBC. 2011a. LEED Rating Systems. http://new.usgbc.org/leed/rating-systems.

USGBC. 2011b. About LEED. https://www.usgbc.org/ShowFile.aspx?DocumentID=4687 (accessed March 13, 2012).

USGBC. 2012. http://www.usgbc.org/ (accessed December 31, 2012).

Van de Ven, Andrew H., and Marshall Scott Poole. 2005. Alternative approaches for studying organizational change. *Organization Studies* 26 (9): 1377–1404.

van Donk, Dirk Pieter, and Eamonn Molloy. 2008. From organising as projects to projects as organisations. *International Journal of Project Management* 26 (2): 129–137.

Van Maanen, John, and Stephen R. Barley. 1984. Occupational communities: Culture and control in organizations. *Research in Organizational Behavior* 6: 287–365.

von Nordenflycht, Andrew. 2007. Is public ownership bad for professional service firms? Ad agency ownership, performance, and creativity. *Academy of Management Journal* 50 (2): 429–445.

von Nordenflycht, Andrew. 2010. What is a professional service firm? Toward a theory and taxonomy of knowledge-intensive firms. *Academy of Management Review* 35 (1): 155–174.

Wallace, Jean E., and Fiona M. Kay. 2008. The professionalism of practising law: A comparison across work contexts. *Journal of Organizational Behavior* 29 (8): 1021–1047.

Walsh, James P., Alan D. Meyer, and Claudia Bird Schoonhoven. 2006. A future for organization theory: Living in and living with changing organizations. *Organization Science* 17 (5): 657–671.

Walsh, James P., Klaus Weber, and Joshua D. Margolis. 2003. Social issues and management: Our lost cause found. *Journal of Management* 29 (6): 859–881.

WCED (World Commission on Environment and Development). 1987. *Our common future*. Oxford: Oxford University Press.

Weick, Karl E. 1987. Organizational culture as a source of high reliability. *California Management Review* 29 (2): 112–127.

Weick, Karl E. 1995. *Sensemaking in organizations*. Thousand Oaks, CA: Sage.

Wilson, Alex. 1999. Daylighting: Energy and productivity benefits. *Environmental Building News* 8 (9): 1–8.

Wilson, Alex. 2005. Making the case for green building. *Environmental Building News* 14 (4): 1–15.

Wilson, Edward O. 1984. *Biophilia*. Cambridge, MA: Harvard University Press.

Zald, Mayer N., and Michael Lounsbury. 2010. The wizards of Oz: Towards an institutional approach to elites, expertise and command posts. *Organization Studies* 31 (7): 963–996.

Zietsma, Charlene, and Thomas B. Lawrence. 2010. Institutional work in the transformation of an organizational field: The interplay of boundary work and practice work. *Administrative Science Quarterly* 55 (2): 189–221.

I

Emerging Professions and Expertise

2

Building Expertise: Renovation as Professional Innovation

Kathryn B. Janda and Gavin Killip

Introduction

The built environment must undergo dramatic changes to meet climate change targets. The World Business Council for Sustainable Development (WBCSD 2009) calls for a worldwide building sector energy reduction of 77 percent below projected 2050 levels. In Britain, the residential sector is the largest consumer of energy and the main emitter of CO_2. Although energy policy in the UK has emphasized energy efficiency in housing (e.g., DTI 2003; DEFRA 2007), the country now recognizes that more radical and transformative changes are needed, particularly for existing homes (DECC 2009). Killip (2008) estimates that transforming the entire UK housing stock by 2050 will require 500,000 refurbishments of older, inefficient properties every year. The sheer scale of these transformations requires radical changes in both technology and work practices.

The large technical potential for improvement in the housing sector has been demonstrated, requiring an integrated combination of ambitious demand reduction strategies (e.g., insulation, improved airtightness, more efficient appliances, behavior modifications) and low and zero carbon (LZC) technologies such as solar technologies and heat pumps (e.g., Boardman et al. 2005; Marchand et al. 2008). Research shows that to reach higher levels of carbon savings in refurbishment (e.g., 50 percent or more) it is not just one technology that needs to be implemented, but a suite of coordinated strategies that treats the dwelling, services it provides, and its occupants as an integrated system (Hermelink 2006; Roudil 2007). We call this the "house as a system" approach. The carbon savings from performing holistic retrofits, even without occupant participation, should not be underestimated. Ürge-Vorsatz, Petrichenko, and Butcher (2011) argue that there is a 79 percent difference between implementing a "state of the art" performance approach in new buildings and renovations worldwide versus a "suboptimal" scenario based on piecemeal individual technologies.

Although optimizing the suite of available technical and social strategies for each existing dwelling will yield the best results in reducing carbon emissions, it is a tremendous challenge to assign this task to a fragmented construction industry. In both the UK and elsewhere, housing refurbishment is the preserve of small and medium-sized enterprises which include general builders, specialist subcontractors (e.g., roofing contractors), plumbers, heating engineers, electricians, architects, design engineers, project managers, building control inspectors and others. These groups are often considered "intermediaries" in the technology adoption process: they occupy an important position between technologies and end-users and as such are expected to provide low carbon refurbishment if their clients demand it. Yet we know that expertise matters, and it is not equally distributed. Quality design and highly skilled installation are essential to the success of low-carbon refurbishment projects, particularly in the areas of insulation, thermal bridging, and airtightness (Lowe and Bell 2000). If some intermediaries are more expert than others, then the supply of low carbon refurbishment is *not* perfectly responsive to the demand. Instead, intermediary groups have their own habits, practices, ways of thinking about problems, and ways of working that affect their ability to provide (and interest in promoting) low carbon refurbishment. How might the need for low carbon refurbishment change the roles of professions, and their interactions? How are existing professions developing to meet the challenge? Which professions will gain control over the new activities involved in low carbon refurbishment?

To address these critical questions, we take up the challenge of discerning which institutions can successfully intervene in the total socio-technical system of the built environment to steer it toward sustainable performance. In doing so, we move from discussions of *what* needs to be done to reduce carbon emissions in the existing housing stock, and draw attention to *who* will do it and *how*. Specifically, we focus on the role of so-called "intermediaries," their expertise, and their ability to enhance (or inhibit) the implementation of sustainable strategies in existing residential buildings. This chapter begins with a review of some recent literature on innovation in the residential sector. Noting that literature on innovation in residential refurbishments is comparatively scarce, we argue that the understanding of this topic needs improvement, particularly with respect to the need for building expertise. To move toward filling this gap, this chapter suggests a socio-technical "system of professions" approach, which addresses the role of experts and expertise in refurbishment. This discussion draws upon the intersection of two theoretical approaches: innovation in socio-technical systems (STSs) and the system of professions (Abbott 1988). The chapter concludes with a snapshot of the UK residential refurbishment industry, approaching existing practices through this theoretical lens. It considers one possible way for the current system of professions to evolve to incorporate low carbon refurbishment, focusing on what it would take to increase the skills and capabilities of small and medium builders in Britain.

Background: Innovation in Housing

Recent work on innovation in construction suggests that influences on multiple levels affect the shape and nature of innovation. Koebel (2008) suggests that there are individual, firm, and industry characteristics of particular importance, including risks associated with innovation, the role of technology champions, and the degree of centralization in decision making between small custom builders and large production builders. With respect to green building, Hoffman and Henn (2008) agree that both individual and organizational factors inhibit innovation, and they add a third level—institutional barriers—which is broader and more pervasive than the structure of the industry itself. In Hoffman and Henn's framing, institutional barriers to green building include regulative, normative, and cognitive aspects of the larger social system in which building occurs. In particular, they assert that social and psychological barriers are in need of greater attention, for they believe that understanding and overcoming these barriers will lead to changes in social structures and in rewards and incentives.

Focusing particularly on passive housing designs, Gentry (2009) argues that it is the building process itself that needs to be changed. Whereas the process employed by large production builders leads to greater fragmentation in the construction of each house and a reduction in labor force skills, Gentry asserts that a design-build approach coupled with integrative design (where structure, systems, and aesthetics act as a team rather than a relay race) should be the way forward. One important aspect of Gentry's proposal is that it reconnects the homeowner with the builder, so that the homeowner (or occupant) becomes more actively engaged with the design and eventual operation of the home. Gentry is one of few authors who treats the resident as an integral part of the housing system.

Taylor and Levitt (2004) also believe that the organizational process of building is important. They delineate the concepts of incremental and systemic innovations in the building industry, arguing on the one hand that incremental innovations happen in the building industry about as readily as they do in manufacturing industries. On the other hand, when it comes to systemic innovations, which require multiple companies to change in a coordinated fashion (e.g., supply chain management), the home building industry is a laggard adopter. Taylor and Levitt hypothesize that systemic innovations will increase when the home builders reduce the number of specialists they use on multiple projects and when the level of interdependence between specialist tasks is decreased. Taylor and Levitt are particularly interested in improving the overall economic efficiency of the industry; they do not mention energy efficiency as a goal of their work.

It is important to note that these studies are all about new housing. For many years, research and policy arenas have ignored renovation and retrofits. The implicit assumption is that because existing housing has already been built, the interesting organizational changes (e.g., integrated vs. sequential design) or radical technical approaches (e.g., passive solar strategies) are not applicable. This orientation, however, is changing in large part due to the carbon reduction agenda. Figure 2.1 shows the projected carbon emissions from the UK domestic sector to 2050 in a 75 percent reduction scenario. The largest block of emissions to be abated is from the existing housing stock. Even if all new homes were "zero carbon" by 2016 in keeping with UK government targets, carbon emissions from an untouched existing stock would swamp the new build improvements. From a demand perspective, 100 percent carbon-free new homes would *at best* leave current carbon emissions unchanged; only retrofitting the existing building stock can actually reduce current emissions.

The refurbishment industry grew substantially in recent years and is poised to grow even faster, in large part because of the emphasis on sustainable development, and due to economic conditions. In central Europe, Kohler and Hassler (2002, 226) claim that these trends have been operating for close to thirty years and believe that they will "oblige the building professions to shift their focus from new construction to maintenance and refurbishment of existing buildings." In the UK, a report commissioned by the Federation of Master Builders presents the poor performance of residences as a business opportunity (Killip 2008). The report claims that building firms, product manufacturers, and suppliers could tap into a new market worth between £3.5 and £6.5 billion ($5.5 to 10 billion) per year if the UK develops policies, skills programs, and financial incentives to upgrade the

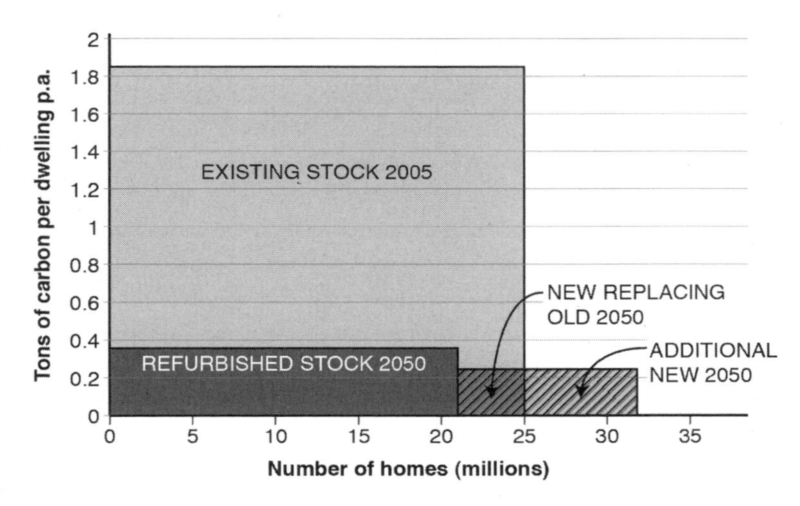

Figure 2.1

Carbon emissions from refurbished and new-build housing in the United Kingdom: 75 percent reduction scenario

existing housing stock to make it greener and more energy efficient. In addition, a refurbished housing stock would help reduce escalating household energy bills while making a real difference to climate change.

Although the authors mentioned recognize the need for changes to the structure of professional practice, this recognition may be a minority view. Michael J. Kelly, chief scientific advisor to the UK Department for Communities and Local Government, recently wrote a commentary in the journal *Building Research & Information* on the importance of retrofitting the existing UK building stock. In this article (Kelly 2009, 198–199), he states that carbon emissions from existing buildings could be tackled in four ways, by

- reengineering the fabric of buildings
- improving the efficiency of appliances
- decarbonizing the sources of energy
- making changes in personal behavior

Kelly states that the first three approaches are "engineering related," and the last one "is a matter of psychology and sociology." In our experience, Kelly's conception of both the nature of the problem and his proposed solution set is fairly common, and carries within it two important assumptions which often go unchallenged. The first is that the problem can in large part be solved by technology rather than people—an assumption based on a determinist view that technology and behavior are easily separable from each other. The second assumption is that social science is limited to investigating how individual people operate as consumers or citizens, not as artisans and professionals engaged in providing services. From this second assumption comes the expectation that social science applies only to changing homeowner behaviors. We challenge these assumptions and their implications in the following section.

A Socio-technical System of Professions Approach

In this section, we introduce a way of thinking about innovation in the refurbishment industry that is informed by theories of socio-technical systems together with the sociology of professions. Our aim is to reconnect the synaptic path that leads policymakers to think that technology is separate from people, and that people only live in houses rather than making their livelihood in them.

Technology Is Inseparable from People

Our first point is that technology and people are intertwined. The science and technology studies (STS) literature provides ample support for this perspective, offering an over-arching framework in which the "seamless web" of social and technological effects of change can be understood. This literature argues that technological change does not come about independently of behavioral change and the development of social norms; rather, the technical and the social coevolve and depend on each other in a complex socio-technical system (Hughes 1983; Bijker, Hughes, and Pinch 1987; Bijker and Law 1992). With regard to energy use in buildings, this means "relating the form, design and specification of more and less energy-efficient buildings to the social processes that underpin their development" (Guy and Shove 2000, 67). The social processes that have been studied in this field often focus on the behavior, habits, and motivations of the individuals who occupy homes (e.g., Wilson and Dowlatabadi 2007) and their interplay with technological attributes of devices within the home (e.g., Gram-Hanssen 2008).

Actor-network theory (ANT)—an approach that originated in the STS literature—further develops the relationship between technologies and their implementation. ANT sees the relationship between technology and people and ideas as a process of translation among multiple actors (both human and nonhuman, including material objects and concepts), all of which can have influence on all of the others in a constant process of (re) creation of customs and norms (Callon 1986; Latour 2005). Three key concepts underpin this idea: first is the idea of "enrollment," by which any actor (including nonhuman agents) successfully assigns a role to other actors in the socio-technical system in question. When all actors are successfully enrolled, the system is said to be "black boxed," meaning that the emergent operation of the entire system is relatively stable. "Black boxing" is the second key concept, portraying habits and customs as just one set of possible outcomes from the processes of translation between actors. In ANT, stability emerges from interacting forces, meaning that new actors (such as a new regulation or profession) or changed objectives can make things unstable again. The third key concept is "dissidence," where any actor (including nonhuman agents) fails to take on the assigned role, with potentially far-reaching consequences for the entire system (see Callon 1986).

Professionals Are People Too

If technologies are inseparable from people, which people matter most in the technology adoption process? In contrast to the research on individuals who use technologies, we concentrate on people who install technologies. Intermediary groups have been shown to have their own habits, practices, ways of thinking about problems, and ways of working that affect their ability to provide (and interest in promoting) energy efficient buildings (Janda 1998). Examples of how intermediaries affect the uptake of energy efficiency include the role of supply chains (Guy and Shove 2000), property agents (Schiellerup and Gwilliam 2009), builders (Killip 2008), heating engineers (Banks 2001), and architects and engineers (Janda 1999). In this section, we focus on two aspects of professionals who renovate homes: the shape and nature of their professional jurisdictions; and their competency in performing various tasks.

System of Professions Approach

A "system of professions" (Abbott 1988) approach fits within the general sociology of professions (Dubar and Tripier 2005). It is concerned with the ways in which different professional or occupational groups define their work and compete for authority, which is linked to their use and appropriation of knowledge. From a system of professions perspective, each work group is linked (neither permanently nor absolutely) to a set of socially accepted tasks considered to be its jurisdiction. Architects, for instance, may see themselves (and be seen by others) as the profession with responsibility for creating quality of place and aesthetic values in the built environment; engineers are more concerned with the technical practicalities of making structures that are safe, healthy, and thermally comfortable. Professional groups compete and develop interdependently, based in part on their ability to perform (and defend) the tasks within their jurisdiction. Jurisdictions and professions change over time and are shaped by a number of social, economic, historical, and institutional factors (Abbott 1988; Bureau and Suquet 2009; Evetts 2006). Abbott focuses mainly on the meso or systems level, investigating relationships between professions, but he also looks at the levels below and above. At the micro level, he considers differentiation *within* professions related to work context, and at the macro level, he discusses the larger social forces that create the "system environment" in which the professions exist.

Abbott admits that his framework explains the shape of existing professional groups better than the development of new groups. However, he posits that growth in general knowledge can create a "new" socially legitimate set of problems and therefore an opportunity for new professional group(s). It is this underexplored element in Abbott's work that most intrigues us. Is growth in knowledge about climate change—its impacts, causes, and opportunities for mitigation—sufficient to challenge the current system of professions operating in the built environment today? Some industry and government organizations believe so. The WBCSD (2009) argues that a new "system integrator" profession is needed to develop the workforce capacity to save energy. The UK is training domestic energy assessors to draw up Energy Performance Certificates (Banks 2008), while the Australian government chose to support the development of a new profession of in-home energy advisors (Berry 2009). Each of these entities asserts that a new profession will help solve the "problem," but each proposed professional solution is different.

Seen from an ANT perspective, many actors will need to be enrolled in the carbon reduction challenge in order to successfully meet it. However, what if one profession is better prepared than another to meet this challenge? Is it possible to have the "wrong" set of actors enrolled in the system? Abbott's perspective allows us to consider what might happen if architects, engineers, builders, solar installers, or some new profession group "takes over" the carbon reduction challenge. It also invites us to consider whether this can be a shared responsibility among the existing professions, based on the established jurisdictions and black-boxed customs and practices of the construction industry in operation today.

Competencies
The traditional focus of training has been on traditionally defined trades (plumbing, plastering, etc.). New LZC technologies and techniques challenge traditional trades to engage in a "multi-skilling" agenda. This agenda represents a shift of emphasis away from trades and specific technologies to an integrated "house as a system" refurbishment focus. A whole-home refurbishment requires multiple skills beyond substituting a more efficient item for a less efficient one. It requires technical understanding of building physics (e.g., how much heat goes out of the roof, wall, windows, and floors). It calls for the integration of demand reduction measures with energy supply technologies (e.g., calculating the heat load for a well-insulated property and sourcing a heating system to match). It involves aspects of project management (e.g., optimal ordering of works on site). Some of the more technical aspects of this work may be best addressed through the development of one or more packages for refurbishment (i.e., an all-inclusive design specification, which can be applied without understanding all of the reasons behind it). Having said that, there is a risk that packages may not work well in practice (or in certain situations), as the assumption that one size fits all is almost certain to be misplaced, given the variety and size of the housing stock. The low-carbon refurbishment agenda therefore presents a series of challenges for training, knowledge, and skills.

Professions + Competencies = ??
Figure 2.2 shows a general conceptual map of the fragmented construction industry, with professional roles arrayed along the horizontal dimension, and skills or competencies stacked along the vertical dimension.

This two-dimensional representation of the "problem space" of our research topic shows how professional jurisdictions are differentiated (or not) from one another in the current system. Notably, gaps appear at the intersections of the professions and competencies, indicating spaces and imperfections in the current system. To this system, we add low carbon refurbishment as a possible new profession or jurisdiction, or both. Existing professional roles (e.g., architect, structural engineer, general builder, roofer) may expand to encompass new competencies (e.g., energy assessment, installation of roof-mounted renewable energy systems, whole-home system integration). Competencies that are well established within one profession may need to be expanded to become the preserve of other roles, for which they have not traditionally been a concern; also, new roles and new competencies may be needed.

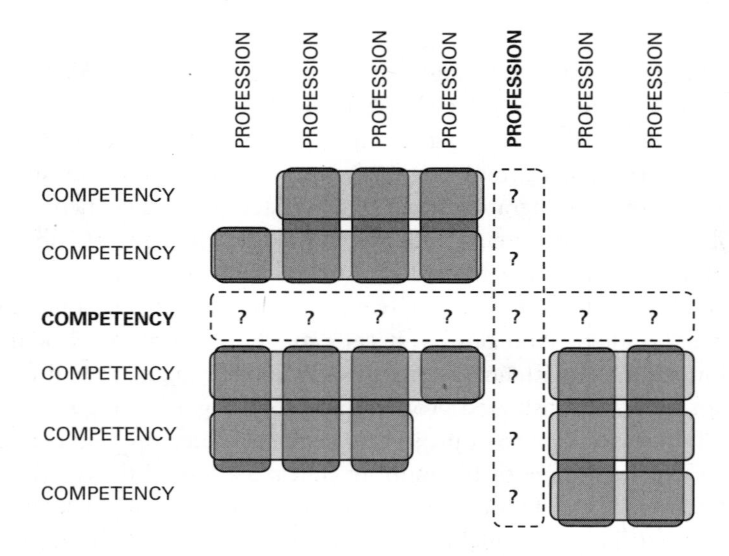

Figure 2.2

Roles and competencies for the integration of low-carbon refurbishment into a system of professions

So what does the current system of professions for housing refurbishment look like in Britain, and how might it change (or need to be changed) to mainstream low-carbon housing refurbishment? We explore this question in the following section.

Toward a UK Low-Carbon Refurbishment Industry

As figure 2.2 suggests, there are many possible configurations for a low-carbon refurbishment industry. Some have suggested that new entrants are needed. One oft-discussed option is for large corporations (e.g., the supermarket chain Tesco, or the home improvement chain B&Q) to lead the way forward by providing one-stop shopping for energy improvements. In this section we take up the question of mainstreaming low carbon refurbishment by developing the skills and capabilities of existing firms rather than new entrants. This section focuses on small and medium builders in Britain, considering their role as key to improving the sector. Whoever takes on the challenge has their work cut out for them.

The State of the Shelf

Construction is a major employer in the UK economy, providing roughly 1.1 million jobs and generating £105 billion ($164 billion) of contractors' output in 2009, of which £39.8 billion ($62 billion) was spent on housing—38 percent of the total (Office for National Statistics 2010, table 2.4).

New construction is a different market from repair, maintenance, and improvement (RMI), just as construction work in housing is a different market from nonresidential work. Many individuals working within the industry move across these boundaries as their careers progress and as the availability of work and subcontractual arrangements shift in response to wider economic forces. Some firms specialize in one particular type of work, while others are generalists; some concentrate on residential work, some stick to commercial projects, and some do both. SMEs (small and medium-sized enterprises) are predominantly involved in RMI work, although some are developers of new housing, mainly on a small scale.

The breakdown of expenditure on housing in 2009 (as opposed to nonresidential buildings) shows that more money was spent in the private sector than the public sector (new-build and repair and maintenance), while the total output on housing repair and maintenance (in both public and private sectors) was £18.9 billion ($29.4 billion), exceeding the £14.6 billion ($22.7 billion) for new housing (see figure 2.3).

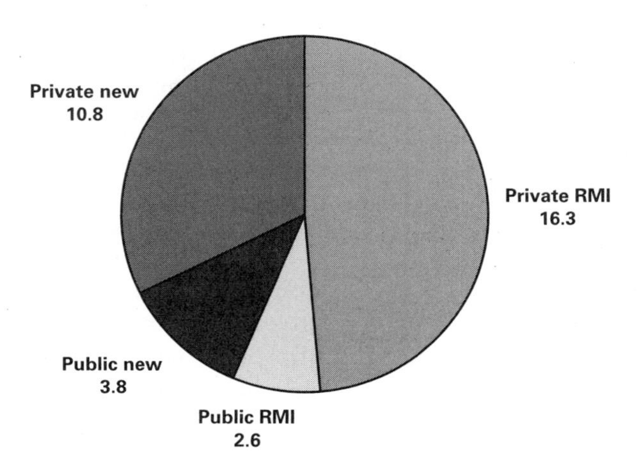

Figure 2.3

Contractors' output in housing in Great Britain, £ Billions (2009)

Construction is multifaceted, ranging from major infrastructure projects, such as the 2012 Olympics, right through to decorating and handyman services in people's homes. The industry is made up of over 194,000 firms: a large number of small businesses, particularly microbusinesses. Firms with one to three employees make up 70 percent of the total number of firms in the industry. Only 35 percent of the total workforce is employed in firms with more than eighty employees (Office for National Statistics 2010, table 3.4).

Innovation in the UK Construction Industry

Sir John Egan's *Rethinking Construction* report (Egan 1998) first set out a clear agenda for improvements to business processes and efficiency in the UK construction sector. This report was followed by a second review, *Accelerating Change*, which described the sector as "a series of sequential and largely separate operations undertaken by individual designers, constructors and suppliers who have no stake in the long term success of the product and no commitment to it" (Egan 2002, 13). A more recent review concluded that progress on the Egan agenda has shown some signs of improvement since 2002, but that these have tended to be "skin-deep" (Wolstenholme et al. 2009). The reality of mainstream practice is that it is dominated by fragmented roles, adversarial rather than collaborative relationships among different professions, wasteful and inefficient management processes, and poor client satisfaction.

Nonetheless, Harris and others (2006) show that innovation does occur in this sector, but that it is hidden from view because of the widespread belief that innovation signifies new technology, and that it can therefore best be measured using conventional metrics that apply to product innovations: research and design spending and patent applications. Instead, Harris and others argue that new regulation and client demand drive innovation in construction, and that it is project based and collaborative. New products are clearly relevant, but so are innovations in practices (ways of doing) and processes (ways of organizing). The metrics that could be used to report on practice and process innovations are not immediately obvious but, given the policy context of an 80 percent reduction in carbon dioxide emissions, they need ultimately to relate to building performance.

The importance and prominence of sustainability issues in construction are much lower on the ground than among policymakers and strategists. Two of the top ten skills issues for Construction Industry Training Board/ConstructionSkills are "making sustainability a reality in construction" and "improving the skills base and competence through client-led demand, enhancing industry's responsiveness to technical change and productivity improvement" (CITB-ConstructionSkills 2008). In the SME sector working on housing refurbishment, there is a long way to go before these aims are met. In a survey of 152 members of three trade federations in Scotland, on average only 10 percent had received any form of training on sustainable development while only 6 percent reported ever having lost business for environmental or social reasons (Brannigan and Tantram 2008). On this evidence, few clients make sustainability a primary objective of the work that they commission. At the same time, 50 percent of all respondents to the Scottish survey believed that the pressure to be more environmentally responsible would grow over the next one to two years, and over 80 percent thought this pressure would definitely increase within five years. The pressure is largely perceived as being policy driven, rather than driven by demand in the market for refurbishment work.

A Socio-technical System of Professions Approach to Building

The usual policy approach to retrofits is to give rebates or assistance for individual qualifying measures, such as cavity wall insulation. The remaining potential for uptake of these measures is still significant—for example, roughly ten million cavity-walled homes do not have their cavity walls insulated (CIGA 2008). At the same time, an acknowledgment is needed that this policy approach is both time limited and insufficient to meet carbon dioxide reduction targets. In the short term this means carrying on with existing measures, while simultaneously preparing for a shift of emphasis in the next few years. Going beyond the existing program requires a new approach, which makes low carbon refurbishment a mainstream "normal" decision for people. Key to managing this transition is to engage more fully with the construction industry, specifically the firms—usually SMEs—that are already involved in housing repair, maintenance, and improvement works. If the opportunities for low carbon refurbishment are to be exploited fully, this kind of renovation needs to be on offer every time a building tradesperson is asked to quote for work. In effect, from an ANT perspective, firms across the entire construction industry need to "enroll" in the low-carbon refurbishment idea, such that this form of refurbishment becomes black boxed. For SME building tradespeople to deliver low carbon refurbishments on a large scale, the sector's capacity to do this kind of work needs to be developed—almost from scratch. In the following sections, we concentrate on three aspects essential to this development: increasing the buildability of innovations, increasing integrated practice and multi-skilling, and developing regional innovation networks.

Increasing Buildability

The work to build capacity in the sector needs to take account of established custom and practice—or the endeavor will result in rejection by most practitioners. This would be called "dissidence" in ANT. This insight is captured in the idea of "buildability," a term intended to capture the reality of how builders operate and the fact that whenever refurbishment is carried out, the contractors have to be confident of their ability to do the work and achieve satisfactory results, both for themselves and for their clients.

The term "buildability" typically applies in the field of construction project management with a focus on the ease of construction, the robustness and quality of what gets built, and a reduced risk of error, delay, or cost overrun. The Construction Industry Research and Information Association defines it as "the extent to which the design of the building facilitates ease of construction, subject to the overall requirements for the completed building" (CIRIA 1983, 26). Crowther (2002, 2) highlights the importance of the conditional clause in the CIRIA definition and argues that "overall project goals may actually restrict the buildability of the project, such that heuristic principles of buildability may not necessarily be appropriate in all cases." In other words, some building projects may be inherently more complicated and carry greater risks for the project manager than other projects.

"Buildability" has come to the fore in recent times as a term used by members of the construction industry to provide a "reality check" to the pursuit of ever tighter energy and carbon dioxide emissions standards for new buildings. An industry task group used buildability to establish a kind of checklist for proposed zero carbon standards for new housing: "Does the specification require overly complicated construction processes? What are the implications for site management and build processes? Will the supply chain be able to deliver circa 200,000 units p.a. to this standard?" (Zero Carbon Hub 2009). These concerns are real enough but they are couched in the language of current incumbents, rather than innovators. What is seen as an "overly complicated" construction process by an incumbent may be something to which others can adapt more readily (see also chapter 6, this volume); site management and build processes may need to change, but that may be because existing techniques, materials, and practices are unsuited to the new standards. Buildability can be seen as an expression of the dilemma of the diffusion of innovation: how can the new and unfamiliar be reconciled to the expectations of the existing mainstream?

If a low-carbon refurbishment strategy can be devised in such a way that it takes account of the need for buildability, then the strategy has the greatest chance of acceptance by the SME construction sector. Without it, it is likely to be ignored or subverted on the ground. Key elements of the buildability idea are that building work needs to comprise products and methods that must be all of the following: practical, replicable, affordable, reliable, sellable, available, guaranteeable, and profitable (Killip 2008, 23–24). Where new products are needed to help meet the low-carbon refurbishment agenda, the key stakeholders (in addition to the SME building tradespeople) are the manufacturers and suppliers. Where new supply chains need to be developed, the key to success is a strong long-term policy commitment from government. This will stimulate investment and strategic business developments, both among existing players in the market and among potential new market entrants.

Increasing Capacity for Integration

The skill sets of traditionally defined tradespeople (e.g., plasterers, electricians, etc.) will need to be expanded so that they understand enough of the low-carbon refurbishment agenda to play their part effectively. This is likely to include a better understanding of how the interaction of different trades on site can lead to loss of overall building performance (e.g., airtightness can be compromised if wet plaster stops at the height of skirting boards instead of reaching floor level; the performance of vapor barriers and insulation materials can be compromised by inaccurate installation and subsequent drilling of holes for pipes, ducts, wires and recessed light fittings). In relation to the installation of low and zero carbon technologies (LZCs), the relevant sector skills council has identified these new technologies as key to the future of mechanical and electrical building services (National Energy Foundation 2007). This council began a process of setting national occupational standards for training on the installation of LZCs, starting with a review of the short courses and other forms of training that have emerged during the early period of market development. This work confirms a widely held observation that innovation in skills training does not start with vocational qualifications, but with short courses. Developing short courses into vocational qualifications is an important part of mainstreaming the capacity to deliver new services. Existing national occupational standards may have to be amended (leading to changes in related vocational qualifications). It is also likely that one or more new sets of occupational standards and vocational qualifications will be needed. Without a perceived need from the industry leaders who guide the development of new skills, none of this work on mainstreaming skills for low carbon refurbishment will come about.

Developing Regional Innovation Networks

Some of the system required to deliver low carbon retrofits already exists and operates at a national scale (e.g., building codes, building energy labeling), while other elements that do not yet exist also imply national-level implementation (e.g., financial incentives or fiscal reform). At the same time there are several reasons for thinking that a regional focus is needed to foster some of the innovation implicit in the low-carbon refurbishment agenda. These mixed scale issues suggest the need for a nationally defined strategy implemented in a series of devolved regional networks. Some of the regional characteristics are summarized as follows.

• Devolved administrations. Both Wales and Scotland have taken different paths to England in terms of the zero carbon new-build agenda, while Scotland's Building Regulations are also significantly divergent. As refurbishment moves up the political agenda, it seems reasonable to assume that the devolved governments will want to define their own strategies for the existing housing stock as well.

• Housing stock variations. Locally and regionally, UK housing has quite different characteristics—from Scottish tenements to back-to-back terraces in northern English cities; from rural houses made of traditional materials (e.g., Devon cob) to inner-city high-rise flats. While some dwelling types are common and ubiquitous, there is also geographical diversity. Tackling these issues at the level of a devolved administration or an English region would allow for more detailed work on the predominant types in that particular part of the country.

• Climate and climate change impacts. Heating demand is typically higher in colder Scottish winters than in Cornwall, while the changing climate may lead to a significant increase in demand for summer cooling in London and the southeast, but not elsewhere.

• Business networks. Most SME construction firms work at a local level, but federations and business-to-business networks typically operate at the slightly larger scale of English regions, nations (Scotland and Wales), and the province of Northern Ireland.

• Regional/devolved development agencies. There is considerable potential for new jobs and new economic development, much of which could benefit from financial assistance and other development services that are available at a regional level.

• Training centers. ConstructionSkills (the agency charged with skills development for the construction industry in England) and the further education (vocational training) college network can be usefully integrated into a regional structure. The involvement of these institutions is key in light of the far-reaching implications that low carbon refurbishment has for skills. Developing the low-carbon refurbishment agenda will require coordination of information, opportunities for networking, and knowledge transfer activities. All of these can usefully take place at a regional level and, in many instances, there are existing partnerships or stakeholder networks in which the low-carbon refurbishment agenda could be accommodated, making good use of existing structures.

Conclusions and Next Steps

This chapter set out some of the issues around low carbon refurbishment and proposed some ideas and recommendations for government bodies and other stakeholders to consider. Much more work clearly is needed to bring about the transformation of the UK housing stock to meet low carbon standards. This amounts to a completely new service provided by the SME construction industry, potentially adding £3.5 to £6.5 billion ($5.5 to 10 billion) to the existing market for housing repair, maintenance, and improvement. A new kind of service is needed, combining new and traditional skilled trades in ways that result in low carbon refurbishment. Many vocational qualifications will need to be amended so that awareness of energy and carbon issues within the SME construction industry is significantly improved and practices changed to meet these new requirements. To increase the chance of success, refurbishment initiatives need to take into account the ways in which building tradespeople operate, making the objectives of policy practically deliverable.

To make these changes, we emphasize the importance of rethinking the ways in which practitioners, policymakers, and academics think, learn, and teach about the built environment. Making significant changes in the built environment is not a matter of reengineering a technical system on paper, it is about reshaping a socio-technical system by redefining established skills, work practices, and professions on the ground.

Although low carbon refurbishment is not currently the norm, we are interested in exploring the ways in which built environment professionals see gaps, opportunities, and challenges for integrating low carbon refurbishment in their work. Work on this topic is ongoing. Nösperger, Killip, and Janda (2011) recently introduced a comparative study of Britain and France that takes a socio-technical system of professions approach to the topic. Because we are interested in whether a new profession might arise, forthcoming work in this vein will focus particularly on the work practices of innovators as providing a key to understanding the social construction of new competencies or roles, or both, that may alter the current system of professions. This focus on innovation will be set against a backdrop of more general work practices and policy context.

References

Abbott, Andrew. 1988. *The system of professions*. Chicago: University of Chicago Press.

Banks, Nick. 2001. Socio-technical networks and the sad case of the condensing boiler. In *Energy efficiency in household appliances and lighting*, ed. Paolo Bertoldi, Andrea Ricci, and Anibal de Almeida, 141–155. Berlin: Springer.

Banks, Nick. 2008. Implementation of energy performance certificates in the domestic sector. UKERC/WP/DR/2008/001. Working Paper. UKERC: Oxford.

Berry, Stephen. 2009. Overview of the green loans programme. Presentation at Oxford University. August 31.

Bijker, Wiebe E., Thomas P. Hughes, and Trevor Pinch, eds. 1987. *The social construction of technological systems: New directions in the sociology and history of technology*. Cambridge, MA: MIT Press.

Bijker, Wiebe E., and John Law, eds. 1992. *Shaping technology/building society: Studies in sociotechnical change*. Cambridge, MA: MIT Press.

Boardman, Brenda, Sarah Darby, Gavin Killip, Mark Hinnells, Christian N. Jardine, Jane Palmer, and Graham Sinden. 2005. 40% House. Oxford: Environmental Change Institute. http://www.eci.ox.ac.uk/research/energy/downloads/40house/40house.pdf.

Brannigan, Jimmy, and Joss Tantram. 2008. *Building future skills*. Scotland: ConstructionSkills.

Bureau, Sylvain, and Jean-Baptiste Suquet. 2009. A professionalization framework to understand the structuring of work. *European Management Journal* 27 (6): 467–475.

Callon, Michel. 1986. Some elements of a sociology of translation: Domestication of the scallops and the fishermen of St Brieuc Bay. In *Power, action and belief: A new sociology of knowledge?* ed. John Law, 196–233. London: Routledge.

CIGA. 2008. The benefits of insulation. Leighton: Cavity Insulation Guarantee Agency. http://www.ciga.co.uk/benefits.html.

CIRIA. 1983. *Buildability: An assessment*. London: Construction Industry Research and Information Association.

CITB-ConstructionSkills. 2008. *Annual review and accounts 2007. HC 387*. London: Construction Industry Training Board-ConstructionSkills.

Crowther, Philip. 2002. Design for buildability and the deconstruction consequences. CIB Publication 272. Karlsruhe: Chartered Institute of Building. http://eprints.qut.edu.au/2885/1/Crowther-TG39-2002.PDF.

DECC. 2009. Heat and energy saving strategy consultation document. London: Department of Energy and Climate Change. http://hes.decc.gov.uk/consultation/download/index-5469.pdf.

DEFRA. 2007. UK energy efficiency action plan. London: Department for Environment Food and Rural Affairs. http://ec.europa.eu/energy/demand/legislation/doc/neeap/uk_en.pdf.

DTI. 2003. Our energy future—creating a low-carbon economy. Energy white paper. London: Department of Trade and Industry. http://webarchive.nationalarchives.gov.uk/+/http://www.berr.gov.uk/files/file10719.pdf.

Dubar, Claude, and Pierre Tripier. 2005. *Sociologie des professions*. 2nd ed. Paris: Armand Colin.

Egan, John. 1998. *Rethinking construction*. London: Department of Trade and Industry. http://www.constructingexcellence.org.uk/pdf/rethinking%20construction/rethinking_construction_report.pdf.

Egan, John. 2002. *Accelerating change*. London: Strategic Forum for Construction. http://www.strategicforum.org.uk/pdf/report_sept02.pdf.

Evetts, Julia. 2006. Short note: The sociology of professional groups: New directions. *Current Sociology* 54 (1): 133–143.

Gentry, Thomas A. 2009. Passive low energy housing: Paradox of behaviors. In *Proceedings of International Conference on Passive and Low Energy Architecture (PLEA)*, 45–50. Québec: Les Presses de l'Université Laval.

Gram-Hanssen, Kirsten. 2008. Consuming technologies—developing routines. *Journal of Cleaner Production* 16 (11): 1181–1189.

Guy, Simon, and Elizabeth Shove. 2000. *A sociology of energy, buildings, and the environment*. London: Routledge.

Harris, Michael, Paul Nightingale, Virginia Acha, Mike Hobday, Pari Patel, Alister Scott, Mike Hopkins, and Caitriona McLeish. 2006. The innovation gap: Why policy needs to reflect the reality of innovation in the UK. London: National Endowment for Science Technology and the Arts. http://www.nesta.org.uk/library/documents/Nesta%20Report%20TIG.pdf.

Hermelink, Andreas. 2006. A retrofit for sustainability: Meeting occupants' needs within environmental limits. In *Proceedings of ACEEE Summer Study on Energy Efficiency in Buildings*. Washington, DC: American Council for an Energy-Efficient Economy. http://aceee.org/proceedings-paper/ss06/panel07/paper10.

Hoffman, Andrew J., and Rebecca Henn. 2008. Overcoming the social and psychological barriers to green building. *Organization & Environment* 21 (4): 390–419.

Hughes, Thomas P. 1983. *Networks of power: Electrification in Western society, 1880–1930*. Baltimore: John Hopkins University Press.

Janda, Kathryn B. 1998. Building change: Effects of professional culture and organizational context on energy efficiency adoption in buildings. PhD diss., Energy and Resources Group, University of California, Berkeley.

Janda, Kathryn B. 1999. Re-inscribing design work: Architects, engineers, and efficiency advocates. In *ECEEE Summer Study Proceedings*. European Council for an Energy-Efficient Economy. http://www.eceee.org/conference_proceedings/eceee/1999/Panel_3/p3_11/paper.

Kelly, Michael J. 2009. Retrofitting the existing UK building stock. *Building Research and Information* 37 (2): 196–200.

Killip, Gavin. 2008. *Building a greener Britain: Transforming the UK's existing housing stock. A report for the Federation of Master Builders*. Oxford: Environmental Change Institute.

Koebel, C. Theodore. 2008. Innovation in homebuilding and the future of housing. *Journal of the American Planning Association. American Planning Association* 74 (1): 45–58.

Kohler, Niklaus, and Uta Hassler. 2002. The building stock as a research object. *Building Research and Information* 30 (4): 226–236.

Latour, Bruno. 2005. *Reassembling the social: An introduction to Actor-Network Theory*. Oxford: Oxford University Press.

Lowe, Robert, and Malcolm Bell. 2000. Building regulation and sustainable housing. Part 2: Technical issues. *Structural Survey* 18 (2): 77–88.

Marchand, Christophe, Marie-Hélène Laurent, Rouzbeh Rezakhanlou, and Yves Bamberger. 2008. Le bâtiment sans énergie fossile? Les bâtiments pourront-ils se passer des énergies fossiles en France à l'horizon 2050? *Futuribles* 343:79–100.

National Energy Foundation. 2007. Identification of renewable energy training provision, qualification accreditation. Milton Keynes, UK: SummitSkills. http://www.summitskills.org.uk/renewables/371.

Nösperger, Stanislas, Gavin Killip, and Kathryn B. Janda. 2011. Building expertise: A system of professions approach to low-carbon refurbishment in the UK and France. In *ECEEE Summer Study Proceedings*. European Council for an Energy-Efficient Economy. http://proceedings.eceee.org/visabstrakt.php?event=1&doc=5-516-11.

Office for National Statistics. 2010. Construction statistics annual. London: Office of Public Sector Information. http://www.google.co.uk/url?sa=t&rct=j&q=&esrc=s&source=web&cd=1&ved=0CDQQFjAA&url=http%3A%2F%2Fwww.ons.gov.uk%2Fons%2Frel%2Fconstruction%2Fconstruction-statistics%2Fno-11--2010-edition%2Fconstruction-statistics-annual-report-2010.pdf&ei=aoHUUPyhOqm70QX0q4D4AQ&usg=AFQjCNHTqjYB_ezivlc3hnFTH6D8Xv_ZOw&sig2=YNf9TJ_V-f4vQ138Y1-ClA&bvm=bv.1355534169,d.d2k&cad=rja.

Roudil, Nadine. 2007. Artisans et énergies renouvelables: une chaîne d'acteurs au cœur d'une situation d'innovation. *Les Annales de la recherche urbaine* 103:101–111.

Schiellerup, Pernille, and Julie Gwilliam. 2009. Social production of desirable space: An exploration of the practice and role of property agents in the UK commercial property market. *Environment and Planning. C, Government & Policy* 27 (5): 801–814.

Taylor, John E., and Raymond E. Levitt. 2004. Inter-organizational knowledge flow and innovation diffusion in project-based industries. CIFE Working Paper #089. Center for Integrated Facility Engineering, Stanford University. http://cife.stanford.edu/sites/default/files/WP089.pdf.

Ürge-Vorsatz, Diana, Ksenia Petrichenko, and Andrew C. Butcher. 2011. How far can buildings take us in solving climate change? A novel approach to building energy and related emission forecasting. In *ECEEE Summer Study Proceedings*. European Council for an Energy-Efficient Economy. http://proceedings.eceee.org/visabstrakt.php?doc=5-429.

WBCSD. 2009. Energy efficiency in buildings: Transforming the market. World Business Council for Sustainable Development. http://www.wbcsd.org/web/eeb/Energyefficiencyinbuilding.pdf.

Wilson, Charlie, and Hadi Dowlatabadi. 2007. Models of decision making and residential energy use. *Annual Review of Environment and Resources* 32:169–203.

Wolstenholme, Andrew, Simon Austin, Malcolm Bairstow, Adrian Blumenthal, John Lorimer, Steve McGuckin, Sandi Rhys Jones, et al. 2009. Never waste a good crisis: A review of progress since *Rethinking construction* and thoughts for our future. London: Constructing Excellence. http://www.constructingexcellence.org.uk/news/article.jsp?id=10886.

Zero Carbon Hub. 2009. Defining a fabric energy efficiency standard for zero carbon homes—task group recommendations. http://www.zerocarbonhub.org/resourcefiles/ZCH-Defining-A-Fabric-Energy-Efficiency-Standard-Task-Group-Recommendations.pdf.

3

LEED, Collaborative Rationality, and Green Building Public Policy

Nicholas B. Rajkovich, Alison G. Kwok, and Larissa Larsen

Introduction

While collaborative practices predate green building and the LEED (Leadership in Energy and Environmental Design) rating system, contractual relationships have traditionally established a hierarchical and adversarial process among the architect, owner, and contractor. More recently, the introduction of new standard contracts by the American Institute of Architects (AIA), and the emergence of new technologies such as Building Information Modeling[1] have helped to create a shift in practice. Most designers now recognize that they cannot produce a high-performance green building if they do not collaborate with other stakeholders. Collaboration is central to the design of green buildings.

This chapter discusses how use of the U.S. Green Building Council (USGBC) LEED rating system promotes collaboration during design, construction, and operations. This new type of practice—defined less by hierarchical, long-term patterns of routine behavior, or linear decision making—recognizes a diversity of interests, and the interdependence of design decisions, which leads to the innovation necessary to achieve a high-performance green building.

However, the LEED rating system is not a panacea to cure all ills. It is often perceived as an overly simplistic checklist, a limit to creativity, and too costly when imposed on a project as a requirement rather than as a voluntary process. To continue to shift the market to greener practices, the USGBC should engage local policymakers to promote a more transparent and collaborative process to revise building and energy codes in the United States.

We begin by describing the emblematic case of the Adam Joseph Lewis Center at Oberlin College in Oberlin, Ohio. This case illustrates how performance goals may slip if critical stakeholders are not included in early visioning sessions, despite the development of ambitious green goals at the onset. Then, we follow with a discussion of the LEED rating system and charrettes, and how these two processes can promote *authentic dialogue* and the formation of *collaborative rationality*. A second descriptive case example, the Chartwell School in Seaside, California, demonstrates how LEED can keep environmental issues tightly framed and project team members' goals closely aligned through the design, construction, and operations process. In the final two sections of the chapter, we encourage the use of similar collaborative practices as a foundation for state and local green building public policy and suggest related research directions.

Case 1: The Adam Joseph Lewis Center

In the early 1990s (prior to the development of the USGBC LEED standards), Professor David W. Orr was laying the foundation for an environmental studies center at Oberlin College in northeastern Ohio. Over the course of several years, he gathered students, faculty, staff, and members of the community to participate in a series of public planning sessions for a new building. During these meetings, they collectively developed a series of principles to help shape the design. One of their principles, to "utilize sunlight as fully as possible," may have been derived from the Hannover Principles,[2] and quickly became the focal point for the building's design (Orr 2004).

In the fall of 1995, Oberlin College issued a request for qualifications (RFQ) for an architect to lead the design process for the new 14,000-square-foot building (see figure 3.1). The building was to house classrooms, faculty offices, and a large auditorium for the college's environmental studies program at a cost of approximately six million dollars. Twenty-six architectural firms responded, and after interviewing five firms, the firm of William McDonough and Partners received the commission (Orr 2004, 162). As the design process unfolded, the architect and client agreed to use the sunlight goal as a primary focus. To operate within the building's "solar income" in Ohio required high levels of energy efficiency; a computer simulation helped the team to evaluate systems and to finalize design decisions. At the end of the design process, the architect and client were confident that a photovoltaic array would produce more electrical energy than the building needed.

After the project was substantially completed and occupied, the architect began to publicly claim that the building "makes more energy than it needs to operate" (McDonough 2005), a statement that captured the

Figure 3.1

View of the south facade of the Adam Joseph Lewis Center at Oberlin College. In the final design, engineers eliminated a vegetated trellis proposed in early design documents through a process of value engineering. Oak trees recently planted will eventually provide shading during the summer to reduce unwanted heat gain. Image © Nicholas B. Rajkovich 2011

imagination of the architectural community and garnered national attention. Unfortunately, an audit conducted by a member of the physics faculty at the college (Scofield 2002a, 2002b; Scofield and Kaufman 2002) and the National Renewable Energy Laboratory (Pless and Torcellini 2004) revealed that the building did not meet its proclaimed zero net energy goal.[3] This discovery caused negative publicity for the college and led to public attacks on the architect's character (Sacks 2008).

Beyond a simple miscalculation of energy loads, the key reason for the mistaken claim was a lack of integration and communication. Although the architect, client, and engineering team were in alignment about the solar income goal and the use of photovoltaics, the contractor held a "value engineering" exercise with college utilities staff late in the process that resulted in the installation of a low-efficiency electric boiler and the elimination of a vegetated trellis (Janda and von Meier 2005). The contractor had not participated in the initial visioning sessions and was unaware that altering the design of the building systems could challenge the entire "concept" of the building[4] (Orr 2004; Oberlin College and Lucid Design Group 2007).

The story of the Adam Joseph Lewis Center is not an isolated incident; similar discussions of the negative effect of value engineering and poor integration pepper green building magazines. At that time in the evolution of green building, the pressure to build green forced many architects to rethink their traditional ways of practicing and to include nontraditional stakeholders in the design process, and also to understand the interrelationship among design decisions. Meeting green goals that exceed the minimum requirements of building codes requires the integration of multiple systems (and subsequently the integration of their respective professions) and requires the tracking of building performance data after occupancy to demonstrate meeting green goals. While some firms had developed their own rules for such endeavors in the early 1990s, the majority of the market lacked a systemized green building approach. With the creation of the USGBC in 1993 and the development of the first LEED rating system in 1998, a new industry standard offered an alternative.

LEED Rating System

The mission of the USGBC is "to transform the way buildings and communities are designed, built and operated, enabling an environmentally and socially responsible, healthy, and prosperous environment that improves the quality of life" (2009, 2). By 1995, after reviewing existing green building rating systems, the USGBC began work on its own system to define and measure green buildings (Scheuer and Keoleian 2002, 16).

The first LEED Pilot Project Program, also referred to as LEED Version 1.0, was launched at a membership summit in August 1998. The current version of the system, LEED 2009, is described by the USGBC as "voluntary, consensus-based, and market-driven" approach (2008, xi). The system evaluates environmental performance using a checklist, and attempts to standardize what constitutes a green building in each phase of design, construction, and operation. There are several versions of the rating system available; versions have been released for rating new and existing commercial, institutional, and residential buildings. Each rating system is organized into five environmental categories: (1) Sustainable Sites, (2) Water Efficiency, (3) Energy and Atmosphere, (4) Materials and Resources, and (5) Indoor Environmental Quality. An additional category, Innovation in Design, addresses innovative and unique design strategies not covered under the five environmental categories, such as client educational programs or institutional measures or programs that extend the points of a given category. Regional bonus points are a recent addition to LEED and begin to acknowledge local conditions in determining environmental design priorities. Buildings are scored out of a possible total of 110 points; LEED 2009 certifications were awarded according to the following scale: "Certified" for 40–49 points, "Silver" for 50–59 points, "Gold" for 60–79 points, and "Platinum" for any project above 80 points (USGBC 2008).

The development of the LEED rating system as a one-page checklist with different point categories reflected the need to simplify the notion of what constituted a green building and provide a checklist that would encourage more holistic thinking. As Lindsey, Todd, and Hayter state, "Process is key to whole-building design. Sustainable design is most effective when applied at the earliest stages of a design. This philosophy of creating a good building must be maintained throughout design and construction" (2003, ii). Each credit in the LEED rating system is typically assigned to a member of the project team and the LEED checklist is then reviewed at each project meeting to determine progress toward certification.

One method that began to tackle the complexity of the LEED rating system, similar in concept to the visioning session that Orr held at Oberlin College, was to hold a LEED charrette—a gathering of all members of the project team to discuss and select LEED strategies for a proposed building. From this new collaborative process, a new specialization—the green building professional—developed to manage the process and assure LEED certification. Their role (called a LEED "Accredited Professional" or "AP" by the USGBC) was to "use strategic planning to overcome conflict . . . focus on both the big picture and the details of a project to produce collaborative agreement on specific goals, strategies, and project priorities. Charrettes establish trust, build consensus, and help to obtain project approval more quickly by allowing participants to be a part of the decision-making process" (Lindsey, Todd, and Hayter 2003, 1). In other words, the green building design charrette encouraged *authentic dialogue* among project participants. This authentic dialogue required each participant to express their interests in a manner that was comprehensible to others (Innes and Booher 2003).[5]

Comprehension among parties is important because collective deliberation on a problem is necessary for the formation of a collaborative rationality. Innes and Booher (2010) state that three conditions are required for a project team to form a collaborative rationality: a full diversity of interests among participants, interdependence of the participants, and engagement in face-to-face dialogue (35). Collaborative rationality results in reciprocity among agents, stronger interpersonal relationships, collective learning, and shared heuristics that transcend disciplinary boundaries (37).

For example, on a LEED project, a lighting designer might take responsibility for achieving credits for daylighting, and a mechanical engineer might manage cooling equipment specification, but both professions may align to negotiate a collective solution as they recognize their interdependence to lower the building's annual energy use. Rather than the owner or architect dictating solutions, other members of the design or construction team or both could actively control aspects of the design and construction process.

While the formation of collaborative rationality may have occurred in previous construction projects, the traditional contractual relationship recommended by the American Institute of Architects (AIA) between the architect and owner (AIA Contract B-141) and owner and contractor (AIA Contracts A-101 and A-201) largely imposed a hierarchical and potentially adversarial relationship among participants (AIA 1997, 2007a, 2007b). In recognition of this, the AIA began work in 2007 to alter the standard contractual agreements provided to their members to address what they called "the status quo of fragmented processes yielding outcomes below expectations to a collaborative, value-based process delivering high-outcome results to the entire building team" (AIA National and AIACC 2007, 1). The result was a new series of integrated project delivery (IPD) standard contracts to supplement B-141 and A-101, and a guide to assist in the delivery of a green building (AIA Document D-503) (AIA 2011).

The goals of the integrated project delivery approach are found in table 3.1. Although never stated explicitly, the practice of green building had embraced the concepts of collaborative rationality to drive the design and construction process. While there were many reasons for the emergence of this collaborative approach, part of which were the LEED rating system and design charrettes, the role of new forms of information technology, such as building energy simulation software, Internet collaboration tools, and the building information model (BIM), also played a crucial role. Together, these factors changed the creation and control of information, created new methodologies and rationales, and produced new power relationships among team members. In the end, the practice of "building" was altered as individuals realized "greener" buildings required more integration.

Table 3.1

Comparison of project delivery approaches (based on AIA National and AIACC 2007, 1)

	Traditional project delivery	Integrated project delivery
Teams	Fragmented; assembled on an "as-needed" basis; strongly hierarchical; controlled	Integrated; assembled early in the process; open; collaborative
Process	Linear, siloed; knowledge gathered "as-needed"; information hoarded; professions guard expertise	Concurrent and multilevel; early contribution of knowledge and expertise; open source
Risk	Individually managed; transferred if possible	Collectively managed; shared risk
Compensation	Individually negotiated; minimum effort for maximum return	Collectively negotiated; team compensation often tied to project performance
Communications	Paper-based; two-dimensional	Digitally-based; three- and four-dimensional Building Information Model
Reward	Encourages unilateral effort; limited learning from process; lack of innovation	Encourages multilateral collaboration; group learning; new heuristics and innovative design

Case 2: Chartwell School

The Chartwell School in Seaside, California, is an example of how the formation of collaborative rationality can lead to a high-performance green building (see figure 3.2). Douglas Atkins, the executive director of this small private grade school for students with learning disabilities began to investigate green building rating systems in 2002. The new school was to be constructed on a former U.S. Army base. The stated goal for the 55,000 square foot project was to create a green, healthy, and high performance school that would support a productive learning environment.

Figure 3.2

View of the main entrance to the Chartwell School. Large windows throughout the facility provide daylight to the majority of spaces, reducing heat gain from lighting and reducing electrical use. Image © Josh Partee 2011

As Atkins began the process, he sought advice from a friend who had completed a number of green buildings in the Northeast. He recounts that a friend suggested that he should "engage all the constituents at every level: students, parents, faculty, administrators, trustees, donors, and community leaders. He said we should interview everybody in order to create a vision about what constitutes a good school. He suggested we pose questions in such a way that they don't presume what the outcome is going to be" (Kwok et al. 2009, 8). In addition, the programming report for the building was created through a discursive process that valued input from people not generally included in the process. Atkins said, "We had the students draw pictures of what they thought a school should be. Some of those were whimsical, and some of them were very analytical; it was really amazing to see what they prioritized. We brought all this information together, and what came out of this picture was a programming document. It allowed us to establish a process of sending out the programming document as part of a broad RFQ" (ibid.).

According to Atkins, several dozen firms from across the United States responded to the RFQ because the project represented an opportunity to design what was intended to be one of the highest performing green buildings in the United States. From the responses to the RFQ, a committee shortlisted three firms, one of which was led by EHDD Architecture of San Francisco. At the interview, this team won the project because according to Atkins, "they worked collaboratively, from the very beginning of their presentation, in a way I hadn't seen in other presentations by other design firms" (Kwok et al. 2009, 9). In addition, when the lead architect Scott Shell presented to the school he brought the engineers with him. To Atkins, openly including the engineers "was another paradigm shift. I saw that architecture isn't, as it might have been a couple of generations ago, just about aesthetics, massing, and color selection. You need engineers to be part of the team as early as possible, in order to achieve those sustainability goals and show that your investment is going to perform in energy efficiency, air quality, and daylighting. They have to pencil out the solutions that can't be intuited, and they have to check that the theoretical performance is actually what you get when you walk into the building and flip the switches" (ibid.).

After the design team was under contract, they engaged in a series of design charrettes to determine the direction for the project. As with any project, power struggles emerged. These were generally resolved by members of the architectural firm EHDD. One of the principals of EHDD, Scott Shell, recounts:

> In design charrettes, there are a lot of different opinions expressed, and they can be expressed in a very animated and passionate manner. When trying to reconcile different views on how to accomplish something, egos can emerge. Then people get bruised, and things can happen that unravel the process . . . through these design charrettes, [we] were able to hear a cacophony of input. Where most people would melt down, or get frustrated, they would be very congenial and fun. They would challenge themselves by looking for ways to do what we were asking them to do. They showed us they could think out loud. They shared their thoughts. (Kwok et al. 2009, 12)

The design team consisted of a group of people who had worked together in the past, valued a nonhierarchical working relationship, and shared a common vision for the green features of the school. By having "an empathetic understanding of why another stakeholder would take a particular view" they were able to build up "reciprocal relationships that become the glue for their continuing work" (Innes and Booher 2000, 11). They learned that it is in their self-interest to work together (Innes and Booher 2003, 42).

As the team considered LEED certification, they realized that their alignment on issues had allowed them to achieve the high level of the Gold rating. According to Atkins, this allowed them to relax about meeting the minimum requirements of LEED, and "some of the more subtle benefits of sustainability started to creep into the process. At a certain point, by being able to stop thinking about the technical aspects of accruing points in a protocol, a wall goes down. Then, you have a permeable relationship with things that can happen which are not based on accruing points. You actually have an opportunity to go into some new territory, explore things, and come up with some solutions that may not have been tried before" (Kwok et al. 2009, 13). As with the Oberlin College case, the creation of a zero net energy building was one ambitious goal beyond the LEED requirements.

While the project was considered successful by all members of the design team and won several local, state, and national design awards, not all issues were resolved at the conclusion of construction. According to architect Scott Shell, "the occupant training on Chartwell was an area where we could have done better. It's easy for the design committee to get excited about the project, because they know all the details. We spend a lot of time together and they're all excited when they move in. Meanwhile, people move in who haven't been involved in that process, and they don't know what all this stuff does or why it was done a certain way" (Kwok et al. 2009, 28). One such person was the facilities manager, Roy Williams. Williams was not hired by the school until 2005 when the project had already broken ground. According to Williams, he "was disappointed in the 'as-built' drawings; they left a lot to be desired. . . . I also would like to have been able to edge in a few things a little earlier, before the design was locked down" (ibid., 73). However, Williams noted that energy bills from the facility indicate that the systems are largely performing as designed, and anecdotal evidence he has gathered indicates that the teaching staff is very pleased with how the building supports the educational mission.

Although the team exceeded initial expectations, the daylighting consultant to the Chartwell School project, George Loisos, cautioned that LEED may not be a solution for every site. "We need to create a situation where we identify what is important to the site. As we all know, each site is different, and LEED is struggling with that. Light pollution in New York is different from light pollution in a national park, and that's not the end of it" (Kwok et al. 2009, 69–70). In addition, he notes that the LEED system may not be sufficiently diverse to incorporate other issues such as social equity, and he laments that because these are not included in the system that project teams may not discuss them. But as someone who had also participated in the Oberlin College project, he states that although he complains about LEED, it has helped green building, has focused the dialogues on relevant issues, and has helped to advance the green building market. He states, "I complain about [LEED] to make it better" (ibid.).

In the Chartwell School example, collaborative rationality appears to have emerged, enabling authentic dialogue among team members with different interests, backgrounds, or professions. As with collaborative planning efforts, there must also be a diversity of stakeholders present early in the process who recognize the interdependence of their actions. Both conditions are necessary to take full advantage of the creativity that comes from collaboration (Innes and Booher 2010, 35). Finally, the process outcome depends on the group being able to follow a discussion where it leads rather than being artificially constrained by rules about what can be discussed or what cannot be changed (Innes and Booher 2003). The group needs to be able to challenge assumptions and question the status quo, a feature of LEED lost when it becomes a mandated building code.

LEED as Public Policy

As described by the USGBC, the LEED rating system is "voluntary, consensus-based, and market-driven" (USGBC 2009, xi). However, many public entities are beginning to encourage the use of LEED as an incentive for faster project review, a prerequisite for additional energy efficiency rebates, and a basic requirement for all new construction. In the move from a voluntary activity to a mandatory building code, the perception of LEED has significantly shifted and resulted in new criticisms of the rating system.

Part of the frustration voiced by the building industry is linked to what True, Jones, and Baumgartner (2007) call "the punctuated equilibrium nature of the building code process." This means that where building codes are required and enforced in the United States, it generally occurs on a three-year cycle. In many rural locations, there may not be regular meetings of a code committee or any local enforcement; revisions are tackled only as the need arises or if here has been an egregious violation of the health, safety, and welfare of the public. Because this political process is characterized by slow progress, stability, and incrementalism, any sudden departure from past practices—such as the sudden adoption of a green building standard—is likely to cause friction and resistance. Borrowing from Udall and Schendler (2005), the resistance to LEED takes several common forms, with two arguments that intensify if LEED standards become mandatory:

1. LEED costs too much. This argument states that the use of the LEED rating system is an unfair burden on building developers and contractors because the technologies required for green building are too expensive, are not widely available, and are often costly to install. This additional cost is above and beyond "standard practice," and therefore represents an unfunded mandate. Several studies (USGBC 2000; Kats et al. 2003; Eichholtz et al. 2010) offer a legitimate rebuttal, citing data from projects that show little to no incremental cost for LEED certification and significant economic benefits from certification. Hoffman and Henn (2008) discuss the technical and economic achievements that have reduced the complexity and cost of green building. Because it is difficult to conduct an "apples-to-apples" comparison of projects from location to location, building type to building type, and contractor to contractor, the likelihood of ever determining a universally applicable first cost or operating cost for all LEED building types is low. Quantifying what a structure would have cost without green features is possible, but it is not possible to verify the costs of a building that never existed. (For a survey of green building economic studies, see chapter 11, this volume.)
2. Overblown claims of green building benefits are misleading. This argument states that the health and productivity benefits of green building have been inflated, and should not be counted as a credit toward the first cost of a building. The difficulty in resolving this argument is similar to the cost debate.

While these first two arguments are valid criticisms of the LEED rating system, they concern the limits of economic modeling. One could argue that all positive externalities of green building should be credited toward the first cost of the building while another could argue that these externalities cannot be quantified and therefore do not relate to the cost of construction. It is unlikely that this issue can be resolved. In states such as California indicate where energy code revisions have occurred, the building industry continues to use a high first cost argument to stall code changes.

Bureaucracy of the LEED Rating System

Udall and Schendler's final argument against the LEED rating system focuses on the bureaucracy of the USGBC. It warrants further discussion because it relates to the punctuated equilibrium nature of the building code process and represents a negative reaction to the power of the USGBC and the sudden adoption of LEED by many jurisdictions. For example, take the following quote from the Pritzker Prize winning architect Thom Mayne. In the last decade he has completed a number of LEED certified buildings for the General Services Administration: "LEED should give performance requirements and let the architect solve the problem. . . . I'd much rather see BTU and CO_2 requirements and let the professional community solve the problem. If you give [prescriptive] requirements, it stagnates new development and research. . . . Because architects deal in creative problem solving, some of that will be curtailed by [prescriptive] systems" (Bowen 2009).

Part of Mayne's statement is a reaction to the loss of the architect's power over the design process, and part of his statement is a criticism of prescriptive requirements being imposed on architectural design. Under the LEED rating system, the power to decide how to meet green building performance requirements is transferred away from the architect. To Mayne, when LEED is no longer a voluntary process, it becomes a regulation like a building code, and puts an unnecessary constraint on the design process.

Next Steps for LEED

We believe that the USGBC should continue to work with practitioners to define the boundaries of green building. Part of the solution is to work on green building standards that cross-disciplinary boundaries such as ANSI/ASHRAE/USGBC/IES Standard 189, "Standard for the Design of High-Performance Green Buildings Except Low-Rise Residential Buildings." This standard, essentially a reformulation of the LEED rating system in language ready for adoption by code agencies, is intended to provide minimum requirements for the design of sustainable buildings "to balance environmental responsibility, resource efficiency, occupant comfort and well-being, and community sensitivity" (Dunlop 2009; ANSI, ASHRAE, USGBC, and IESNA 2009). By engaging the two main building engineering professional societies in the United States (ASHRAE, the American Society of Heating, Refrigerating and Air-Conditioning Engineers; and IESNA, the Illuminating Engineering Society of North America) the proposed standard and requirements will appeal to each of these professions because their respective fields collaborated in green building policymaking.

With these new formulations of green building requirements in hand, the USGBC can tap local volunteers to work with local, city, and state code officials and code councils to include these standards into state and local building codes. By engaging the local design community in a collaborative planning process, and modifying the standard to address local issues, the USGBC can engage more of the professional community in green building and receive valuable feedback to inform new versions of the system or green building codes. One caution from the collaborative planning literature is that this may take a lot of time and resources, but the net benefit may be greater engagement in policy formation and lasting, defensible results in the form of greener buildings.

Additionally, the USGBC should continue to develop supporting materials that describe the theory behind each credit, and the models used to support the achievement of each. Guy, Farmer, and Moore, in their review of the sustainable architecture literature, found that participants' diverse and divergent sets of ideas may challenge the development of a universally acceptable standard (Guy and Farmer 2001; Guy and Moore 2007). They suggest that rather than create a singular approach to green building, we should engage diverse approaches to build green. Recent initiatives by the USGBC to improve green building post-secondary and professional education, and efforts to go beyond teaching to the LEED rating system may help professionals build green buildings even if they do not pursue LEED certification.

This reflective engagement on the issues of sustainability would be sympathetic to the pragmatist tradition, promote social learning, and allow for a greater set of discourses to be considered. All of these approaches fit the mission of the USGBC, and will allow the USGBC to be critical of its own power and its singular approach to green building. One could imagine multiple formats of the LEED green building rating system, with equivalent performance metrics that do not specify credits or prescriptive requirements. While these approaches will not necessarily mesh with the original USGBC concept for a green building rating system, they can appease critics, reduce fragmentation in the field, and lead to a greater adoption of green building practices across the industry.

Conclusions and Directions for Future Research

This chapter began with the case of Oberlin College, where we argued that performance goals may slip if critical stakeholders are not included in early visioning sessions and a collaborative process. The chapter continued with a discussion of the LEED rating system and charrettes, and how these two processes can promote *authentic dialogue* and the formation of *collaborative rationality*. A second descriptive case example, Chartwell School, demonstrated how LEED keeps environmental issues tightly framed and project teams aligned on goals through the design, construction, and operations process. In the last section of the chapter, we suggested how green building design may influence adoption of new building codes and requirements, and offered potential new directions for the development and evolution of green building practices.

Green building practice is still early in its evolution. The unintended effects of governments adopting voluntary standards such as LEED may not be straightforward and constitute a fertile area for future research. With the introduction of the USGBC Green Building Information Gateway (http://www.gbig.org/), the performance of LEED buildings may be evaluated among other LEED projects as well as against projects where the certification was required as a local code or owner mandate (USGBC 2013). After controlling for variables such as the building type, climate, team chemistry, and delivery method, comparisons of key metrics such as energy use might reveal that those project teams that pursued LEED voluntarily had a higher level of performance. In addition, teams with better collaborative practices or chemistry may also achieve higher levels of performance. (See Korkmaz, Riley, and Horman 2010 for a list of variables that might be considered in the analysis.) Qualitative analysis of project documents and interviews with project participants might also reveal differences in framing environmental issues or approaches to LEED certification. This research could help to inform future versions of the LEED system, approaches to public policy, or improvements in the certification process.

Another research area is suggested by the work of Meyer and colleagues (see chapter 10, this volume). While their research followed the progress of one school district in pursuing high-performance building, it would be valuable to follow the development of green building advocates at the individual, organizational, and institutional levels (as suggested by Hoffman and Henn 2008). Combined with insights on the formation of collaborative rationalities, one could ask such questions as: If an individual gains experience working on green building projects, does her decision-making processes improve or does building performance increase? How do such advocates interact with and modify their organizations and institutions as they gain experience? Does alignment on environmental issues coalesce over time at each level of an organization? And does the construction of green buildings lead to changes in the culture of the organization as a whole? (Regarding the latter, see chapter 9, this volume.) Results from this type of research could help design and construction teams to improve management of green building projects, or help grassroots efforts to "go green" within companies.

Finally, questions emerge over the need for a new "systems integrator" profession that may be necessary to advance green buildings, such as low-carbon refurbishment of homes in the United Kingdom (see chapter 2, this volume). Since the introduction of the LEED program, the USGBC has promoted the development of LEED APs in a similar role for green building. But their knowledge and skills must go beyond technical aspects of green buildings and administrative aspects of the LEED process. As facilitators, these integrators must develop skills to engage in collaborative practices. This calls for a need to reshape post-secondary design and construction education. The development of such skills is critical to the development of future green building professionals and future green buildings.

Acknowledgments

We would like to thank Dick Norton, Rebecca Henn, and Andrew Hoffman for reviewing earlier drafts of this chapter and providing valuable feedback. We would also like to thank Karol Kaiser, the former Director of Education at the USGBC for ongoing research support and access to the Chartwell School narratives before official publication. Graduate students Britni L. Jessup, Kristen B. DiStefano, Amanda M. Rhodes, Rachel B. Auerbach, and Thomas D. Collins in the Case Study Lab at the University of Oregon helped with the collection of stories from practitioners and the transcription of the interviews used in this chapter. Josh Partee kindly provided permission to use an image of the Chartwell School. Finally, we appreciate the feedback early in the research process from John Forester and Richard L. Hayes. Without their support and the support of the Upjohn Research Initiative at the AIA, this chapter would not have been possible.

This work is supported by the National Science Foundation Graduate Research Fellowship under Grant No. DGE 0718128.

Notes

1. A building information model (BIM) is an object-oriented database that represents the physical and functional characteristics of a facility. BIM promotes collaboration by different stakeholders at different phases of the life cycle of a facility as they insert, extract, update, or modify information in the BIM (AIACC 2007).

2. The Hannover Principles were a series of sustainable development principles developed by William McDonough and Partners for the 2000 World Expo. The purpose of the document was to "insure that the design and construction related to the fair will represent sustainable development for the city, region, and world." One of the nine Hannover Principles stated that "human designs should, like the living world, derive their creative forces from perpetual solar income" (McDonough 1992, 2, 9).

3. A zero net energy building is "a residential or commercial building with greatly reduced energy needs through efficiency gains such that the balance of energy needs can be supplied with renewable technologies" (Torcellini et al. 2006, 1).

4. Although the Adam Joseph Lewis Center did not initially meet its solar income goal, an energy audit conducted by the National Renewable Energy Lab (Pless and Torcellini 2004) helped the college identify energy savings opportunities in the building to help achieve its zero net energy goal. Several subsequent retrofit projects have replaced inefficient building equipment, trained occupants, and additional solar panels have doubled total photovoltaic output. As a teaching tool, commissioning aid, and example for other zero net energy efforts, the college posts building performance data online: http://oberlin.edu/ajlc/. The building is one of eight buildings recognized by the U.S. Department of Energy (2008) as a zero net energy building.

5. The concept of authentic dialogue is borrowed from the Frankfurt School of critical theorists (Habermas 1981) and the notion of collaboration that underlies authentic public participation process is also applicable to green building.

References

AIA. 1997. *B141–2007: Standard form of agreement between owner and architect.* American Institute of Architects. Washington, DC: AIA.

AIA. 2007a. *A101–2007: Standard form of agreement between owner and contractor where the basis of payment is a stipulated sum.* American Institute of Architects. Washington, DC: AIA.

AIA. 2007b. *A201–2007: General conditions of the contract for construction.* American Institute of Architects. Washington, DC: AIA.

AIA. 2011. *D503–2011: Guide for sustainable projects, including agreement amendments and supplementary conditions.* American Institute of Architects. Washington, DC: AIA.

AIACC. 2007. A working definition: Integrated project delivery. AIA California Council. http://www.ipd-ca.net/IPD%20Definition.htm.

AIA National and AIACC. 2007. *Integrated project delivery: A guide*. Washington, DC: AIA, AIACC.

ANSI, ASHRAE, USGBC, and IESNA. 2009. Standard for the design of high-performance green buildings except low-rise residential buildings. American National Standards Institute, American Society of Heating Refrigerating and Air Conditioning Engineers, U.S. Green Building Council, and Illuminating Engineering Society of North America. Atlanta, GA: ASHRAE.

Bowen, Ted Smalley. 2009. The ArchRecord Interview: Thom Mayne on green design. *Architectural Record*. http://archrecord.construction.com/features/interviews/0711thommayne/0711thommayne-1.asp.

Dunlop, Jodi. 2009. ASHRAE, USGBC, IESNA partner on baseline standard for green building. ASHRAE. http://www.ashrae.org/news/.

Eichholtz, Piet, Nils Kok, and John M. Quigley. 2010. Doing well by doing good? Green office buildings. *American Economic Review* 100 (5): 2492–2509. doi:10.1257/aer.100.5.2492.

Guy, Simon, and Graham Farmer. 2001. Reinterpreting sustainable architecture: The place of technology. *Journal of Architectural Education* 54 (3): 140–148. doi:10.1162/10464880152632451.

Guy, Simon, and Steven A. Moore. 2007. Sustainable architecture and the pluralist imagination. *Journal of Architectural Education* 60 (4): 15–23. doi:10.1111/j.1531-314X.2007.00104.x.

Habermas, Jurgen. 1981. *The theory of communicative action: Reason and the rationalization of society*. Boston: Beacon Press.

Hoffman, Andrew J., and Rebecca Henn. 2008. Overcoming the social and psychological barriers to green building. *Organization & Environment* 21 (4): 390–419. doi:10.1177/1086026608326129.

Innes, Judith E., and David E. Booher. 2000. Collaborative dialogue as a policy making strategy. IURD Working Paper 2000–05, Institute of Urban and Regional Development, University of California, Berkeley. http://www.escholarship.org/uc/item/8523r5zt.

Innes, Judith E., and David E. Booher. 2003. Collaborative policymaking: Governance through dialogue. In *Deliberative policy analysis: Understanding governance in the network society*, ed. Maarten Hajer and Hendrik Wagenaar, 33–59. New York: Cambridge University Press. doi:10.1017/CBO9780511490934.003.

Innes, Judith E., and David E. Booher. 2010. *Planning with complexity: An introduction to collaborative rationality for public policy*. New York: Routledge.

Janda, Kathryn, and Alexandra von Meier. 2005. Theory, practice and proof: Learning from buildings that teach. In *Sustainable architectures: Cultures and natures in Europe and North America*, ed. Simon Guy and Steven A. Moore, 31–50. New York: Spon Press. doi:10.4324/9780203412800_chapter_3.

Kats, Greg, Leon Alevantis, Adam Berman, Evan Mills, and Jeff Perlman. 2003. *The costs and financial benefits of green buildings*. Sacramento: California's Sustainable Building Task Force.

Korkmaz, Sinem, David Riley, and Michael Horman. 2010. Piloting evaluation metrics for sustainable high-performance building project delivery. *Journal of Construction Engineering and Management* 136 (8): 877–885. doi:10.1061/(ASCE)CO.1943-7862.0000195.

Kwok, Alison G., Britni L. Jessup, Kristen B. DiStefano, Amanda M. Rhodes, and Rachel B. Auerbach. 2009. Case study: Chartwell School—stories of practice. In *Stories of practice*, ed. Alison G. Kwok. Eugene, OR: University of Oregon and USGBC.

Lindsey, Gail, Joel Ann Todd, and Sheila J. Hayter. 2003. A handbook for planning and conducting charrettes for high-performance projects. National Renewable Energy Laboratory Report NREL/BK-710–33425. http://www.nrel.gov/docs/fy03osti/33425.pdf.

McDonough, William. 1992. *The Hannover principles: Design for sustainability*. 5th ed. New York: William McDonough Architects.

McDonough, William. 2005. Ted Talks: William McDonough on cradle to cradle design. TED, http://www.ted.com/talks/william_mcdonough_on_cradle_to_cradle_design.html.

Oberlin College, and the Lucid Design Group. 2007. Adam Joseph Lewis Center for Environmental Studies at Oberlin College. http://buildingdashboard.net/oberlin/ajlc/.

Orr, David W. 2004. Can educational institutions learn? The creation of the Adam Joseph Lewis Center at Oberlin College. In *Sustainability on campus: Stories and strategies for change*, ed. Peggy F. Barlett and Geoffrey W. Chase, 159–176. Cambridge, MA: MIT Press.

Pless, S. D., and P. A. Torcellini. 2004. *Energy performance evaluation of an educational facility: The Adam Joseph Lewis Center for Environmental Studies, Oberlin College, Oberlin, Ohio*. National Renewable Energy Laboratory Report NREL/TP-550–33180. http://www.nrel.gov/docs/fy05osti/33180.pdf.

Sacks, Danielle. 2008. Green guru gone wrong: William McDonough. *Fast Company*. http://www.fastcompany.com/1042475/green-guru-gone-wrong-william-mcdonough.

Scheuer, Chris W., and Gregory A. Keoleian. 2002. Evaluation of LEED using life cycle assessment methods. National Institute of Standards and Technology Report NIST GCR 02-836. http://www.fire.nist.gov/bfrlpubs/build02/PDF/b02170.pdf.

Scofield, John H. 2002a. Early performance of a green academic building. *ASHRAE Transactions: Symposium* (June): 1–17.

Scofield, John H. 2002b. The environment and Oberlin: An update—one scientist's perspective on the Lewis Center. *Oberlin Alumni Magazine*. http://www.oberlin.edu/alummag/oamcurrent/oam_summer2002/feat_enviro4.htm.

Scofield, John H., and David Kaufman. 2002. First year performance for the roof-mounted, 45-KW PV-array on Oberlin College's Adam Joseph Lewis Center. In *29th IEEE Photovoltaic Specialists Conference*. New Orleans, LA.

Torcellini, Paul, Shanti Pless, Michael Deru, and Drury Crawley. 2006. Zero energy buildings: A critical look at the definition. In *ACEEE summer study*. Pacific Grove, CA.

True, James L., Bryan D. Jones, and Frank R. Baumgartner. 2007. Punctuated Equilibrium Theory. In *Theories of the policy process*, ed. Paul A. Sabatier, 97–115. Cambridge, MA: Westview Press.

Udall, Randy, and Auden Schendler. 2005. LEED is broken—let's fix it. *iGreenbuild.com*. http://www.igreenbuild.com/cd_1706.aspx.

U.S. Department of Energy. 2008. Zero energy buildings. U.S. Department of Energy and BuildingGreen, Inc. http://zeb.buildinggreen.com/index.cfm.

USGBC. 2013. The green building information gateway. U.S. Green Building Council. http://www.gbig.org/.

USGBC. 2000. *Making the business case for high performance green buildings*. U.S. Green Building Council. Washington, DC: USGBC.

USGBC. 2008. *LEED 2009 for new construction and major renovations rating system*. U.S. Green Building Council. Washington, DC: USGBC.

USGBC. 2009. *U.S. Green Building Council Strategic Plan*. U.S. Green Building Council. Washington, DC: USGBC.

4

Beyond Platinum: Making the Case for Titanium Buildings

Jock Herron, Amy C. Edmondson, and Robert G. Eccles

Introduction

In the first half of the twenty-first century, smarter, more sustainable buildings and cities have the potential to play a key role in shaping the global environment. In the years since the first Earth Day in 1970, the world's attentiveness to and technical grasp of environmental dynamics have both expanded considerably. Environmental progress in industrial countries has been made on a number of important fronts: rivers are cleaner; air quality is better; and cars, furnaces, and appliances are more energy efficient. Yet, critical gaps remain. In particular, progress is slow in the realm of buildings and construction, where there are major opportunities for substantial reductions in carbon emissions and other forms of pollution and waste.

One reason that so much unrealized potential remains in buildings is that practitioners in the fields of design and construction tend to conceive of buildings as objects rather than as dynamic entities created through a social process that endures over time. Buildings are seen as nouns, not verbs—things, not processes. The relationship between the design and construction process, on the one hand, and the operating life cycle of the occupied building, on the other, is thus typically misunderstood or ignored. Although both collaborative building practices and post-delivery performance auditing are filtering into the benchmarking standards of green buildings, we believe they deserve more attention because of the transformational contribution these practices can make toward improving both how buildings are built and how buildings perform over time.

Taking this broader perspective, we argue for a life-cycle-oriented framework for evaluating sustainable construction. To make this approach more tangible, we introduce the term "Titanium" as a rating category that could be awarded to buildings whose *actual* performance over time matches or exceeds the *potential* expected from the LEED Platinum rating. To begin with, we advocate evaluation of both the construction process and the audited, real-time performance of individual buildings over time. We also argue that the evaluation of new buildings must consider their relationships with other buildings and with the surrounding city. Building on existing concepts and prior research, we suggest that a building could earn the Titanium rating if it met three primary conditions:

1. Qualify for the leading category of building sustainability, currently the Leadership in Energy and Environmental Design (LEED) Platinum rating sponsored by the United States Green Building Council (USGBC).
2. Be constructed using a cross-disciplinary, cross-firm collaborative process, such as integrated project delivery (IPD), complemented by building information modeling (BIM) software that contractually and substantively incorporate diverse vendors into a more organic whole.
3. Have accurate tracking systems that make it feasible to monitor sustainability performance over the life of the building.[1]

As important as the design and engineering advances promoted by the LEED ratings have been, we believe that the key hurdles to green construction are primarily social and managerial, rather than technical. Examples of technical innovations in the industry include more efficient materials, better supply chain tracking, more sophisticated daylighting systems, and more robust building metrics and modeling tools. Social innovations such as better management techniques, more thoughtfully aligned incentives, and better and more integrative training lag behind the technical developments. In contrast, the automotive and aerospace industries provide informative examples of engineering-driven industries that have achieved substantial productivity gains through innovations in social and organizational practices to complement a variety of technical advancements. In building, however, engineering advances have outstripped organizational ones. We argue that a greater commitment to social innovation will be a prerequisite to establishing sustainability as a prevailing paradigm in construction and building performance.

The aim of our chapter is threefold: (1) to remind readers why *Constructing Green* matters to society as a whole, not just to those in the design and construction sectors; (2) to suggest that construction efficiency and actual operating performance of the occupied building are underemphasized in current sustainability rating methodologies; and (3) to promote a dynamic, longitudinal perspective on sustainable buildings. To be clear, promoting collaborative building processes and better building performance auditing are not by themselves novel ideas (see chapter 10, this volume). A number of promising initiatives promote the more rigorous monitoring of actual building performance through life-cycle assessments (LCA) and the development of dashboards like the Lucid Design Group's Building Dashboard, which visually monitors resource use in real time (see chapter 8, this volume). But relative to the scale of opportunity, the awareness of and progress in using such systems has been modest. The primary focus of current initiatives is on the energy-based retrofitting of existing buildings (Deutsche Bank 2008; RMI 2009) (see also chapter 2, this volume), a sector that plays a vital role in demonstrating the potential of more sophisticated resource management.

In the case of more collaborative design and construction processes, the American Institute of Architects (AIA) and other groups have promoted new contracting techniques to facilitate greater cooperation across disciplines (see chapter 3, this volume). However, implementation has gained limited traction. A more systematically urgent approach is needed to leverage technological developments into broadly scaled applications (see chapter 7, this volume). The central premise of our argument is that more collaborative building processes and more rigorous auditing of actual building performance should be positioned as key requirements rather than peripheral dimensions of green building. Despite good intentions and some promising, but voluntary, performance monitoring initiatives such as the USGBC's Green Building Information Gateway pilot, neither practice is widespread today.

Green Construction and the Social Influence of Ratings

Using energy consumption as a proxy for environmental footprint, buildings in the United States accounted for 40 percent of energy use in 2008, an increase of 50 percent from 1980 (Kelso 2011). A recent McKinsey study on global resource productivity placed improved building energy efficiency first in the top 15 of 130 opportunities for improving how scarce resources are used. Rethinking how buildings are designed, constructed and operated will yield new opportunities to reduce waste and improve sustainability. Construction is a logical focal point, because the industry, fraught with waste and inefficiency, remains a productivity laggard.

A number of factors contribute to construction inefficiencies. The industry is highly fragmented (U.S. Department of Energy 2008).[2] Designers, general contractors, engineers, and developers are the most visible participants, but on large projects there are dozens, even hundreds, of subcontractors, suppliers, and other specialists who are integral to the building process. For the most part, contractors typically move from site to site, modifying supply chains, working with new subcontractors, and adapting as necessary to the idiosyncratic demands of each location. Although large-scale residential developments and replicable retail and hotel facilities often take advantage of efficiencies provided by manufacturing-like production processes conducted on a common site, this approach remains an exception in the industry.

When LEED was first established by the USGBC in 1998, it was properly and understandably focused on potential building *outcomes*, not the *process* of construction itself or the *actual* performance of the building. Although the USGBC has advanced its LEED program on numerous fronts, including existing buildings, infrastructure, neighborhoods, and the voluntary monitoring of performance, it remains less focused on the process of construction and the mandatory post-delivery auditing of building performance. These two aspects of green building are critical to improving the overall sustainability of the built environment.

The USGBC's LEED rating system has contributed to a profound shift in awareness among architects, engineers, contractors, real estate developers, building managers, building occupants, manufacturers, government officials, and investors of the potential for more sustainable buildings (Turner Construction 2011).[3] Ratings matter! Ratings, along with associated third party certification, can motivate owners and the teams that work with them to create buildings that aspire to ambitious performance levels they would otherwise have failed to consider. Once widely adopted, rating systems like LEED help build a positive reputation for those who adopt them. Building on the positive impact of LEED, we believe that a higher level of sustainability can be achieved through a more ambitious, socially innovative, and dynamic approach to rating—one that explicitly rather than tangentially promotes collaborative design and more integrative construction processes as well as the actual, not simply the potential, sustainability performance of new buildings.

Beginning in 1990 with the Building Research Establishment's Environmental Assessment Method (BREEAM) in the United Kingdom, an array of environmental rating systems for buildings has been developed and deployed globally (Fowler and Rauch 2006; Kane and Allen 2010).[4] Although each of the certification protocols has benefits and deficits, we focus on LEED for several reasons. First, in little more than a decade, it has become the de facto standard in the United States for green building, increasing in stature and influence throughout the world. Second, with 170,000 accredited professionals (APs) around the world, LEED has had a striking impact on professional training and development. Third, there are now enough extant LEED buildings—in 2010 there were 32,210 registered and 7,552 certified LEED projects in the United States (Passive House Institute 2011; USGBC 2010)—that credible empirical research on the costs and benefits is increasingly feasible (see chapter 11, this volume). Fourth, LEED has developed into a highly regarded brand that has gained broad and increasingly global acceptance among developers, occupants, and investors, as well as engineers, architects, and other key service providers.

LEED certification has proven to be an impressively adaptive process that since its inception has successfully incorporated new ideas and increasingly broadened in scope. Committed to methodological transparency, LEED is constantly being improved. Criteria are regularly refined, the nature of different building types is being factored into the certification process, and efforts have been made to evaluate clusters of buildings (e.g., neighborhoods and building portfolios), not just individual buildings. LEED's shortcoming remains that the USGBC has focused primarily on the performance goals and attributes of the building upon completion, rather than the fuller life cycle of the building, perhaps because USGBC's core constituency includes but does not feature the ultimate owners and occupants of buildings.[5] In time, this may change, but for now we view it as a shortcoming.

Titanium: A New Category for Sustainable Building

To capture the spirit of our argument, we propose the category "Titanium" to symbolize a new and aspirational level of sustainability for buildings. In our framework, Titanium status would be assessed by evaluating the design and construction process, the sustainability potential of the completed building on opening day (which LEED does very well), and the actual (not just the anticipated) performance of the building over time. We chose the term "Titanium"—which refers to a material that is abundant, strong, inexpensive, and light—to symbolize ambitious performance goals. Indeed, titanium has a higher strength-to-weight ratio than any metal on earth and—consistent with our emphasis on collaboration—alloys easily with other metals. The term titanium also references the metal-based LEED ratings of Silver, Gold, and Platinum, but shifts the implicit emphasis from "better as more rare and expensive," to "better as lighter, stronger, and more efficient." In so doing, our intention is to reframe green building from a frame rooted in expensive objects to one rooted in robust process and performance.

In focusing on the full life cycle of a building, Titanium can accommodate the interests of its long-term owners, property managers, and occupants, as well as the designers, engineers, and contractors who are key players in the earlier life of a building. As a result, we believe our proposed Titanium standard is more than an extension of LEED Platinum: it is a fundamentally new category.

Before exploring the three principal criteria for achieving Titanium, it may be helpful to frame our proposal with some preliminary comments about the life cycle of buildings and the interests of different stakeholders over time. To do so, we consider the phases of a building's life from concept to construction, through occupancy, and ending in demolition.

In promoting green construction, it is essential to appreciate that one size does not suit all. As the evolution of LEED reflects, different building types warrant different assessment techniques and standards. A climate-sensitive hospital has fundamentally different performance characteristics than an unrefrigerated warehouse. It is easier to benchmark the relative performance of a retrofitted existing building than an entirely new building, because the baseline data of the existing building is "real" while that of the new building is "theoretical."

We focus on new commercial construction, such as an office building, because it encompasses the full life cycle of a building from original concept to ultimate demolition. However, our advocacy for more collaborative production processes and the post-delivery performance auditing of buildings applies to retrofitting existing buildings as well. Indeed, a compelling case can be made that retrofitting will have a far greater sustainability impact than efficient new construction (see chapter 13, this volume).

Sustainability practices must take a building's entire life cycle into consideration. An overview of a building's life cycle and the principal stakeholders is shown in table 4.1. In most cases, stakeholders differ from phase to phase, especially once the building is "delivered." Within phases, different stakeholders have different objectives (e.g., financial return, reputation, quality, and risk mitigation), different time horizons (e.g., from very short to, in relatively rare cases, the full life cycle of the building), and different levels of authority (e.g., owner/high authority, small subcontractor/low authority). Traditional contracting practices and associated liability concerns tend to re-enforce stakeholder silos, although a tendency

Table 4.1

The life cycle of new buildings and the key stakeholders within each stage

	Concept	Construction	Occupancy	Demolition
Time frame	1–5 years	2–3 years	30–50 years	1 year
Eco-footprint	Very low	Substantial	Very substantial	Modest
Stakeholders	Developer Architect Engineer	Developer Architect Engineer Contractor Subcontractors Supply chain Construction lender Permitting authority Neighbors	Owner(s) Lenders Occupants Building manager Neighbors	Owner(s) Contractor Reverse supply chain Permitting authority Neighbors

toward repeat partnering by particular architects, engineers, and contractors softens the boundaries between them (Eccles 1981). The phases in a building's life are typically loosely coupled, creating little "institutional" memory or accountability across the full life cycle of the building. We propose that an effective and enduring approach to green construction is one that creates resilient links between each of the phases and, to the greatest extent possible, reconciles the various interests and time horizons of different stakeholders.

Because of their critical influence on a building's long-run sustainability, the *design* and *construction* (or building) phases warrant particular attention. In general, the construction industry is fragmented, craft-based, guild-oriented, and markedly inefficient. On most projects, authority is shared and exercised through influence across group boundaries, rather than through centralized control. Because the interests of collaborating parties are often not well aligned, actions that might prevent process failures are rarely identified in advance. Unpredicted but inevitable bottlenecks—created by supply chain, transportation, labor, and weather problems, as well as by ambiguous building code interpretations—are common. Although studies differ on how to calculate productivity in the construction industry, there is a general consensus that construction has enjoyed considerably fewer productivity gains than other similar industries (Chapman and Butry 2008). As elaborated later, new software tools, along with process innovations to improve collaboration across project phases, have tightened links between architectural design and construction engineering, planting the seeds of future productivity improvements (Eastman 2011).

The occupancy phase, typically decades but up to hundreds of years, constitutes the bulk of a building's life. This is where the building can be seen to perform its basic function. As important as it is to minimize waste during construction, managing the cumulative use of resources over the life of a building is thus an even greater determinant of sustainability. One challenge is stakeholder turnover. Developers typically hand new buildings over to owners or occupiers, who then powerfully influence the building's sustainability performance. Sophisticated control technologies, environmentally sound materials, and shrewd siting of the building decided during the design and construction phase are essential to a building's sustainability, but they do not ensure the intelligent management of a building over its life. Designing a building to accommodate unanticipated new technologies and creating incentives for occupiers of a building to monitor and use resources efficiently are thus illustrative priorities. Finally, the *demolition* phase, although brief in duration, is also vital to sustainability performance. Attention must be paid to the recyclability of building materials and to reducing toxic chemicals and the resources required to prudently dismantle the building.

Titanium's Three Elements

The Titanium classification represents a next generation for sustainable construction and building management certification. As summarized earlier and discussed in more detail in this section, we propose three defining "elements" for achieving Titanium: (1) a LEED Platinum rating; (2) formalized collaborative project management and building information modeling (BIM); and (3) life-cycle auditing of building performance. This systemic approach to assessment would make green construction an increasingly robust proposition (see table 4.2).

1 LEED Platinum Rating

Clearly, a Titanium building should first satisfy the requirements of a broadly accepted benchmarking standard for building sustainability. Although there are other candidates, LEED's highest certification category, Platinum, is broadly accepted as the current "state of the art" for

Table 4.2

Three elements of Titanium buildings

Phase	Building process	Building product	Building performance
Element	CPM/BIM	LEED Platinum	Auditing
Key goals	Efficiency Waste reduction Time compression Sustainability Improved design	Performance potential Sustainability Reputational value	Validation of potential Enable ongoing remediation
Core attributes	Collaborative decision making IT-supported designing Risk sharing	Transparent rating Point-in-time estimate Up-front costs	IT-enabled diagnostics Ongoing/iterative Enabled remediation
Core activity	Teaming	Certifying	Learning

assessing the sustainability potential of a new building. We view this as a necessary but insufficient condition for assessing a building's sustainability performance.

Completed in 2000, the Chesapeake Bay Foundation Headquarters in Annapolis, Maryland, was the first building to achieve Platinum status. The project demonstrated the impact that an enlightened, mission-driven client can have in elevating the standard of sustainable building performance. Ironically, the operational problems of the building—a roof allegedly weakened by poorly treated "green" lumber—highlight the importance of full life-cycle performance assessment, given that the building has failed to perform as anticipated (*Chesapeake Bay Foundation vs. Weyerhaeuser Company* 2011, 1–22). Taking a broader approach to assessment, our concept of Titanium would, therefore, require an ambitious client dedicated to playing a leadership role. A pioneering client seeking to obtain Titanium status would help elevate the goals and standards of green construction into a new, more comprehensive realm.

2 Collaborative Project Management and Building Information Modeling

Next, we propose that a form of deeply collaborative project management (CPM) become an added requirement for the sustainable building. Integrated project delivery is an increasingly well-recognized example of this kind of intensive collaboration, which involves all players in a design and building project from the outset, including the signing of a single legal contract (see chapter 3, this volume) (Edmondson and Rashid 2011; Yi, Ramirez, and Bendewald 2010). This collaborative approach must also be supported by an acceptable version of building information modeling (BIM) software. The combination of IPD and BIM constitute a social "forcing mechanism" that tightens relationships and more fully aligns incentives among the diverse stakeholders involved in developing, designing, constructing, and, in some cases, even occupying a new building.

IPD[6] has emerged over the past decade in the United States as a promising, albeit infrequently used multidisciplinary approach to linking, monitoring, and compensating diverse participants on a building project based on the overall performance of the project itself and not just the slice of work performed by separate experts (Mathews and Howell 2005). As the AIA formally defines it, "IPD is a project delivery approach that integrates people, systems, business structures and practices into a process that collaboratively harnesses the talents and insights of all participants to reduce waste and optimize efficiency through all phases of design, fabrication and construction" (Cohen 2010). The operating principles of integrated project delivery include:

1. Early and substantive participation by co-located principals—in particular the architect, engineer, builder, and project owner/developer.
2. Joint project management with risk sharing between principals and incentives crafted to align the interests of each party toward common goals including performance standards.
3. Contractual clarity about roles and responsibilities.
4. Non-litigious dispute resolution (Cohen 2010).

The British Defense Ministry produced a study of construction procurement estimating that the use of IPD principles could save as much as 30 percent of the initial cost of a project by integrating supply chain management directly into the design process (UK Office of Government Commerce 2003, 1–24). Despite that potential, the adoption of IPD has been slow, in part due to the initial learning curve required of interested parties. Interest is growing, however. In 2010, almost twenty thousand sample IPD contracts were downloaded from the AIA website and in a follow-up survey, nearly a quarter of those downloading them noted that they were using some form of IPD (Roberts 2011). Unlike LEED, IPD is not an off-the-shelf set of protocols. It is a contractually flexible process that incorporates the silo-softening attributes identified earlier (American Institute of Architects 2007).[7] The objective is to integrate the obligations and contributions of diverse parties by explicitly linking their roles to a clear common purpose—less wasteful and, hence, greener construction. The scope and complexity of each project will determine the scope and complexity of the IPD contract.

Qualitative case research has found that IPD promotes efficiencies in project duration and reduced waste because it codifies the rules of interaction among participants, makes procedures transparent, facilitates mutual understanding of other disciplines, and builds trust (Edmondson and Rashid 2012). Moreover, signing a single contract that precludes suing collaborating companies on the project is integral to how IPD has been conceptualized. By mitigating the threat of litigation, a primary source of mistrust in high-risk projects, IPD encourages participants to solve problems rather than identify culprits to blame for them. As a result, the design and building process is likely to be characterized by swift corrective action, collaboration, and mutual learning that can also be transferred to subsequent projects. While IPD will not cure the inherent fragmentation of the construction industry, it can be a source of cost savings that will lower the barrier to achieving LEED Platinum status.

These benefits are offset by challenges to implementing IPD. In particular, IPD is an unfamiliar way of working—one that involves new contractual relationships among participating firms. Another challenge is ensuring that net gains in efficiency are properly shared, which is made more complicated due to new responsibilities and the altered costs of working that accompany IPD arrangements.

While IPD contracts may be tailored to particular projects, they are supported by BIM, an emerging software application that provides a standard but flexible platform to track developments across both disciplines and time. As such, BIM helps project professionals to integrate design, engineering, and construction activities in a transparent, team-oriented way by offering a robust set of tools that allow specialists with different perspectives—design, engineering, scheduling, costing, and performance tracking—to work collaboratively within a framework that is simultaneously dynamic (e.g., can accommodate change) and stable (e.g., tracks and updates change).

A number of BIM benefits have been identified that support and complement IPD. Examples include (Eastman 2011):

- The development of an original building model that broadly relates design, cost, and functionality—including sustainability features—and ultimately informs the detailed design.
- Tighter, real-time links between design and cost projection and procurement.
- Improved modeling of sustainability factors such as energy efficiency, water usage, and other resource considerations.
- Early detection of potential design errors and omissions, code violations, and obstacles to long-run sustainability performance.
- Better coordination among design, construction, and procurement to minimize waste, supply chain kinks, inventory control, and people management.
- More efficient commissioning of the building due to more systematically catalogued data on the building itself (including benchmarking to similar buildings).
- Improved operational efficiency of the completed building as BIM provides a natural starting point for monitoring ongoing building performance.

It is important to recognize that these benefits will be more surely realized if BIM methodologies are broadly accepted across the construction industry, the allied professions of architecture and engineering, and the many links along the procurement supply chain (Bernstein 2010, 16).[8] There are network effects to BIM adoption. A broader acceptance of BIM as a shared platform with a commonly understood language would extend the user network further, thus increasing the likelihood that the construction process itself—along with the completed building—will become more efficient, and consequently, more "green." Establishing the use of CPM (for example, IPD) and BIM as an element of the Titanium standard could accelerate and broaden the adoption of both in the design and building industry.

3 Life-Cycle Auditing of Building Performance

The third element we include in the proposed approach to Titanium certification is an assessment of ongoing environmental performance. This assessment would be generated by a dashboard-friendly operating system that facilitates credible, sensor-based, cost-efficient, real-time sustainability audits of actual building performance. A building would earn Titanium status with the inclusion of such an operating system, but it would only retain the status if it actually performed according to plan over its full life. In other words, Titanium status could be rescinded if a building did not perform as expected. It is important to emphasize that although technology is an enabler, a set of necessary tools, it is insufficient without the socially engaged and even self-interested participation of occupants and building managers (see chapter 8, this volume).

Over the past several decades, companies like Honeywell and Johnson Controls have made major improvements in controlling both energy and water use within buildings and in tracking "building performance." The combination of wireless computing and the prospect of comprehensive sensor-based tracking of building performance could catalyze a step change in the real-time management of building environments. It also creates an opportunity for a "reality check" on buildings certified as sustainable by LEED and other groups. Do these buildings perform as expected? Good intentions aside, do they deserve their ratings over time? Who is on the hook if the building fails to perform as expected? The old adage that "you can't manage what you can't measure" applies (Lucid Design Group 2011).

For a social rather than technical tool to make performance measurement over the life cycle of the building more consequential, we would add some version of performance-based contracting for the key principals involved in the design and construction of the building. As appropriate, service providers (especially architects, contractors, and engineers) should have some of their fees/profits contingent on the sustainability performance of the building over some initial period of occupancy. This presents challenges as actual building performance would depend in no small part on how the building was used, which would be largely outside the control of the original architects, engineers, and contractors. However, if parameters were chosen that aligned well with the interests of the owners and occupants of the building (e.g., energy usage) then there would be incentives to operate the building as efficiently as the original design intended.

The retrofitting of existing buildings has generated what may be the most innovative developments in performance-based sustainability contracting. For example, the Empire State Building was renovated using energy-based contract contingencies (Navarro 2009). While performance-based contracting is typical in industries like investment management and manufacturing, it remains rare in new building construction. However, outcome-based risk sharing and commonly aligned incentives would be a natural extension of IPD and BIM and, as such, they would keep collaborators focused on what matters most in sustainable construction: how the building actually performs over its entire life.

The longer-term orientation through a life-cycle approach will also change the frame of reference for how owners and developers view the economics of the building, another factor which can lower the barriers to seeking a LEED Platinum rating. Higher initial construction costs would be offset by verifiably lower operating costs over the useful life of the building. The sooner the net present value of these cost savings exceeds the construction premium, the greater the incentive for pursuing Titanium.[9] In addition to direct cost savings from operating performance efficiencies, sustainability goals might be promoted by government tax or other incentives designed to accelerate the adoption of the Titanium rating in building construction.

Although LEED Platinum certification establishes the sustainability *potential* of the building, neither the efficiency of the construction process nor the audited performance of the completed building are addressed directly. By explicitly setting sustainability as a project goal and by increasing the transparency of collaborative decision making, the use of IPD and BIM increases the likelihood that the construction *process* itself will be "green." That said, it takes a robust, user-friendly monitoring system to establish that the ongoing *performance* of the building meets the sustainability criteria set for it, and to provide remediation clues as necessary.

Translating ideas to practice is never straightforward. We focus next on implementation strategies and the related question of professional development. We conclude with some thoughts about extending our framework to the broader landscape of the city in which buildings are situated.

Toward Titanium Certification: Issues of Implementation

Relying, as Titanium does, on social innovations to realize what are largely, but not exclusively, technical outcomes, successful implementation requires that particular attention be paid to the organizational side of green construction. Outlined below are six steps that could accelerate and sustain the adoption of Titanium:

1 Institutional Sponsorship

Titanium needs an institutional sponsor. One obvious candidate is the USGBC, which has been successful in establishing LEED as the premier environmental rating system for buildings. The USGBC has thus demonstrated how to translate a compelling idea into a constantly improving and broadly scaled reality. It has the capabilities and credentials to adopt the LEED rating system to incorporate more explicit links between the design, construction, and operating phases of a building's life cycle. As mentioned earlier, however, the shortcoming of the USGBC as a Titanium sponsor is that its focus has been on the "building" of buildings, rather than on operating or owning them. However, that more narrow perspective might be addressed by an alliance with the Building Owners and Managers Association (BOMA), an arrangement that would encompass the fuller range of constituencies and capabilities in a building's life cycle. There are other sponsorship candidates as well. McGraw-Hill,[10] which owns S&P, is a longstanding leader in building information and either has or could easily acquire the expertise to independently rate buildings (Bernstein 2010). The U.S. government has had successful award-based initiatives, like the Malcolm Baldrige National Quality Award and the Energy Star program for efficient appliances, and could do the same in promoting Titanium. The AIA, perhaps in alliance with the American Society of Landscape Architects (ASLA), or the American Society of Heating, Refrigerating and Air Conditioning Engineers (ASHRAE), or both, could sponsor the Titanium rating. The same is true for a large and well-respected foundation (e.g., Pew, Gates, MacArthur, or the Energy Foundation in San Francisco) with a long time horizon and a commitment to sustainability.

2 A Pilot Project

The Titanium rating needs a pilot project with an enlightened and influential developer and owner willing to take the risk of building and then assessing the building in this way, and sharing what is learned in the process. Ideally, subsequent projects, reflecting learnings from the pilot, would follow. Several such examples—the Genzyme Corporation, Autodesk, the Chesapeake Bay Foundation, and the Bullitt Foundation (Nelson 2011)—have earned coverage in the popular media. An environmentally oriented foundation, a serial builder such as a large, prominent university, or a corporation committed to a sustainable strategy all present natural candidates to lead a pilot project.

3 The Scaling Strategy

Truly significant sustainability gains in the construction and operation of buildings can only be achieved at a large scale. However inspiring individual projects might be, their impact will be marginal unless there is a clear path to leverage good ideas and individual experiences into generalized solutions that can be easily applied to all buildings. We do not wish to imply that solutions for all of the anticipated challenges should be in place before moving forward. In fact, the more complicated the early steps are for rolling out such an initiative, the less likely the early Titanium projects are to be successful. The potential for long-term success will be increased by keeping conditions as simple as possible at first (e.g., limiting the scope of CPM agreements; giving leeway in how BIM is used; deferring the use of performance-based fees) and then refining them over time as more is learned about the process of achieving Titanium status. Another option for promoting widespread adoption is government-mandated building regulations (see chapter 15, this volume). While the logic of regulation is compelling, especially in achieving energy independence, the politics of environmental regulation have proven to be daunting.

4 Continuous Learning

Uniformly supported in principle but too often underemphasized in practice, a commitment to "continuous learning" is a defining attribute of concepts that successfully scale into broadly applied solutions. The elements of Titanium are linked sequentially, and move across three phases of activity, which we simplify in figure 4.1, as a shift from *teaming* to *certifying* to *learning*, which, in concert, facilitate continuous improvement.

Over time the learning that takes place in the occupancy phase of the building will inform and improve the design and construction phases of future buildings as well as the operation of existing buildings, thereby improving teaming across disciplines—a process of collaborating in the absence of stable teams—which will, in turn, improve the certification process and building performance.[11] Thus, we can modify figure 4.1 to depict a continuous learning model as shown in figure 4.2.

In such a system, continuous learning would take place in forums that include conferences, community charrettes, academic journals, the so-called "invisible colleges" of professional practice, case studies, and the informal transfers of knowledge between close colleagues and less familiar acquaintances.

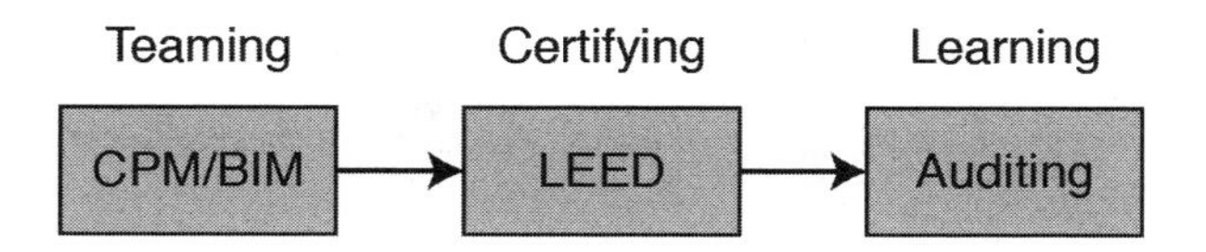

Figure 4.1

Linear connection between elements of the Titanium building standard

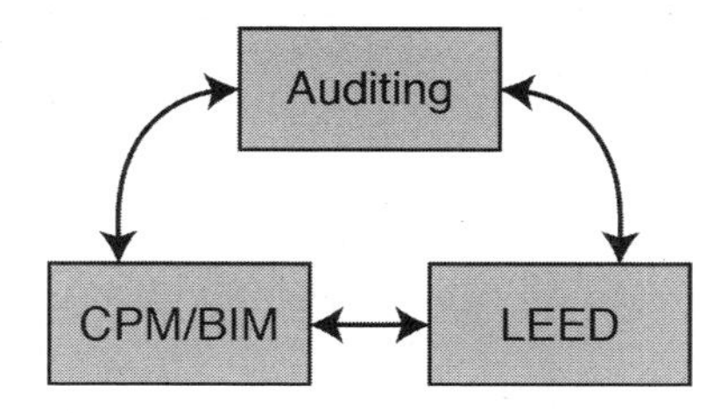

Figure 4.2

Dynamic relationship between elements of the Titanium building standard

5 Updating Required Capabilities

Although it is useful to begin with a set of specific capabilities to get the Titanium rating launched, it is essential to have feedback loops in place to update those capabilities in light of hard-earned experience. It is difficult to anticipate what new capabilities might emerge, but it is clear that a collaborative, transparent, flexible process can ensure that refinements are made continually.

6 Professional Development and Training

A key driver of LEED's growth has been the striking increase in LEED professional accreditation primarily among architects, engineers, and the allied professions, giving rise to a mutually reinforcing cycle propelled by the increase of LEED-accredited professionals (Roberts 2010). A decade ago LEED professional accreditation was comparatively rare. In 2012, it has become prevalent. What LEED professional accreditation has lost as a professional differentiator, it has more than gained as an expected accomplishment, a "must have" rather than a "nice to have." Titanium adds the elements of building process (CPM/BIM) and building performance (LBPA), each of which requires new domains of professional training.

In particular, Titanium APs would need a more comprehensive knowledge base than LEED APs. In addition to being LEED accredited, they would have to understand the contractual complexities of CPM, the software capabilities of BIM, and the monitoring requirements, options, and remedial implications of building performance auditing, including sensor-based climate control systems. They would need a deep knowledge of building commissioning, building life-cycle analysis, organizational management, and—more broadly—systems thinking. Similar to commissioning under LEED but involving far wider scope, Titanium certification puts a premium on team building and management skills. Teaming thus becomes an essential aspect of designing and building for sustainability and an essential aspect of the Titanium certification.

The Relationship between Building Titanium and Sustainable Cities

Ultimately the concept of Titanium may even be applied at the level of the city. The first aspect, the use of both process and product in rating buildings, explicitly recognizes that buildings are verbs, not nouns. Buildings are dynamic systems. The Titanium category could be extended over time to assess how fluidly a building connects to other buildings, the white space between them, the infrastructure supporting them, and, ultimately, the hinterland servicing them (Forman 2008).[12] Is the building designed, or redesigned (retrofitted), to make it easier to operate efficiently in partnership with other buildings and urban management systems? If a smart grid is introduced, is the building wired to benefit from it? Are rents and other charges contractually aligned to share benefits offered by smarter urban infrastructure? By evaluating real-time processes, as well as point-in-time conditions, the Titanium standard offers a richer set of evaluation tools to assess particular buildings and makes it easier to learn how buildings actually operate (Brand 1995).

The second aspect of Titanium, collaborative project management, incorporates techniques such as transparent, real-time decision making, risk sharing, and IT-supported project costing and tracking, which are designed to help diverse groups work collaboratively with a common purpose on complex projects. At the moment, implementing these attributes in construction projects remains a challenge. Over time, however, the successful use of CPM on particular buildings is likely to make it easier to incorporate a progressively broader design scope into the process. In this way, the professions will learn to consider how a building relates to other buildings and to the city as a whole.

The third aspect—robust, real-time auditing of actual building performance—is especially critical as new technologies, especially sensors and wireless communications, create a phase change in the quality, cost, and availability of environmental monitoring. Titanium promotes these technologies as both a diagnostic tool providing a reality check on the promise of building performance and as a remediation tool to improve building performance over time. These new technologies will contribute in important ways to more efficient cities. Infrastructure—bridges, streets and highways, interchanges, parking facilities, utilities, and reservoirs—equipped with sensors will make it easier to track the urban flow of people, cars, energy, water, and other resources and to design incentive systems to optimize those flows. Integrated software platforms—for example, the Living PlanIT Urban Operating System—will incorporate a wide range of real-time data and facilitate the comprehensive monitoring and management of cities (Living PlanIT 2011).

Over time, research is needed to evaluate the performance interdependencies that exist among buildings, some of which are positive and should be leveraged further and some of which are negative and need to be mitigated. An example of complementary interdependencies would be buildings that use electrical power from the same grid but do so at different times. Much like the diversification benefits that come from negatively correlated securities in a portfolio, a building whose energy use is countercyclical to other buildings (i.e., off peak) improves the overall energy efficiency of the city as a whole. An example of a negative interdependency would be a building designed to reflect light in order to lower air conditioning costs and that reflects that light onto another building, thereby making it hotter and *increasing* its air conditioning costs. In this case, the first building has created an externality on the second building and an assessment must be made of how to reduce the air conditioning costs of both buildings taken together. Recognizing interdependencies could serve as the basis for providing different incentives—such as tax abatements, infrastructure commitments, and zoning concessions—to create a more sustainable city.

Conclusion

The promise of a fuller, lifecycle approach to rating a building's sustainability is predicated on our fundamental premise that social and organizational innovations are central to the next generation of green construction. Relying on technical progress alone is inadequate and shortsighted.

Developing a new sustainability category, such as our proposed category of Titanium, provides a concrete example of a social innovation that we believe could contribute meaningfully to the design and management of smarter buildings, and even promote the larger prize of more sustainable cities. The concepts of green construction and smarter, greener cities are closely related and can be developed further in an integrated way. To date, very different professional and academic communities have focused on these two topics (buildings and cities), with limited dialogue among them. Green construction has tended to be the domain of designers, environmental and civil engineers, and BIM software firms. The locus of interest in "smarter" cities has tended to be hardware firms that develop sensors and controls, as well as in software firms focused on urban management, while those focused on "greener" cities have tended to be landscape architects and ecologically oriented urban designers and architects. These groups need a social bridge constructed with well-aligned incentives, a unifying common language, and shared goals and time frames to realize the potential of particular technical innovations. Cross-disciplinary concepts like Titanium incorporate organizational and social strategies to address sustainability opportunities that are framed too often and too narrowly in purely technical terms. Broader and more inclusive perspectives are essential to the green construction of genuinely green buildings and ultimately to the development of smarter, greener cities in the future.

Acknowledgments

We gratefully acknowledge Sydney Ribot for superb research and editing that contributed immensely to the development of this chapter.

Notes

1. It is important to note that sustainability calculations (e.g., average kilowatt hours used per square foot for a discrete period) are easier to calculate than interpret (e.g., exogenous events such as unusual weather patterns could distort calculations) so judgment is needed to assess data. Further design and research are needed on performance-based contracting that rewards and penalizes key stakeholders (e.g., designers, engineers, lead contractors, and developers) based on whether the building performs as expected. This, too, presents managerial, legal, and organizational challenges as results may well be contested despite initial efforts to smoothly align stakeholder interests with overall building performance.

2. Although the United States does not track industry concentration data for the construction industry, the United Kingdom calculated that the UK construction

industry is among the least concentrated with the top five firms accounting for only 5 percent of total output (UK Office of Government Commerce 2003). For example, because the top five residential builders comprise only 20 percent of the market and the remaining 79 percent have little market reach, it is difficult to influence the industry to address productivity concerns.

3. For example, Turner Construction now has 11,000 LEED APs on staff, over ten times the number of LEED APs it employed in 2005.

4. A representative list of environmental ratings systems for buildings would include the Building Research Establishment's Environmental Assessment Method (BREEAM) which originated in 1990 in the United Kingdom, followed by LEED in 1998 in the United States and the Comprehensive Assessment System for Building Environmental Efficiency (CASBEE) in 2001 in Japan.

5. An analogy from another industry would be the Chartered Financial Analysts program, which is the leading global brand in certifying investment professionals. The origin of the program in the portfolio analysis of investment securities would logically position the CFA Institute, the organization that certifies investment professionals through its eponymous program, to be a leader in the area of enterprise risk management, but the strengths of CFAs remain in financial analysis rather than the more comprehensive analysis of operational and other nonfinancial risks.

6. The first building to be constructed using IPD was the Orlando Utilities Commission North Chiller Plant in 2004. Owen Mathews, CEO of Westbrook Air Conditioning & Plumbing, baptized the collaborative process "Integrated Project Delivery" after successfully getting several parties to collaborate to share risk and reward in the construction of the plant. The American Institute of Architects further solidified IPD as a process to aspire to in 2007 when it created *Integrated Project Delivery: A Guide*. This guide is summarized in Cohen 2010.

7. Silo-based stakeholders with narrowly defined interests are a major impediment to project collaboration. IPD aligns the interests of different stakeholders with the success of the project as whole. The behavioral differences between silo-based and collaborative stakeholders are most clearly illustrated by their different approaches to risk management. In what the AIA calls traditional project delivery, risk is "individually managed; transferred to the greatest extent possible" while in IPD it is "collectively managed and appropriately shared."

8. Some 78 percent of contractors and 74 percent of architects surveyed cite saving time and money in the pursuit of LEED certification as factors contributing to their use of BIM. A 2010 SmartMarket Report by McGraw-Hill Construction entitled *Green BIM* further addresses how building information modeling (BIM) is contributing to green design and construction.

9. As noted in note 8, in *Green BIM* the drivers for adopting BIM and IPD include cost reduction.

10. Through a combination of services and sources, including *Engineering-News Record*, *GreenSource*, eleven regional publications, and *Dodge* and *Sweets*, McGraw-Hill Construction serves over one million customers.

11. To read more about the way that teams' ability to learn and adapt impacts organizational success, see Edmondson 2012.

12. The growing literature on smart cities tends to underplay—or even ignore—the critical importance of regional context: access to fresh water, locally grown food, efficient transportation, recreational amenities, and so on.

References

Bernstein, Harvey M. 2010. *Green BIM: How building information modeling is contributing to green design and construction*. McGraw-Hill Construction Smart-Market Report. Bedford, MA: McGraw-Hill Research & Analytics.

Brand, Stewart. 1995. *How buildings learn: What happens after they're built*. New York: Penguin Books.

Chapman, Robert E., and David T. Butry. 2008. *Measuring and improving productivity of the US construction industry: Issues, challenges, and opportunities*. Gaithersburg, MD: National Institute of Standards and Technology.

Chesapeake Bay Foundation vs. Weyerhaeuser Company. 2011. Circuit Court Civil Action No. 341442, Montgomery County, MD. Case 8:11-cv-00047-AW Document 2. http://www.greenbuildinglawupdate.com/uploads/file/2-main.pdf (accessed January 30, 2012).

Cohen, J. 2010. *Integrated Project Delivery: Case Studies*. Report under A Joint Project of AIA California Council, Integrated Project Delivery Steering Committee, and AIA National Integrated Project Delivery Interest Group, Sacramento, CA: AIA California Council.

Deutsche Bank. 2008. Deutsche Bank Americas Foundation contributes $25,000 to Enterprise Community Partners, New York. http://www.db.com/us/docs/Enterprise_Contribution_FINAL.pdf (accessed December 20, 2012).

Eastman, Chuck, ed. 2011. *BIM handbook: A guide to building information modeling for owners, managers, designers, engineers and contractors*. Hoboken, NJ: Wiley.

Eccles, Robert G. 1981. The quasifirm in the construction industry. *Journal of Economic Behavior & Organization* 2 (4): 335–357.

Edmondson, Amy C. 2012. *Teaming: How organizations learn, innovate and compete in the knowledge economy*. San Francisco, CA: Jossey-Bass.

Edmondson, Amy, and Faaiza Rashid. 2011. *Integrated project delivery at Autodesk, Inc. (A)*. Harvard Business School Case. Cambridge, MA: Harvard Business Publishing.

Edmondson, Amy, and Faaiza Rashid. 2012. Risky trust: How multi-entity teams build trust despite high risk. In *Restoring trust in organizations and leaders*, ed. Roderick Kramer and Todd Pittinsky, 129–150. New York: Oxford University Press.

Forman, Richard T. T. 2008. *Urban regions: Ecology and planning beyond the city*. Cambridge, UK: Cambridge University Press.

Fowler, Kimberly, and Emily Rauch. 2006. *Sustainable building rating systems summary. Pacific Northwest National Laboratory*. Washington, DC: U.S. Department of Energy.

Kane, Anthony, and Jill Allen. 2010. *The Zofnass rating system for sustainable infrastructure*. Cambridge, MA: Harvard University Graduate School of Design.

Kelso, Jordan D., ed. 2011. Buildings sector. In *2010 buildings energy data book*. U.S. Department of Energy, 1–28. Silver Spring, MD: D&R International.

Living PlanIT. 2011. *Cities in the cloud: A Living PlanIT introduction to future city technologies*. Living PlanIT, London. http://living-planit.com/resources.htm (accessed January 30, 2012).

Lucid Design Group. 2011. Building Dashboard® network. http://www.lucid designgroup.com/products.php (accessed January 30, 2012).

Mathews, Owen, and Gregory A. Howell. 2005. Integrated project delivery: An example of relational contracting. *Lean Construction Journal* 2 (1): 46–61.

Navarro, Mireya. 2009. Empire State Building plans environmental retrofit. *New York Times*, April 7. http://www.nytimes.com/2009/04/07/science/earth/07empire .html?_r=0 (accessed December 21, 2012).

Nelson, Bryn. 2011. Seattle's Bullitt Center aims to be energy self-sufficient. *New York Times*, October 5. http://www.nytimes.com/2011/10/05/realestate/commercial/ seattles-bullitt-center-aims-to-be-energy-self-sufficient.html?pagewanted=all (accessed January 30, 2012).

Passive House Institute. 2011. What is a passive house? Passivhaus Institut. http:// passiv.de/en/02_informations/01_whatisapassivehouse/01_whatisapassivehouse .htm (accessed December 21, 2012).

RMI. 2009. Project Case Study: Empire State Building. Rocky Mountain Institute. http://www.rmi.org/retrofit_depot_get_connected_true_retrofit_stories (accessed December 21, 2012).

Roberts, Tristan. 2010. LEED AP credential maintenance: Cracking the code. *Building Green*. http://www2.buildinggreen.com/blogs/leed-ap-credential-maintenance -cracking-code (accessed December 21, 2012).

Roberts, Tristan. 2011. Insurance breakthrough for integrated project delivery. *GreenSource: The Magazine of Sustainable Design*. http://greensource.construction .com/news/2011/04/110427_IPD.asp (accessed December 21, 2012).

Turner Construction. 2011. Turner's Commitment to Green Buildings Results in Achieving LEED Certification of 200th Project. http://www.turnerconstruction.com/ news/item/763/Turners-Commitment-to-Green-Buildings-Results-in-Achieving -LEED-Certification-of-200th-Project (accessed December 21, 2012).

UK Office of Government Commerce. 2003. *Achieving excellence in construction procurement guidance 05*. London.

United Nations. 2010. *World urbanization prospects: The 2009 revision*. New York: United Nations.

U.S. Department of Energy. 2008. *Energy efficiency trends in residential and buildings, vol. 13*. Washington, DC: U.S. Department of Energy.

USGBC. 2010. LEED statistics. North Carolina Triangle Chapter, Raleigh, NC. http://www.triangleusgbc.org/leed/leed-statistics (accessed January 30, 2012).

Yi, Aris, Samuel Ramirez, and Michael Bendewald. 2010. *Lovins GreenHome 1.0. factor ten engineering case study*. Boulder, CO: Rocky Mountain Institute.

II

Market Structures and Strategies

5

Why Multinational Corporations Still Need to Keep It Local: Environment, Operations, and Ownership in the Hospitality Industry

Jie J. Zhang, Nitin R. Joglekar, and Rohit Verma

Introduction

While the goal of environmental sustainability offers positive global impact, its implementation must consider the inherent variations in the local environmental, social, and economic contexts. For instance, EnerNOC, a U.S. firm engaged in demand-side management of energy consumption, employs a business model that reduces aggregate energy demand during peak loads, and passes the savings to clients who respond to its conservation calls by reducing their energy demand (Healy 2007). EnerNOC has been successful in signing up key accounts including large corporations, governmental agencies, hotels, hospitals, and universities, yet its growth is limited by the ability of its sales force to engage individual building and facility managers to implement the demand-response practice (Tuttelman 2008). EnerNOC's experience shows that local variations in both the built environment and operational issues are central to the diffusion of sustainability practices. A wide community of users and environmental leaders recognize this connection, such as the USGBC (U.S. Green Building Council) through its LEED (Leadership in Energy and Environmental Design) rating system. Specifically, the LEED Existing Buildings: Operations & Maintenance certification system (LEED EBOM) underlines how operations can stand in the way of realizing the potential of sustainable design and construction. We recognize this critical lens of local operations, and use it to assess the effectiveness of design parameters, construction standards, and operations in terms of consumption outcomes.

In the built environment, operations managers of individual sites face two major challenges in implementing sustainability practices. The first challenge involves difficulties in tracking environmental performance, because sites often lack consistent ways of measuring resource consumption or monitoring energy consumption patterns (Schleich and Gruber 2008).

To this end, we develop operating measurements that reflect the contingent nature of environmental sustainability at the site level. With these measurements in place, corporations and facility managers can monitor outcomes of sustainability practices, thus providing feedback necessary for further improvement.

The second challenge lies in the conventional firm-centric view that overlooks the effect of diverging interests and performance-evaluation criteria of collaborating parties such as the building owner and the building operator. Delmas and Toffel (2008) show that within an organization, the difference in adopting environmental management practices relates to the receptivity of market and nonmarket pressures across functional departments such as corporate legal affairs and marketing departments. We argue that such varying receptivity is even more pronounced across firm boundaries. For example, in a hotel service chain, the central players, collaborating through franchise or management contracts or both, typically include the owner, who focuses on the development and management of the hotel real estate property, and the operator, who focuses on managing the daily hotel operations (i.e., guest services, facility management operations, and reservation systems). Consequently, the owner prefers practices that enhance the valuation of the physical assets, while the operator prefers practices that increase revenue—usually the base of royalty and management fees. In other words, due to their specialization and distinct reward structures, each service supply chain partner strives for a unique subset of business benefits (see chapter 11, this volume). Recognizing that supply chain partners do not share uniform interests in adopting sustainability practices, this chapter develops a theoretical framework to explicate the diverging interests along the chain and suggests potential economic and policy measures to align them.

We inform our arguments through an exploratory study of sustainability measures in a typical built environment—hotels—which is an ideal lab to research sustainability for three reasons:

1. Data availability—the contractual arrangement between hotel owner and hotel operator requires cost reporting on resources consumed, resulting in data that are difficult to gather in other built environments such as residences or commercial office buildings.

2. Industry commitment—major hotel chains identify sustainability as a strategic issue (Sherwyn 2010) for its potential to enhance brand image and reduce costs, creating strong momentum for sustainability innovations.

3. Demonstration effect—guests satisfied with their green lodging experience might adopt some sustainability practices in their homes and workplaces, yielding positive externalities.

To address the first challenge of measuring and comparing the performance of sustainability practices, we track 984 hotel sites across the United States over an eight-year period and construct a standardized measure of environmental sustainability based on resource efficiency. This sustainability measure is driven by two factors: an operating cost factor (OCF) and a consumer behavior cost factor (CBCF). We offer a comparative analysis of these two factors based on variations in five basic hotel characteristics: ownership structure, operating structure, level of urban development, type of guests, and regional ambient temperature. We illustrate that the values of these two factors vary systematically for each of the five characteristics.

We use these differences to argue that it is possible—in fact desirable—to incorporate local variations into environmental policy changes aimed at improving sustainability in the built environment. For instance, effective environmental policy needs to address both the investment and operational issues related to ownership structure: the asset owner may choose to invest in solar panels, but this investment requires the operator to change related energy and facility management routines to fully realize the return. However, the return on investment often only accrues to one party—utility savings for the owner in the case of solar panels. So there is little incentive for the operator to participate. A reverse situation is also possible. The asset owner's building architecture and energy equipment constrains how far the operator's efforts can go toward establishing a genuinely green brand reputation. Such situations are best described by the Prisoner's Dilemma with two players (Pruitt 1967). If both players commit to improvement, a win-win solution emerges; if both the players defect (i.e., cheat in the hope of private gain through the partner's commitment), a lose-lose scenario results. The threat to the cooperative efforts, hence the dilemma, arises from uncertainty of the partners' actions, in the presence of diverging performance goals. Unequal cooperation results in one side reaping rewards and the other becoming a "sucker." Clearly, no side wants to be the sucker.

Having provided an overview of our study, this chapter is organized as follows. We first describe the industry setting for our empirical study, then define and calculate the resource-efficiency-based environmental sustainability measures. Drawing from prior research on corporate motivations to adopt sustainability, we then develop a framework to identify the factors that influence the multiple agents in the built environment who have diverging goals, constraints, and planning horizons. We flesh out specific tradeoffs within this multi-agent commitment framework, suggest contracting and negotiation tools to bridge the incentive needs of various agents, and identify and close behavioral loopholes to accelerate the diffusion of sustainability practices in the built environment.

The U.S. Hospitality Industry

The U.S. hospitality industry, with buildings situated in diverse settings across the country, and with ongoing sustainability efforts of varying types and degrees, lends itself to study supply chain coordination and incentive structures. Each hotel site is both a real estate investment for hotel owners and a service and brand building opportunity for hotel operators. Multiple ownership and operating structure configurations are possible in this industry: owner operated (they own or lease the property), major chain operated (branded management), or nonbranded hotel management company operated. Further, the hospitality industry measures its performance in terms of return on assets (owner) and operating performance (operator), and both metrics are intimately linked with environmental sustainability practices. Table 5.1 illustrates the hospitality service supply chain for several common ownership structures and operating structures.

We conducted our empirical study on 984 U.S. hotel sites' annual operating statements from 2001 to 2008. These hotel sites are located across the United States, and represent all major U.S. hotel chains. We collected comparable information on all aspects of the hotel operations with an emphasis on operating expenses including consumption of fundamental resources such as electricity, water, and materials used as operating supplies. PKF Hospitality Research, an industry trend research firm that has tracked thousands of hotels for nearly seventy years, provided the data.

To resolve the first challenge listed previously of comparable data, we developed a measure to benchmark environmental sustainability in the hospitality industry based on the efficiency of hotels' use of fundamental resources to generate revenue (measured in revenue per available room—RevPAR). A wide range of resources are included: electricity; water and sewer; maintenance supplies; and laundry, linen, and supplies used in the rooms and food and beverage departments. This measure stemmed from analyzing the preceding panel dataset using exploratory factor analysis—a statistical method that analyzes the varying patterns of resource

Table 5.1

Hospitality service supply chain players: Ownership and operating structure choices in the hotel industry

Types of ownership and operating structure	Responsibilities		
	Manage property asset	Manage operations	Set standards and manage distribution
Vertically integrated	Owner	Owner	Owner
Franchisor/brand operated	Owner	Brand	Brand
Professionally (non-branded) managed	Owner	Operator	Brand

consumption and identifies primary drivers of resource efficiency in the sample. This analysis yielded one measure of environmental sustainability consisting of OCF that assigns large weights to electricity, water, sewer, and maintenance expenses, and CBCF that assigns large weights to expenses from laundry, linen, and supplies used in the rooms and food and beverage departments. Although all expenses are subject to influences from both management and guests, the expenses measured by the operating cost factor are more affected by managerial decisions, and the expenses measured by the consumer behavior cost factor are largely driven by guests' choices. We generate normalized factor scores to indicate how efficient the hotels are in using the resources for every unit of revenue generated—hotels that score below zero are more resource efficient than the average, and hotels that score above zero are less resource efficient. In summary, this two-factor measure of environmental sustainability is a consistent industry-wide measure for benchmarking sustainability performance by considering both internal environmental management choices and the guests' resource consumption behavior. We compare the factor scores across five basic hotel characteristics: ownership structure, operating structure, level of urban development, type of guests, and ambient temperature.

Corporate Motivations to Adopt Sustainability

A hotelier can expect several possible outcomes when she decides to build and operate sustainably. However, prior literature made conflicting predictions regarding the outcomes. Some of the earliest studies on corporate sustainability issues argue that environmental concerns reduce profit maximization (Friedman 1970). Later, Porter and van der Linde (1995) argued that a firm's pollution is often associated with wasted resources (material, energy, etc.), and that more stringent environmental regulation can stimulate green innovations that may offset pollution prevention costs. These debates generated a series of arguments around why a firm ought to invest in environmental protection efforts, and the possible outcomes. For instance, some scholars cite incentives such as near-term profitability gains through toxic waste prevention (King and Lenox 2002) and financial markets' positive response to environmental awards announced and confirmed by third parties (Klassen and McLaughlin 1996). Drawing on institutional theory, early work in this arena predicts an inevitable shift in corporate environmentalism in which sustainability goals become a way of organizational life (Hoffman and Ehrenfeld 1998, 73).

Various stakeholders (Freeman 1984) may also provide the motivation for building and operating green. Ambec and Lanoie (2008) conducted an extensive literature review and summarized the mechanisms through which firm characteristics may interact with various internal and external factors to influence economic performance. For instance, a firm may comply with government regulations, such as green procurement programs for EPA-designated products and services, and consequently enjoy increased revenue.

In spite of this extensive knowledge about a firm's motivation for sustainability, we have just begun to understand the sustainability motivations of organizations that are hybrid (i.e., neither market nor hierarchy) or extend across firm boundaries (e.g., supply chains) but connected through contractual "bridges" (Baker, Gibbons, and Murphy 2008). Prior research on the challenges of implementing strategic choices in individual operating units demonstrates the need for considering the multi-agent aspect of our study. For example, in the manufacturing context, researchers (Boyer and Lewis 2002; Boyer and McDermott 1999) found that a firm's personnel at different organizational levels can substantially disagree on strategic decisions. For instance, operations-level employees tended to rate investments in technology disproportionately higher than plant managers. Further, Boyer and Lewis (ibid.) present case evidence of significant inconsistencies in competitive priorities (i.e., cost, delivery, flexibility, and quality) across all levels of a firm's organizational chart. It is conceivable that similar inconsistencies exist across a firm's operating units and supply chain, with each unit optimizing its environmental performance according to its own set of priorities, constraints, and reward structure.

Drawing on the research streams discussed previously, we examine the environmental sustainability outcomes at individual hotel sites. Each site is considered a standalone operating unit with a unique profile consisting of internal resources (e.g., physical assets, brand equity, management expertise) and local conditions (e.g., location, ambient temperature). The supply chain partners at individual hotel sites interact differently with the external environment, such as conflicting responses to changes in the economy. For example, the performance goals of the hotel owner and the hotel operator can diverge, and tension often intensifies in tough economic times, especially in the higher-tier luxury market segment. Hotel owners, facing debt service obligations on their multimillion-dollar investments, seek to cut costs. But high-end, branded hotel operators seek to maintain their revenues and uphold their premier brand image, which requires sustained operating cost levels, and leads to higher cost pressure on the hotel owner. Indeed, during the recent economic downtown, several highly contentious hospitality industry cases surfaced where the owner ousted the operator over operating costs disputes (Segal 2009) or poor management performance (Berzon and Hudson 2011). Therefore, the diverging economic objectives and collaborative requirement of sustainability practices at individual hotel sites necessitates a closer look at local variations. In particular, we make comparisons across five key local variations in the U.S. hotel industry and try to understand the causes and potential relief of the tensions that afflict the owners and operators in pursuit of environmental sustainability.

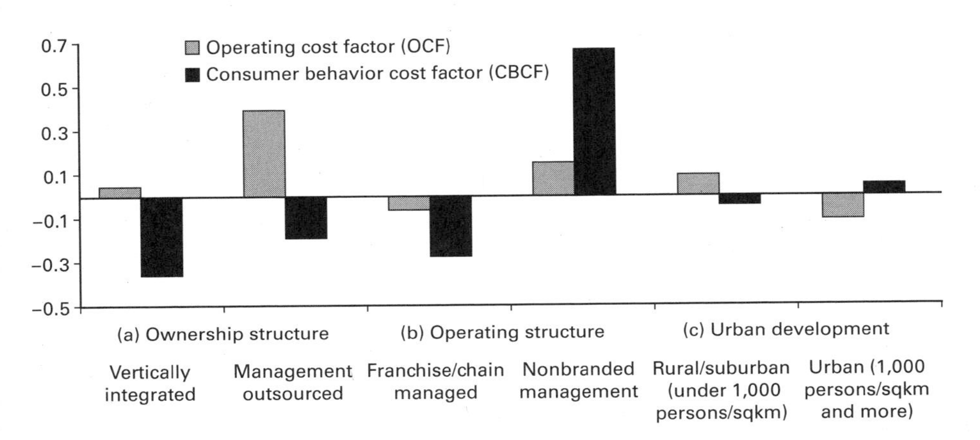

Figure 5.1

Environmental sustainability factor core comparisons in the hotel sector.

Ownership Structure: Vertically Integrated vs. Management Outsourced
Typically, the land or the building, or both, of a hotel may be owned or leased by the hotel owner, who then outsources the hotel operation to a management company via a long-term contract. Under this ownership structure, the hotel's operator often identifies capital expenditures aimed at increasing energy efficiency, but the owner must approve and fund the efforts, resulting in a split-incentive problem (Schleich and Gruber 2008). Evidence from the commercial real estate market (Fisher and Rothkopf 1989; Jaffe and Stavins 1994) suggests that neither the landlord (building owner), nor the tenant (the operating company) may have an incentive to invest in energy efficient equipment or services, although both parties may benefit from building value appreciation or the energy cost savings associated with increased energy efficiency. The landlord will not invest in energy efficiency if the investment costs cannot be passed on to the tenant. The tenant will not invest if the lease expires or terminates before recouping the energy efficiency investment costs. Sometimes, the hotel owner may choose to manage self-owned assets, resulting in a vertically integrated hotel where the owner is singly responsible for the costs and benefits of sustainability investment. Such integrated ownership structure requires the owner to have expertise in both real estate asset management and hotel service operations, but attenuates the split-incentive problem.

Figure 5.1a shows the two factor scores OCF and CBCF of the environmental sustainability measure for two groups of hotels based on the ownership structure distinction (vertically integrated vs. management outsourced). The figure suggests that the hotels that are vertically integrated score lower values in both OCF and CBCF, indicating a lower level of resource consumption per unit of revenue (i.e., higher resource efficiency). The resource efficiency lead of the vertically integrated hotels in figure 5.1a confirms the need to consider ownership structure and associated issues including split incentive when designing sustainability strategy and evaluating sustainability performance.

Operating Structure: Franchisor/Chain Managed vs. Nonbranded Management Company
In the United States, about 70 percent of hotels are affiliated with branded chains such as Marriott, Hilton, or Wyndham. The benefits of branding include instant access to an established customer base, extensive reservation and marketing systems, and participation in dynamic processes such as strategy making (Bradach 1997). These benefits lead to synergy that enhances service uniformity and performance through innovation

and learning across the sites (Barlow 2000). A branded hotel may be operated by the brand franchisor, a branded chain offering management services, or a nonbranded management company. On the one hand, in the franchisor/chain-managed hotels, by committing to the chain business model and investing in the mandated brand standards, the hotel owner faces potential hold-up problems that give the chain increased bargaining power (Unsal and Taylor 2009). On the other hand, the nonbranded hotel management company builds its core competences in hotel operations, and possesses no stake in the hotel's brand image. Because the choice of operating structure relates to the type of asset invested, the core competencies of agents, and the reward system, it affects a wide range of strategic decisions, including sustainability. Figure 5.1b shows the factor scores—OCF and CBCF—for two groups of hotels: those managed by a franchisor or hotel chain company vs. those managed by nonbranded hotel management companies. The results show that franchisor/chain-managed hotels are more resource efficient in both factors. This may indicate that the franchisor and the chain operators have stronger incentive to increase resource efficiency due to their larger stake (e.g., in terms of brand equity) in the operation. Therefore, operating structure and its associated incentive issues are important considerations in the decisions to build and operate green.

Location: Urban vs. Rural/Suburban

The process of urbanization intertwines with people moving into cities to seek economic opportunities. So we draw the distinction between urban and rural/suburban areas by the population density—areas that have 1,000 and more persons per square kilometer are considered urban, while areas that have less than 1,000 persons per square kilometer are considered rural/suburban. Higher population density accompanies increased per capita energy consumption and extensive modification of the natural environment (McDonnell and Pickett 1990). Consequently, hotels in various locations may face varying resource and environmental pressures. For instance, high prices for gas and electricity and resource shortage (such as brownouts and water bans) are more likely to affect hotels located in urban areas. Sustainable development strategies need to reflect the varying resource constraints across locations. Figure 5.1c shows a comparison of the two environmental sustainability factor scores for hotels located in urban areas and those located in rural or suburban areas.

The graph shows that relative to their rural/suburban counterparts, the hotels in an urban environment consume fewer resources measured by operating cost factor (e.g., maintenance and utilities) but consume more measured by the consumer behavior cost factor (e.g., supplies for rooms and food and beverage departments) per unit of revenue generated. Because the factor scores reflect cost-based resource efficiency, either higher revenue or lower total cost of consumption may contribute to these differences. On the revenue side, urban hotels tend to charge higher prices for their services than their rural/suburban peers, resulting in a larger revenue base for urban hotels. On the consumption cost side, opposite forces are at play: for OCF-driven decisions, higher resource scarcity in the urban areas may provide a stronger incentive for hotel managers to improve their resource efficiency, while perceived abundant resources may support wasteful processes and result in lower efficiency in the rural/suburban hotel sites. As for the CBCF, since the hotels have limited control over their guests' consumption behavior, the higher resource prices in urban areas dominate and result in lower consumer behavior-related resource efficiency in the urban areas. Therefore, the varying effects of location-based constraints must be considered in the decision-making process for sustainability.

Type of Guests Served: Transient vs. Group

Some hotel guests travel for leisure (creating a "transient" income stream in hospitality industry terminology) and others for business purposes (creating a "group" income stream). Although almost identical in terms of the service delivery processes, these two types of income streams—transient and group—may involve different decision-making processes by the guests. For example, a tourist is likely to choose a hotel based on her budget, personal experience, and family consideration, while the same person on a business trip must consider her company travel policy in choosing a hotel. In other words, while transient income results from the individual guest's personal value system, group income is jointly determined by personal and institutional choices. One industry trend serves as a good indication of how these institutional and personal choices may interact. Increasingly, corporations are favoring hotels that promote efforts toward environmental sustainability. A recent survey shows that 65 percent of corporate travel executives responsible for over $10 million in annual travel budgets are in various stages of implementing green business travel guidelines (HSPI 2011). It stands to reason that hotels that derive more income from group business and conference activities have stronger incentive to adopt sustainability initiatives. The types of guests served closely relate to the hotel property type, that is, the facilities

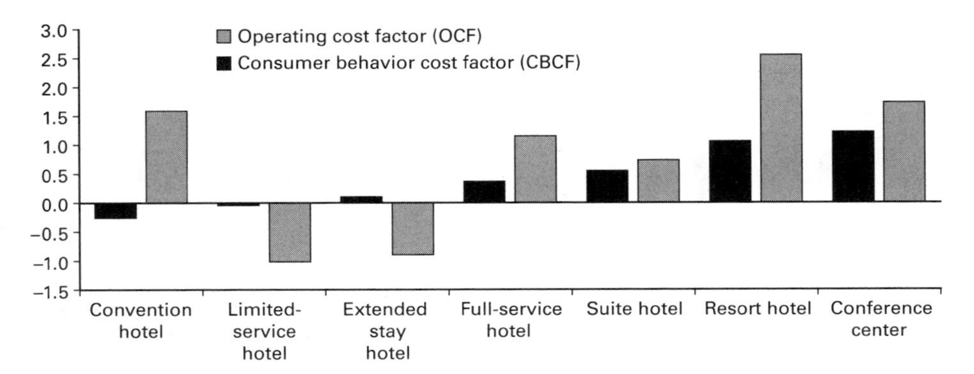

Figure 5.2

Property type and environmental sustainability factor scores in the hotel sector

and amenities available to support certain guest activities. Our dataset contains seven types of hotel properties: conference center, resort hotel, full-service hotel, limited-service hotel, suite hotel, convention hotel, and extended-stay hotel. (The glossary details the definition for each hotel type.) Figure 5.2 depicts how the two factor scores of the environmental sustainability measure vary across different hotel types.

The results show that resource consumption correlates positively with the level of service the hotels provide—resort and conference hotels consume more resources along OCF and CBCF, while the limited-service and extended-stay hotels consume much less in both factors. At one end, resort and conference hotels provide extensive amenities such as pools/saunas/health clubs, restaurants, shops, and ballrooms, and may be a destination in their own right. At the other extreme are limited-service hotels that provide only essential lodging services and where guests tend to engage in activities outside the hotel property boundary, and thus use fewer resources on the hotel premises.

Regional Ambient Temperature: Hot vs. Cold

The aggregate warming at the global level coupled with increasingly volatile local weather conditions pose great challenges for hotels trying to predict future energy needs and balance guest comfort and energy requirements. Both the energy sources currently used to provide cooling and heating and potential or newly developed alternative energy sources are closely connected with the local natural resources and environment. For

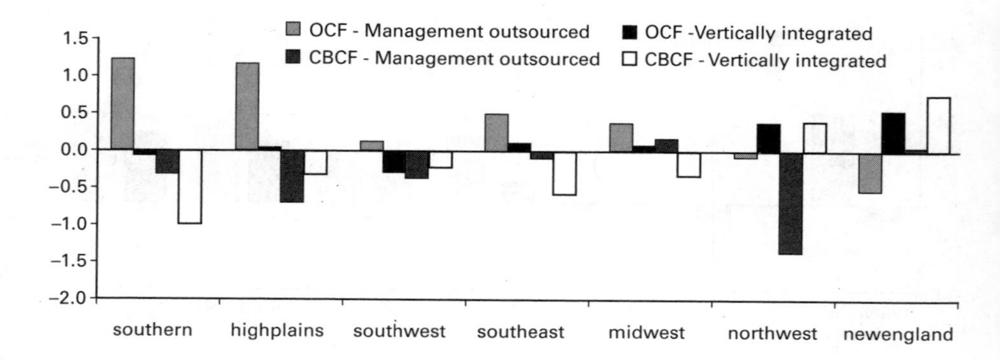

Figure 5.3

Environmental sustainability factor scores for the hotel sector (OCF and CBCF) by climate zone: owner managed vs. other managed

instance, hotels in the U.S. Southwest face increasing pressure from water shortage problems due to the dry climate, but also are situated in an ideal climate to take advantage of solar energy innovations. Forming strategic relationships with appropriate energy and energy solution suppliers then becomes an important strategy for hotels. A sound environmental sustainability strategy for a hotel site must therefore consider the markedly different long-term resource use patterns and potential scenarios. Figure 5.3 depicts the operating cost factor scores and consumer behavior cost factor scores across different climate zones of the vertically integrated hotels and management outsourced hotels in our sample.

Figure 5.3 identifies large variations in environmental sustainability as measured by OCF and CBCF across climate zones. Further, these two factors exhibit different variation patterns: specifically, vertically integrated hotels display a much smaller variability in OCF (dark solid bars in the graph) across climate zones.

Paired-Case Comparisons

We then conducted a series of *t*-tests to validate the intuitions gleaned from the graphs presented earlier (i.e., figures 5.1–5.3). The *t*-test assesses whether the means of two groups are statistically different from each other. For each of the five hotel characteristics, we divide the sample into two groups and compare the means of the OCF score and the CBCF score. For example, we first divide the sample into two groups based on the ownership structure and use “O” to represent the group of hotels under owner management and use "T" to represent the group of hotels where management is outsourced. In addition, we compared subgroups by considering combinations of multiple characteristics, such as hotels in urban locations under owner management (UO) vs. urban hotels where management is outsourced (UT). Table 5.2 summarizes these paired-case *t*-test results. All the comparisons in table 5.2 are statistically significant at the 0.05 level.

To summarize, we provide evidence on the contingent nature of sustainability practices by comparing the measured environmental sustainability of the hotels in our sample along local site variations including ownership structure (vertically integrated vs. management outsourced), operating structure (franchisor/chain managed vs. non-branded management), type of guests (transient vs. group), location (urban vs. rural/suburban), and ambient temperature (hot vs. cold). The graphs (figures 5.1–5.3) illustrate

Table 5.2

The *t*-test summary of environmental sustainability factor score comparisons across basic hotel characteristics[1]

	OCF	CBCF
Owner-managed (O) vs. Outsourced (T)	O>T[2]	O>T
Franchisor/chain (F) vs. Nonbranded (M)	F>M	F>M
Franchisor/chain O vs. T	FO>FT	FO>FT
Nonbranded O vs. T	MO<MT	MO<MT
Rural/suburban (R) vs. Urban (U)	R<U	R>U
Rural/suburban O vs. T	RO>RT	RO>RT
Urban O vs. T	UO>UT	UO=UT
Ambient temperature Hot (H) vs. Cold (C)	H<C	H>C
Hot O vs. T	HO>HT	HO>HT
Cold O vs. T	CO<CT	CO<CT
Full Service (S) vs. Limited (L)	S<L	S<L
Full Service O vs. T	SO<ST	SO<ST
Limited Service O vs. T	LO>LT	LO>LT

1. The letters in the parentheses are shorthand for the hotel site characteristics. For example, FO represents a hotel that is owned and managed by a franchisor/chain; while FT indicates a hotel for which the owner hires a franchisor/chain as the operator.

2. The ">" sign indicates better environmental sustainability as measured by the factor scores.

the distinct patterns in the measured environmental sustainability. The *t*-test comparisons statistically validate the intuitions from the graphs. (Readers may refer to Zhang, Joglekar, and Verma (2012) for statistical analyses that quantify the individual effects in the presence of other sources of variability such as scale of operations, market segment, and the overall economic cycle.) These analyses show that a multilevel theoretical approach is necessary to illuminate the various modes of inter-firm collaborations and intra-firm coordination, especially for organizations with complex distributed operations carrying distinct local characteristics.

Managerial and Policy Implications

A simple conclusion is that we observe considerable heterogeneity in the operating cost and consumer behavior cost components of environmental sustainability regarding local variations. Thus, a uniform policy aimed at coordinating multi-agent commitment may not be the most effective way to improve resource efficiency. That is, even if we draw upon the firm-centric view of the stakeholder-driven strategy formation process, we cannot attempt a unified implementation of strategies. The variation in sustainability performance associated with differences in asset ownership, operating structure, location, and so on, requires a contingent approach that explicitly considers the diverging interests of multiple agents and varying contextual factors.

Multi-agent Collaboration

A second and equally important implication of our research is about how hotel owners and operators decide on increasing resource efficiency. We focus on the dyadic relationship surrounding an asset: the hotel property owner and hotel operator. Nearly one third of the built environment in our hospitality industry dataset is owned by real estate investors (individuals and groups—e.g., corporations, LLCs). Generally, they maximize a combination of their rents and their assets' long-term value. Organizationally, these owners may be absentee landlords with little time or inclination to understand and manage the day-to-day operations and the challenges. These properties typically are managed by large hotel chains that are experts in service operations. These operators understand and track the operating cost factor, as well as the guest behavior that drives resource consumption.

It is clear that these two parties need to work collaboratively toward increasing resource efficiency. For instance, the installation of solar panels to heat water requires approval and funding from the property owner.

Yet merely adding solar panels usually is not enough to realize cost savings, because the peak generation of energy mid-day does not coincide with the peak demand for hot water, typically in the morning. The hotel operator may decide to change the housekeeping and laundry schedule to take advantage of the abundant hot water during the middle of the day. This is easier to achieve when the asset ownership and operating tasks are governed by a single entity such as in vertically integrated hotels.

Incentive Design

The ownership of a hospitality physical asset increasingly is separated from its operator, and the economic and social interests of the asset owner and the operator are rarely aligned. The Prisoner's Dilemma is a well-known metaphor used in psychological, sociological, and economic research to model situations of social conflict between two or more interdependent actors (Luce and Raiffa 1957; Janssen 2008; Pruitt 1967). The essence of the dilemma is that each individual actor has an incentive to act according to competitive, narrow self-interests even though, in general, all actors receive greater payoffs if they collectively cooperate. We follow the structure outlined by Cable and Shane (1997) to explain the underlying tensions in the Prisoner's Dilemma in figure 5.4.

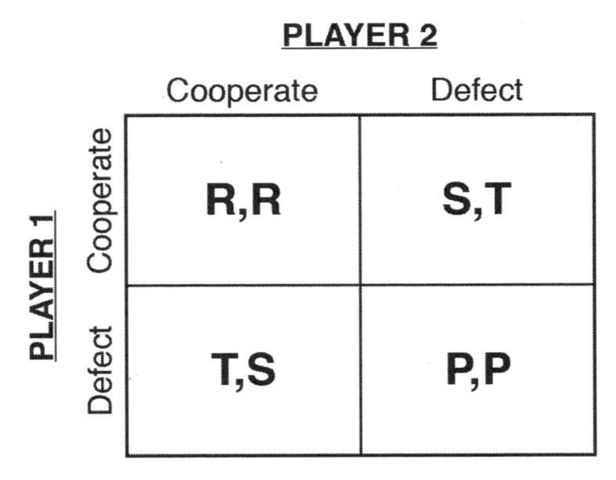

Figure 5.4

General 2 × 2 payoff matrix for the classical two-person Prisoner's Dilemma (Cable and Shane 1997)

Cable and Shane (1997) point out that "consistent with past research (e.g., Axelrod and Dion 1988), the payoffs for each actor are dictated by the strategy adopted by the other actor and follow the payoff structure $T > R > P > S$, in which T represents the temptation of extra payoff from defection, R represents the reward for mutual cooperation, P signifies the penalty for mutual defection, and S represents the 'sucker's' payoff (the penalty for cooperating while the other actor furtively defects, unknown to the sucker). This payoff structure illustrates the Dilemma's conceptual value by highlighting the conflict between individual and collective rationality." In the built environment, such dilemmas have been examined for problems such as demand-and-supply mismatch driven by cyclical markets (Kummerow 1999). Prisoner's dilemmas have also been identified in the diffusion of environmental practices in settings such as coastal zone and watershed management (Reddy 2000; O'Riordan 1993, 3–36). In figure 5.5, we outline the key factors that drive the commitment of the asset owner and operator based on the hospitality site database.

We follow Cable and Shane (1997) in identifying these factors:

• *Site-specific alignment and information sharing:* Site-specific variation is evident in table 5.2 based on differences in ownership structure, operating alignment between actors drives the nature and extent of their information exchange.

• *Time horizon:* The various actors have different planning horizon needs. A real estate investor's time horizon is typically longer than that of a corporate group trying to make operational improvements. The investor's time horizon is driven by tax laws, and the nature of investors in real estate investment trusts (REITs)—typically three to thirty years. A corporate group tends to set planning horizons based on organizational needs that generally range between three and thirty months. In the hospitality industry, the prevailing contract types are revenue-based franchising contracts and profit-based management contracts, whose inherent short-term performance bias discourages sustainability investments in the delayed-payoff and high-upfront-cost scenarios. Assigning the sustainability investment ownership to the contractual party that has most to gain is one way to address the split incentive (Mathewson and Winter 1985). For example, new contracting mechanisms can be designed to tie the cost and payoff of environmental sustainability investments with the life of the investment instead of the contractual relationship. Government guarantees or third-party investors willing to take the risks (and share the profits) from uncertainty may be the broker or even hold the ownership to the investment—an arrangement analogous to home mortgages.

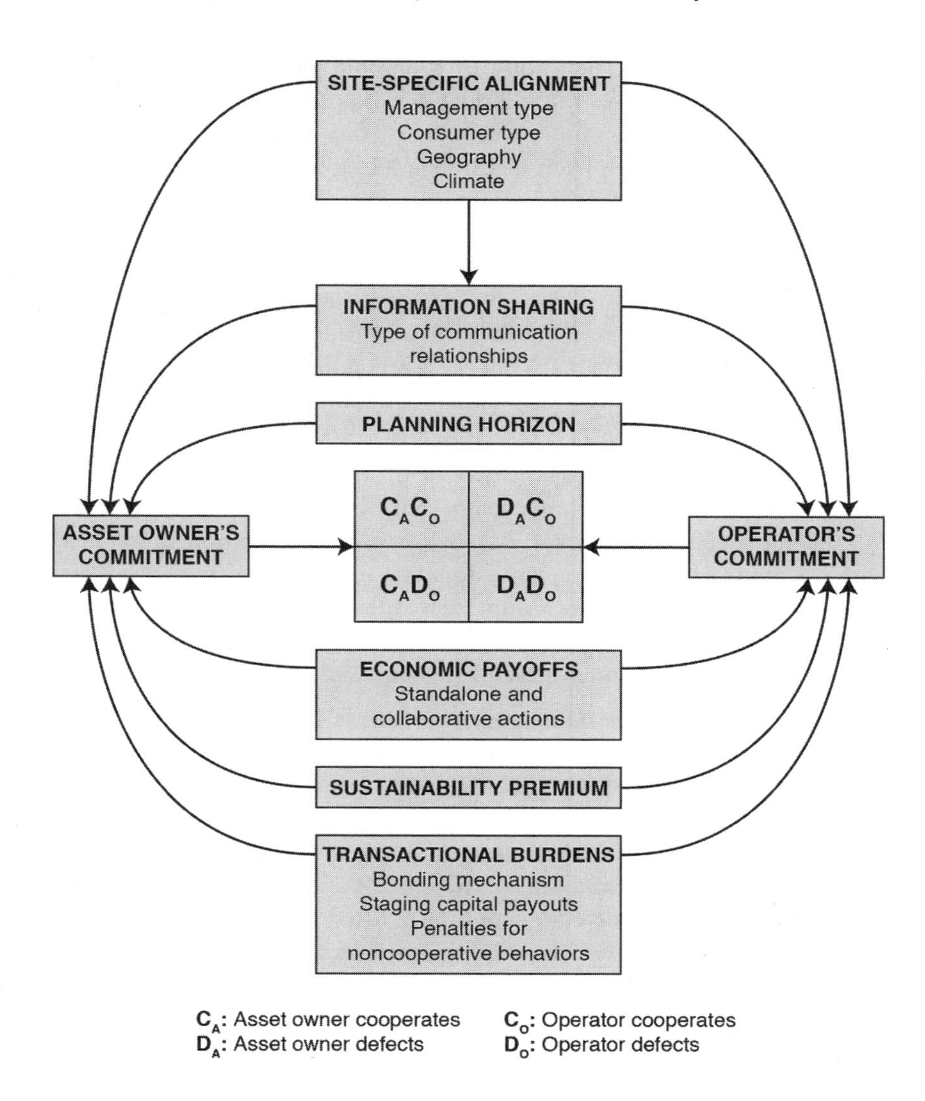

Figure 5.5

A framework comparing asset owner and operator commitment in the hotel sector

• *Economic payoffs:* The nature of economic payoff depends on the resource-savings technologies and the business processes affected by the owner's and operator's commitment to increasing resource efficiency. Often the payoff requires upfront investment along with ongoing operating commitments such as investment in terms of staff and working capital. Certain technologies, such as insulation, geothermal, and solar units may be eligible for government tax credits based on their sustainability impact. Other technologies such as an environmental management system may also contribute to sustainability, but are often ineligible for the tax credits. Life-cycle analysis (LCA) of these technologies evaluates the cost and payoff comprehensively and therefore should be the basis of policy or tax incentives.

• *Transactional burdens:* The behavior of individual actors is affected not only by contractual arrangements that bind the actors, but also by the size and schedule of payoff as well as penalties for noncompliance with the contractual arrangements. It is important to realize that parties outside the contractual relationship, such as the hotel guests, may actively affect the payoff of the sustainability investment through their influence on the consumer behavior cost factor (Zhang, Joglekar, and Verma 2011).

We recognize that our framework holds many limitations. It considers only two key players—owner and operator—out of the many stakeholders in the hotel service supply chain. There is significant difficulty in measuring in practice the key constructs depicted in figure 5.5. We assume a single-shot game set-up, yet the agents often need to make repeated decisions, such that their brand image influences the observers' assessments. Fleshing out such a framework should offer useful solutions for built environment settings where sustainability practices are difficult to instill.

Policy Implications for the Built Environment

We posit that instituting public policy and program changes may prompt the hotel property owner and operator to commit to sustainability practices and stimulate the diffusion of environmental sustainability practices in this industry. First, to facilitate information exchange and comparison, it may be more effective for government agencies such as the EPA to recognize the contingent nature of performance across sites in future standards and programs. For example, the EPA's Portfolio Manager in its current version, which ranks all U.S. hotels in the same group, misses the opportunity to identify local sustainability champions and to help spread green practices. The EPA could better achieve its goals by customizing its approach and creating programs that serve specific types of hotel sites based on ownership, operating structure, guest type, location, and ambient weather conditions. Moreover, two kinds of monitoring mechanisms are needed. One would ensure that the contractual parties fulfill their commitments. A system similar to the financial accounting reporting system would likely work for environmental reporting. The other monitoring mechanism would ensure that the full potential of the design and construction standards is actually realized in operation. This monitoring system will rely on the "smart" technologies emphasizing sensing and real-time energy management being deployed in the built environment. Movement in this positive direction is also reflected in the changes proposed in the 2012 LEED standards, which include resource metering and data sharing.

Conclusion

As the world we live in becomes flatter and hotter, it is increasingly important to identify sustainability solutions sensitive to local variations and for policies to support those solutions. In this chapter, we offer preliminary evidence on how these local variations may affect sustainability performance in a typical built environment—the hospitality industry—and present the policy implications of these local variations.

By focusing on two primary drivers of resource consumption—operational choices and customer behavior—we obtain an industry-wide measure of sustainability performance. On the operational side, information for evaluating and selecting sustainability initiatives is lacking due to the paucity of metrics for resource consumption and efficiency. However, the advent of the "smart hotel" that utilizes sensor and automation technologies to acquire real-time information promises more detailed and sensitive metrics to be constructed using an approach similar to ours. On the consumer behavior side, the sociological development in green marketing (Prakash 2002) and consumer receptivity to sustainable products and services (Shrum, McCarty, and Lowrey 1995) are key issues. Consumer behavior research experiments coupled with discrete choice analysis (McFadden 1974; Verma, Thompson, and Louviere 1999) can be designed to study the factors influencing customer choice.

We show that systematic differences in sustainability performance exist across basic hotel characteristics ranging from ownership structure to local weather conditions. These differences informed our framework for understanding the manner in which multiple agents can cooperate and commit to sustainability practices. Future research can address the complexity involved in this framework in two complementary directions: game theoretical analysis followed by empirical validation highlights this framework's salient effects and relationships, which are briefly explored in the "incentive design" section of this chapter; simulation using system dynamics (Forrester 1971) tools is another fruitful research direction. A multi-agent approach (for example, Swaminathan, Smith, and Sadeh 1998) could be especially useful in elucidating the relationships among supply chain partners, and significantly enrich contexts where key effects unfold.

Acknowledgments

The authors sincerely thank the Cornell Center for Hospitality Research (CHR) and PKF Hospitality Research (PKF-HR) for their invaluable industry insights and data support. The discussions at the CHR 2009 and 2010 Sustainability Roundtables greatly enriched the ideas in this chapter. Any opinions, findings, and conclusions or recommendations expressed in this chapter are those of the authors and do not necessarily reflect the views of CHR or PKF-HR. An initial version of this research was presented at the 2010 Constructing Green Conference hosted by Erb Institute at the University of Michigan, Ann Arbor. We gratefully acknowledge feedback from the conference participants.

References

Ambec, Stefan, and Paul Lanoie. 2008. Does it pay to be green? A systematic overview. *Academy of Management Perspectives* 22 (4): 45–62.

Axelrod, Robert, and Douglas Dion. 1988. The further evolution of cooperation. *Science* 242 (4884): 1385–1390.

Baker, George P., Robert Gibbons, and Kevin J. Murphy. 2008. Strategic alliances: Bridges between "islands of conscious power." *Journal of the Japanese and International Economies* 22 (2): 146–163.

Barlow, James. 2000. Innovation and learning in complex offshore construction projects. *Research Policy* 29 (7–8): 973–989.

Berzon, Alexandra, and Kris Hudson. 2011. Marriott loses trendy Waikiki hotel as owner changes locks overnight. *The Wall Street Journal*, August 29, Business.

Boyer, Kenneth K., and Marianne W. Lewis. 2002. Competitive priorities: Investigating the need for trade-offs in operations strategy. *Production and Operations Management* 11 (1): 9–20.

Boyer, Kenneth K., and Christopher McDermott. 1999. Strategic consensus in operations strategy. *Journal of Operations Management* 17 (3): 289–305.

Bradach, Jeffrey L. 1997. Using the plural form in the management of restaurant chains. *Administrative Science Quarterly* 42 (2): 276–303.

Cable, Daniel M., and Scott Shane. 1997. A prisoner's dilemma approach to entrepreneur-venture capitalist relationships. *Academy of Management Review* 22 (1): 142–176.

Delmas, Magali A., and Michael W. Toffel. 2008. Organizational responses to environmental demands: Opening the black box. *Strategic Management Journal* 29 (10): 1027–1055.

Fisher, Anthony C., and Michael H. Rothkopf. 1989. Market failure and energy policy: A rationale for selective conservation. *Energy Policy* 17 (4): 397–406.

Forrester, Jay W. 1971. *World dynamics*. Cambridge, MA: Wright-Allen Press.

Freeman, R. Edward. 1984. Strategic management: A stakeholder approach. Lanham, MD: Pitman.

Friedman, Milton. 1970. The social responsibility of business is to increase its profits. *The New York Times Magazine*. September 13: 32.

Healy, Tim. 2007. EnerNOC chairman's letter to shareholders. *Annual Report*: 5–9. http://www.shareholder.com/visitors/dynamicdoc/document.cfm?documentid=2424&companyid=ENOC&page=1&pin=537039828&language=EN&resizethree=yes&scale=100&zid= (accessed April 17, 2010).

Hoffman, Andrew J., and John R. Ehrenfeld. 1998. Corporate environmentalism, sustainability, and management studies. In *Sustainability: Strategies for industry. The future of corporate practice*, ed. Nigel J. Roome, 55–73. Washington, DC: Island Press.

HSPI. 2011. *MindClick SGM Study: Green corporate business travel to make strong impact on hotel brand selection*. Hospitality Sustainable Purchasing Index. http://www.hspiconsortium.com/press-releases/mindclick-sgm-study-green-corporate-business-travel-to-make-strong-impact-on-hotel-brand-selection/ (accessed September 8, 2011).

Jaffe, A. B., and R. N. Stavins. 1994. The energy-efficiency gap. *Energy Policy* 22 (10): 804–810.

Janssen, Marco A. 2008. Evolution of cooperation in a one-shot Prisoner's Dilemma based on recognition of trustworthy and untrustworthy agents. *Journal of Economic Behavior & Organization* 65 (3–4): 458–471.

King, Andrew A., and Michael J. Lenox. 2002. Exploring the locus of profitable pollution reduction. *Management Science* 48 (2): 289–299.

Klassen, Robert D., and Curtis P. McLaughlin. 1996. The impact of environmental management on firm performance. *Management Science* 42 (8): 1199–1214.

Kummerow, Max. 1999. A system dynamics model of cyclical office oversupply. *Journal of Real Estate Research* 18 (1): 233–256.

Luce, R. Duncan, and Howard Raiffa. 1957. *Games and decisions: Introduction and critical survey*. New York: Wiley.

Mathewson, G. Frank, and Ralph A. Winter. 1985. The economics of franchise contracts. *Journal of Law & Economics* 28 (3): 503–526.

McDonnell, Mark J., and Steward T. A. Pickett. 1990. Ecosystem structure and function along urban-rural gradients: An unexploited opportunity for ecology. *Ecology* 71 (4): 1232–1237.

McFadden, Daniel. 1974. Conditional logit analysis of qualitative choice behavior. In *Frontiers in econometrics*, ed. Paul Zarembka, 105–142. New York: Academic Press.

O'Riordan, T. 1993. The politics of sustainability. In *Sustainable environmental economics and management: Principles and practice*, ed. R. Kerry Turner, 3–36. Hoboken, NJ: John Wiley & Sons.

Porter, Michael E., and Claas van der Linde. 1995. Toward a new conception of the environment-competitiveness relationship. *Journal of Economic Perspectives* 9 (4): 97–118.

Prakash, Aseem. 2002. Green marketing, public policy and managerial strategies. *Business Strategy and the Environment* 11 (5): 285–297.

Pruitt, Dean G. 1967. Reward structure and cooperation: The decomposed prisoner's dilemma game. *Journal of Personality and Social Psychology* 7 (1): 21–27.

Reddy, V. Ratna. 2000. Sustainable watershed management: Institutional approach. *Economic and Political Weekly* 35 (38): 3435–3444.

Schleich, Joachim, and Edelgard Gruber. 2008. Beyond case studies: Barriers to energy efficiency in commerce and the services sector. *Energy Economics* 30 (2): 449–464.

Segal, David. 2009. Pillow fights at the Four Seasons. *New York Times*, June 28, 2009, Business: 1.

Sherwyn, David. 2010. The hotel industry seeks the elusive "green bullet." *Cornell Hospitality Roundtable Proceedings* 2 (1). http://www.hotelschool.cornell.edu/chr/pdf/showpdf/chr/roundtableproceedings/sustainabilitytopost.pdf?my_path_info=chr/roundtableproceedings/sustainabilitytopost.pdf (accessed December 7, 2011).

Shrum, L. J., John A. McCarty, and Tina M. Lowrey. 1995. Buyer characteristics of the green consumer and their implications for advertising strategy. *Journal of Advertising* 24 (2): 71–82.

Swaminathan, Jayashankar M., Stephen F. Smith, and Norman M. Sadeh. 1998. Modeling supply chain dynamics: A multiagent approach. *Decision Sciences* 29 (3): 607–632.

Tuttelman, Mathew. 2008. EnerNOC energy start-up presentation at Boston University School of Management. November 19.

Unsal, Hakan I., and John E. Taylor. 2009. Integrating agent-based simulation with game theory to investigate the hold-up problem in project networks. In *Construction Research Congress proceedings: Building a sustainable future*, ed.

Samuel T. Ariaratnam and Eddy M. Rojas, 1280–1289. Reston, VA: American Society of Civil Engineers.

Verma, Rohit, Gary M. Thompson, and Jordan J. Louviere. 1999. Configuring service operations in accordance with customer needs and preferences. *Journal of Service Research* 1 (3): 262–274.

Zhang, Jie J., Nitin Joglekar, and Rohit Verma. 2011. Contract incentive design for environmental sustainability in service operations: A principal-agent analysis in the U.S. hospitality industry. Working paper. University of Vermont.

Zhang, Jie J., Nitin Joglekar, and Rohit Verma. 2012. Pushing the frontier of sustainable service operations management: Evidence from U.S. hospitality industry. *Journal of Service Management* 23 (3): 377–399.

6

The Evolution of the Green Building Supply Industry: Entrepreneurial Entrants and Diversifying Incumbents

Michael Conger and Jeffrey G. York

Introduction

The pursuit of ecological sustainability by entrepreneurs, existing corporations, and nonprofit organizations has triggered the emergence of diverse new industries (Sine and Lee 2009; Weber, Heinze, and DeSoucey 2008); green building, renewable energy, biofuel, and organic foods are a few well-known examples. These emerging, ecologically relevant industries are important drivers of the replacement of environmentally damaging practices, and raise questions for traditional theories of economics. Better understanding of these new sectors may help us build on theories of industry emergence, see how the social and economic focus of firms affects the evolution and legitimacy of the industrial field, and, specifically, help us explain the emergence of green building.

Scholars have long explored the people, institutions, and events that lead to the emergence of new markets and industries that provide spillover to the public good (Lounsbury 1998, 2003; Haveman, Rao, and Paruchuri 2007; Hiatt, Sine, and Tolbert 2009; Sine and Lee 2009; Weber, Heinze, and DeSoucey 2008), benefits that accrue to society and the natural environment as a consequence of the economic activities. Prior research shows that social movement actors play an important role in shaping these industries as they form and evolve (Hoffman 1997; Sine and Lee 2009). In fact, ecologically relevant industries remain dependent on social movements even as they start to take on the characteristics of a commercial market (Lounsbury 2003; Sine and Lee 2009; Urbano, Toledano, and Ribeiro Soriano 2010; Weber, Heinze, and Desoucey 2008). However, scholars often overlook the role that *for-profit* firms play in influencing the norms and institutional structures that develop as these industries mature. Examining these firms and their actions is particularly important if we want to understand the differences among industries that provide spillover to the public good and those that do not. What determines when and why a particular firm will enter an ecologically relevant industry? How does this differ for new firms vs. existing firms that already operate in another industry? And, what role do firms play in shaping the industry once they become part of it?

In this chapter, we address these questions by offering a theoretical explanation of the emergence and evolution of the green building supply industry that grew in tandem with the emergence of the green building construction industry following the introduction of the LEED (Leadership in Energy and Environmental Design) standard in 1998. This chapter first outlines existing scholarly theories that could explain who would enter new industries—such as the green building industry—during their emergence. In doing so, we outline what we would expect to find "on the ground." Finally, we illustrate how these theories bore out in the emergence of the green building industry.

We view the green building sector, encompassing the new markets and businesses that emerged around the introduction of LEED, as an integral part of an ongoing social movement aimed at changing construction practices in the United States and beyond. The firms in this sector, such as green building construction firms (i.e., contractors constructing LEED certified buildings) and the green building supply firms (i.e., firms producing materials and systems for use in LEED projects) compete in economic markets. However, they differ from firms in the conventional building sector in that they intentionally create a specific spillover of public goods—the reduction of environmental damage in the construction process and through the life cycle of the building—in addition to the value appropriated by industry firms. Because environmentalism has traditionally been the realm of social movements, we expect that firms seeking to succeed in the green building supply sector will be cognizant of, and may even partner with, social movement organizations. Using the green building supply industry as a context, we theorize that these two industry characteristics—the presence of public goods spillover and the influence of social movement actors—will affect the kind of firms that will enter the industry and the timing of their entry. We also argue that the firms that populate the industry will, out of necessity, work together to establish common practices and industry norms. Through this collective action, and in cooperation with other social movement actors, the industry's firms will significantly influence the economic and institutional characteristics of the industry.

We suggest that the early life of the newly emerging green building supply industry was laced with uncertainty about the potential economic benefits of opportunities and the permanence of the institutional structure that would ultimately define the new field. This uncertainty was compounded by the unique characteristics of the industry as part of an environmentalism ethos based on the concept that the niche of green building could change the broader field of the built environment (see chapter 9, this volume). Unlike a typical market, in which there is uncertainty about the potential of opportunities to create value, the presence of public goods spillover also caused uncertainty about whether that value could or *should* be kept as profit by the firms that created it. Would customers be willing to pay a premium for green building? How could one be sure that materials were truly "green"? Finally, how would local and state-level policies evolve to recognize, or even encourage, the nascent LEED standards?

The public goods aspect of green building amplified uncertainty about the future institutional characteristics of the industry because it was not clear how economic and ecological concerns could or should be simultaneously pursued. Firms choosing to enter at this early stage would be those that did not see this uncertainty as a concern because they believed that the value they would be able to appropriate would be sufficient and, more important, that society and the natural environment *should* benefit directly from their pursing the opportunity. In fact, uncertainty may have been an advantage for these firms because it allowed them to enter without having to compete with diversifying firms. In other words, their ethical commitment to the ecological goals of green building may have reduced the impact of economic uncertainty on these firms' decisions to enter the nascent green building supply industry. We expect that entrepreneurs whose personal values aligned with the goals of the environmental social movement often founded these new firms.

Meanwhile, most existing building industry firms, having already found alignment with the economic and institutional logics of the existing commercial building industry, saw early opportunities as unattractive due to this market uncertainty, regardless of public good spillover, and were less likely to enter. As the industry matured, it developed and gained sufficient density to reduce economic uncertainty. The practical and ethical consequences of the presence of public goods spillover were now clearer. At the same time, institutional uncertainty was also reduced since influential institutional actors had determined the norms of the emerging green building industry and established its legitimacy. These included social movement actors such as the USGBC (U.S. Green Building Council) and, especially important, the firms that had entered earlier and engaged the institutional development of the industry. At this later stage, diversifying incumbent firms were more likely to enter since it was possible to do so with greater certainty.

We see the development of the green building supply industry as an ideal context in which to develop a theory that explains how ecologically and/or socially relevant industries evolve. While the mechanisms, such as the effects of uncertainty, are similar to those in any emerging industry, the presence of public goods spillover, powerful social movement actors, and socially minded entrepreneurs affects the development of these industries and makes them distinct from other, similar industries. To understand these processes in more detail, we look next to entrepreneurship and institutional theory literature and show how these theories can explain the actions of individual firms as they entered and helped shape the green building supply industry.

How New Industries Evolve

Management scholars have long sought to understand the evolution of markets and the factors that lead to the formation of new industries. Their primary questions ask: How do new industries emerge and change over time? What determines which firms will enter an industry at different stages in its development? How do the actions of firms determine the characteristics of the industry as it evolves? Many prominent theories of the firm emphasize rationality and efficiency as the driving forces behind new firm formation, horizontal and vertical integration decisions, forms of governance, and new market entry. Prominent examples include transaction cost economics (Coase 1937; Williamson 1975, 1985) and agency theory (Jensen and Meckling 1976).

While these theories can help explain how firms come into existence, they tell us little about how a new industry is shaped by the pattern of entry and the subsequent actions of these entrants. Even less understood is how this process may be different in an industry where environmental benefits are a central concern. In this chapter, we follow the tradition of defining incumbents as those firms that existed before entering the new industry, perhaps operating in a related existing industry. Studies of firm entry examine multiple factors that impact the likelihood that firms will enter a newly emerging market. New goods and services emerge as industries evolve; products may be extensions of existing products (Abernathy and Utterback 1978) or they may emerge following a shock to the organizational environment such as a technological innovation (Tushman and Anderson 1986) or the development of new knowledge such as improved business processes or shifting industry standards (Nelson and Winter 1982). When new industries emerge (whether complementary or competing), managers in incumbent firms must decide whether to enter the new field. Prior literature has argued both for and against incumbent advantage in new market entry.

The Entrepreneurial Theory of the Firm

Why and how, then, does any *particular* market or firm come into being and how do they evolve over time? Recent work in entrepreneurship research attempts to address this question with a more finely pointed theory of the firm. These scholars describe firm creation as resulting from the nexus of the individual and an opportunity (Shane and Venkataraman 2000; Venkataraman 1997). Rooting their argument in this tradition, Dew, Velamuri, and Venkataraman (2004) propose an entrepreneurial theory of the firm based on Hayek's (1945) concept of knowledge dispersion and uncertainty. According to this theory, entrepreneurial opportunities exist because unique knowledge about any particular time and place is dispersed among the individual people who inhabit that time and place. This dispersion of knowledge, combined with unavoidable uncertainty about the future (Knight 1957/1921) leads to different expectations about the potential value of an opportunity. Since this potential value is not only unknown, but also unknowable, it is not possible for everyone to have the same beliefs about the opportunity (Davidson 2001). Instead, the entrepreneur acts on her own, individual beliefs about the opportunity's potential. Dew, Velamuri, and Venkataraman (2004) suggest that this process also explains the emergence of new markets and why new rather than existing firms will enter them. They argue that the agreement about market opportunities among firms is different from the agreement among individuals within an individual firm. When agreement among firms is low, as would be the case when uncertainty is high, only firms with higher internal agreement will see entrepreneurial opportunity. Central to these processes of decision making are considerations for both economic and institutional uncertainty.

Economic Uncertainty

Novelty is key to making a new field attractive to entrepreneurial firms, which will see an opportunity to create a market, and unattractive to incumbent firms, which will see either a small, unattractive market with questionable ability to support value creation or, possibly, see no market at all. Of course, some established firms will also see value in early entry, perhaps being uniquely well positioned to recognize and pursue opportunities in the face of uncertainty, or seeking a first-mover advantage. However, these early diversifying incumbents should be a relatively small subset of the incumbent firms that will eventually enter the field. In both cases, these early entrants see potential value despite a lack of agreement about the potential of the newly emerging field.

Newly emerging industries, and particularly those with ecological relevance, represent a novel case of industry emergence. Green industries represent markets with public goods spillover that had not existed previously. At the outset, the potential for economic profit and the landscape of customers, suppliers, and competitors are all unknown. Although uncertainty about opportunities is present in any new market, it is compounded in industries with public goods spillover because it is unclear how the value of the public good may be captured (Olson 1971). Central to the emergence of the green building sector is the fundamental idea that the goals of green building as a social movement—reducing the negative impact of building construction and improving the health and well-being of building occupants—can and should be addressed through the market rather than through government policy alone.

Thus, the inherent ideology that public goods and private economic profit can and *should* be pursued simultaneously compounds uncertainty about market opportunities in the green building sector. Not only is the potential for creating value unknown, but the way in which the created value can and should be distributed between the firm and the public good is unclear. A potential entrant to the industry, therefore, has increased uncertainty regarding the potential to create value or how much of it they could expect to appropriate to make an informed calculation of risk. Firms can only enter through entrepreneurial action, judging the opportunity to be viable and attractive in the face of irreducible uncertainty (McMullen and Shepherd 2006). According to entrepreneurial theory, we expect that new firms will more likely enter during this period of uncertainty. However, the unique, additional goal of addressing ecological degradation may act to push would-be entrepreneurs *into entry when they would not otherwise do so* based on economic calculations alone. In the green building sector, this phase of socially minded entrepreneurship occurred roughly from 2000 to 2005, as the LEED standards were introduced and supported through the USGBC's efforts to build legitimacy for the practice.

As the industry grows and the number of overall entrants increases, the viability of economic opportunities in the sector becomes more obvious. Also, the way in which value can and should be appropriated by the private firm becomes clear. At this point, the level of agreement between firms about market potential and the nature of the competitive landscape peaks and theories of transaction costs become relevant. Firms will now enter based on their ability to estimate the likelihood of success in the new market since it is also now much easier to attain agreement within the firm. We expect to see incumbent firms enter much more frequently at this point.

Institutional Uncertainty

Institutional theorists also address the uncertainty inherent in new industries. In a newly forming industry, institutional voids are commonplace (Aldrich and Fiol 1994; Scott 1995; Zimmerman and Zeitz 2002). Also, institutional actors are often in disagreement or outright conflict as the norms and shared meanings of institutions are converging (Kraatz and Block 2008; Scott 1995). The outcomes, and even the processes of institutional formation itself, impact the success and the very survival of firms that enter the field (Oliver 1991; Scott 1995; Stinchcombe 1965; Suchman 1995). These initial conditions, by themselves, represent an uncertain institutional environment.

At the same time, entering firms are institutional actors, affecting the success and survival of the field (Hannan and Freeman 1984), thereby subjecting an already unpredictable institutional environment to dynamic change. For these reasons, the ability of any entrant to make objective judgments about the institutional landscape of newly emerging industries in the green building sector is subject to uncertainty. As we explained earlier, uncertainty in these industries is compounded due to the presence of public goods spillover; while social movement organizations are in the early process of establishing legitimacy for the field, it is not possible to know what the institutional structure of the sector and the new industries within it will ultimately look like. In the case of the green building sector, the USGBC quickly became the dominant institutional actor from the very beginning. And, there is little doubt that the LEED certification was a catalyst in the birth of the green building sector. However, it may be that the rapid emergence of these institutional changes also served as a sorting mechanism for potential industry entrants, helping to determine which firms would enter and when.

Incumbent firms are often bound to existing institutional norms for the sectors in which they already operate. Institutional change is usually met with skepticism by actors tied to existing institutions, especially when those institutions are directly challenged, as in the case of green building (Deephouse and Suchman 2008; Kraatz and Block 2008; Oliver 1991). We argue, on the one hand, that incumbent firms will see the institutional landscape of the new sector as uncertain. Moreover, they may even see the emerging institutional norms as lacking legitimacy given their newness and conflict with institutions already deemed legitimate.

New firms, on the other hand, will see opportunity in institutional change. They likely attribute more legitimacy to the USGBC and the LEED standards than they do to the institutions of the conventional building sector. Also, in the same way that entrepreneurs find opportunities attractive in the face of economic uncertainty, they may also find an uncertain institutional environment attractive (York and Venkataraman 2010). Firms that enter new industries, particularly when they enter early, are obliged to participate in the institutional development of that industry (Aldrich and Fiol 1994). Doing so is essential to their survival and the survival of the industry (Zimmerman and Zeitz 2002). But new firms should also see this as an opportunity to help establish the industry norms and have a hand in establishing rules of the game that will favor the outcomes they prefer for themselves and the industry.

And so, institutional uncertainty, like economic uncertainty, inhibits agreement between firms and favors the entrepreneur over the diversifying incumbent firm in the early days of a field like the green building supply industry. Entrepreneurial firms will have greater internal agreement about the institutional forces that will govern the industry in the future. These firms are therefore more likely to see the uncertainty as an opportunity to shape the industry to align with their goals.

Institutional Entrepreneurship and Collective Action

Firms operating in sectors with public goods spillover have a strong tendency to engage in collective action with other firms and institutional actors in the sector (Austin, Stevenson, and Wei-Skillern 2006; Mair and Martí 2006, 2009). Common tactics include increased cooperation with other organizations, participation in, or partnership with, foundations and other social movement groups, and even engaging in lobbying or other political efforts. Ventures with prominent social or environmental goals tend to treat other organizations in the field as potential collaborators rather than potential competitors (Austin, Stevenson, and

Wei-Skillern 2006). Furthermore, because of the positive environmental impact promised by the sector's survival, firms with similar social goals will have a vested interest in filling the sector's institutional voids (Mair and Martí 2009).

In other words, these firms are not only entrepreneurial, but they also act as de facto institutional entrepreneurs, linking the success of their venture with the success of the field. Early in the life of a new field, legitimacy is low since the institutional framework is still being built and may be contested by multiple actors (Aldrich and Fiol 1994; Kraatz and Block 2008). Early entrants will deal with the lack of legitimacy by helping to establish institutions for the industry through collective action and by voluntarily conforming to the norms and expectations of those institutions (Aldrich and Fiol 1994; Oliver 1991; Suchman 1995; Zimmerman and Zeitz 2002). Also, just as the success of the industry affects the firms operating in it, the success or failure of early entrants also has a significant and lasting impact on the legitimacy and survival of the industry (Hannan and Freeman 1984). As the industry matures, legitimacy will grow and agreed upon institutions will emerge (Zucker 1987). As this happens, differences in the institutional actions different firms take will fade, and both new and incumbent firms will become more alike and homogenous, taking symbolic actions that demonstrate they are conforming to the field's institutional norms (DiMaggio and Powell 1983).

We illustrate the theories we propose in the context of green building, and the green building supply industry. In the remainder of this chapter, we describe the emergence and evolution of the industry.

The U.S. Green Building Supply Industry

The USGBC defines green building as "design and construction practices that significantly reduce or eliminate the negative impact of buildings on the environment and occupants" (USGBC 2004). In short, green buildings are designed to reduce waste, negative impact on the surrounding natural environment, and damage to human health. This is accomplished through the use of a variety of practices, technologies, and materials not commonly used in conventional building design and construction. Examples include: choosing sustainable building sites; reusing materials; using water conservation systems, green power, rapidly renewing building materials, and only nontoxic materials; and designing for increased natural ventilation, daylighting, and outdoor views (ibid.).

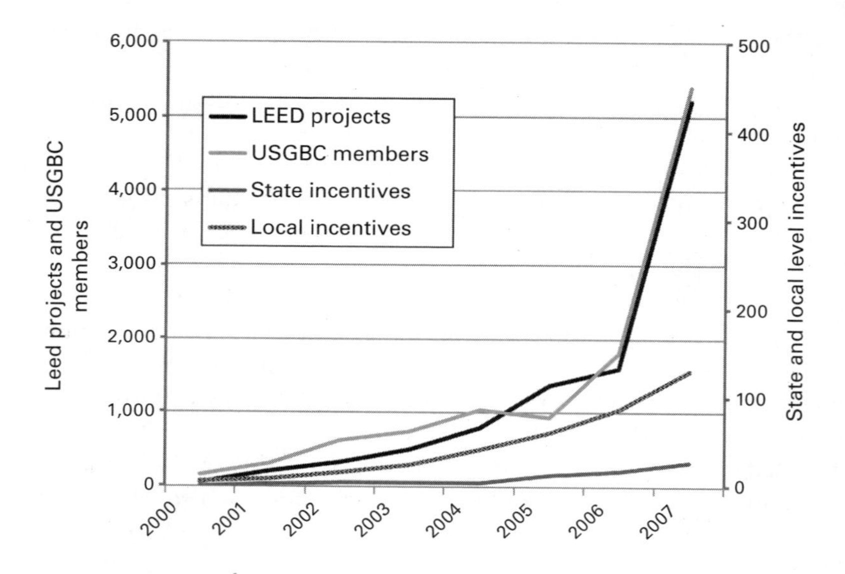

Figure 6.1

Evolution of green building sector through adoption, social movements, and policy (data supplied to authors by USGBC, August 29, 2008)

The green building sector effectively came into being in 2000 with the release of the LEED 2.0 green building certification to the public. The LEED certification defines standards for green building and evaluates project compliance with those standards. In 2000, forty-one commercial building projects were registered with LEED in the United States. By November 2011, there were 32,451 registered commercial projects and 11,082 commercial projects had achieved LEED certification (USGBC 2011b). The USGBC also developed related professional accreditation standards for individuals in the field. In 2010, the sector boasted over 157,000 LEED Accredited Professionals (APs) active across all areas of practice in the building industry (USGBC 2011a).

Often encouraged by the rise in LEED projects, supplier firms emerged, further fueling growth in the sector and broadening the green building field. These firms include both new ventures and incumbent building supply firms diversifying into the green building market. During the early time period, the green building sector was marked by intensive institutional development in the form of legitimacy-building activities in which actors in the sector, including green suppliers, could participate. Examples include the explosion of LEED-certified building projects, growing membership of the USGBC, the introduction of multiple LEED professional accreditation standards, rapid growth in the number of LEED APs, growing attendance at USGBC conferences, the establishment of programs to educate consumers about green building benefits/practices/standards, and the establishment of lobbying committees for state or federal legislation, or both, which promotes the green building sector. As illustrated in figure 6.1, LEED registered projects, USGBC membership, and state-level and local-level policies all rose sharply in the years 2000 to 2007.

The Evolution of the Green Building Supply Industry

When the green building supply industry first emerged in 2000, it was laced with uncertainty in several important respects. First, there was uncertainty about the potential scale and profitability of the green building market. Given the newness of the LEED certification standard and the relatively small number of green building projects underway, it was not at all certain that producing supplies to the exacting (or at the time, esoteric) standards required by LEED would pay off and it was even less certain that entering a market to produce materials and products to supply these projects was a sound decision. Growth in green construction drove the demand for these products. If the number of LEED projects did not grow, it was unlikely that customers in conventional building markets would need or want green supplies.

There was also a question of appropriability—the factors that govern an innovator's ability to capture profits generated by an innovation. Given the newness of the LEED standards and the central role of technological innovation in developing solutions for greener construction, products and services created for this market were very costly. At the same time, firms could not be sure about the market's ability to bear the higher prices that would necessarily follow, at least in the short term. It was clear that these improved products would benefit the natural environment and the people who inhabited LEED buildings, but there was no way to know whether this additional public good value would also generate profits for suppliers. If not, firms in the industry would be doing society and the planet a service but would be bearing the burden alone.

Finally there was fundamental uncertainty about the nature and strength of institutional forces that would develop and define the sector in the future, calling the legitimacy, and even the survival, of the industry into question. For example, one entrepreneur described the early days of her green building consulting business:

> I probably got a little discouraged there, that there wasn't enough activity in the industry. And I'd heard about the USGBC forming. In fact, a very close friend and colleague from the early solar days was one of the founding board members. So I would get reports from him and I'd be kind of . . . paying attention to this U.S. Green Building Council. Like, who is it and what are you doing? And it was very small and slow to take off. So . . . in some ways I think of myself as jumping in a little bit late, which was in . . . oh, probably end of 2003, in terms of the whole leaping at the beginning of 2004 . . . all of this was slow moving and trying to get some traction and find its way and define itself. (Anonymous interview conducted by authors, May 2010)

Even though collective action through the USGBC was remarkably successful given the challenges of initiating institutional change at such a large scale, it was not universally accepted. Not all organizations involved, directly or indirectly, in the construction of buildings would enter the green construction sector. Certainly many powerful players in the incumbent building sector were threatened by the emergence of the USGBC, LEED, and the green building sector as a whole. So, while many early entrepreneurs may have seen the production of products designed to reduce ecological degradation as morally correct, others questioned the efficacy and economic legitimacy of the new industry.

Together, these sources of uncertainty reduced agreement among firms about opportunities in the industry. In the early days of the industry, most established building supply firms did not see the potential market for supplying LEED projects as financially viable. Instead, entrepreneurial firms that could muster sufficient agreement within the organization began to enter the market. Often these early entrepreneurs acted in the role of both business venture owners, and proponents for the logic of green building. These early entrants saw opportunities to create future profits and to shape the institutions that influence the new industry. In doing so, they may have given themselves a better chance to survive and to achieve their goals. Where others saw risk, they saw an opportunity to create a venture that not only would provide them with financial benefit but also would achieve a higher moral purpose and benefit society. Many early entrants acted as institutional entrepreneurs, actively participating in the legitimating of the field through collective action and voluntary conformance to emerging institutional norms and expectations, typically through the USGBC. In this way, they played a significant role in shaping the institutional character of the sector. By their actions through the USGBC and participation in the green building supply industry, these entrepreneurial firms were an active part of green building as a social movement.

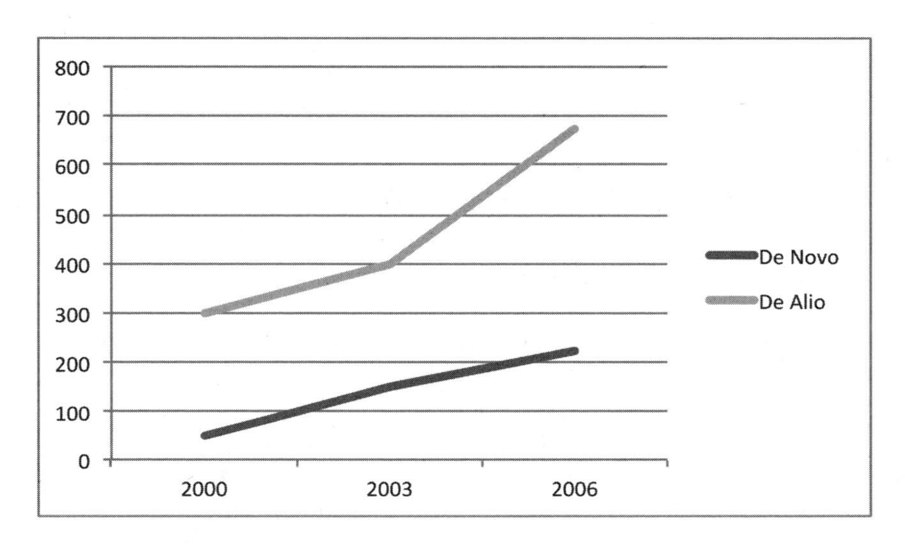

Figure 6.2

Green building sector firm entry rates per year (source: GreenSpec Catalog, 2000–2006)

Over time, as the industry developed, both economic and institutional uncertainty was reduced. LEED projects began to spring up all across the country and it became clear that the green building philosophy was more than a fad; by the year 2003 there were 494 LEED-registered projects vs. 41 in 2000. At this point, we also see an increased rate of entry from incumbent firms (see figure 6.2).

Professionals in the construction industry began to take an interest in incorporating green building techniques and materials in conventional building projects, and the public even began to view green supplies as a viable purchase for home construction and improvement projects. As more firms entered the industry and were successful, the viability of economic opportunities in the sector became more obvious. The mere increase in the number of firms in the industry reduces uncertainty about the market's future and shows that success, or at least survival, is possible. These firms are also models for future entrants, demonstrating the ways in which value appropriation is acceptable while maintaining a commitment to the simultaneous creation of public goods. This also reduces uncertainty and boosts agreement among firms about the attractiveness of the market. Also, as firms persist in the industry, they engage in institution-building activities since early entrants will be obliged act together in order to survive and ensure the survival of the industry. Their collective action is a kind of inter-firm agreement and may also serve as a shortcut to agreement for future entrants.

Also borne out was the collective work of entrepreneurial firms in the green building sector, acting as institutional entrepreneurs alongside and through the USGBC, to establish institutional structures. Clarity regarding the correctness and stability of the characteristics and norms of the industry became more apparent and the legitimacy of the field grew. Professional standards for quality and environmental sustainability were clearly codified and enforced through the LEED AP program. Regulators recognized green building and lent considerable legitimacy to the field by enacting green building support policies in most states. Educational initiatives began to bear fruit, igniting interest among architects, builders, suppliers, and even the public at large. At this stage, around 2004, the growing agreement among firms reached a tipping point, after which diversifying incumbents saw opportunity in the sector and began to enter at an increased rate. An architect involved in the green building sector since its beginning described this convergence in an interview:

> You can look at the LEED statistics, you can look at the number of people who have become LEED APs. . . . All those are the latest indicators that this has become a mainstream idea and people ask for it. People expect it. . . . [Now] everybody's sort of touting their green credentials. And, I think the USGBC's had a substantial part in making green almost always a part of the conversation. And I think that's the biggest accomplishment. (Anonymous interview conducted by authors, January 2010)

Later entrants to the industry, by necessity, conformed to institutional pressures to comply with the new institutional norms. They engaged in activities that represented a kind of membership dues that all firms must pay to participate in the market. These firms would, for example, recognize LEED certifications and become members of the USGBC. At this stage, green building began to be seen as mainstream and more incumbent firms from the conventional building supply industry were willing to change their products and strategies to fit within the institutional context of the now attractive and influential green building supply industry. We even suggest that incumbent firms *were forced* to enter the green building supply industry or risk losing out on a lucrative and established sector.

Conclusion

Understanding how new industries that focus on public goods such as environmental sustainability emerge and evolve over time has important implications for future research and for the firms that enter these industries. It is essential to recognize that these industries emerge under conditions of uncertainty and that this uncertainty is compounded by the presence of spillover to the public good, and the fact that they are market-based extensions of ongoing social movements. Because of this, entrepreneurs play perhaps the most important role in making socially and environmentally relevant markets a reality.

Certainly, the groundwork for these industries is laid by many different actors such as professional and regulative organizations—most notably, the USGBC, government policymakers, developers of new technology, and social or environmental activist groups. But, entrepreneurial firms that create opportunities in the face of uncertainty will be the first to actually participate in the new market. They are both the builders of the new industry and the raw materials from which it is made. In the end, it is their success or failure that will determine the future of the industry and initiate the widespread agreement that is necessary for the industrial field to shift toward more environmentally sensitive practices.

This is an important point because it suggests that entrepreneurs must attend to both their social mission and the economic realities of operating in a new market. It is this dual focus that allows these early entrants to effect real and lasting change. The growth of suppliers in the green building sector indicates a shift in the way we think about how buildings are to be designed, built, and used. Entrepreneurial firms in the industry demonstrate that the practice of green building can be economically viable. They are also uniquely positioned to shape the institutional structure of the industry and establish norms that favor environmental sustainability in the long run. In this way, they simultaneously make the market attractive to established firms and help establish the norms and other institutional characteristics that will define the industry. In this way, entrepreneurs have the potential to fundamentally change or replace existing business practices with socially/environmentally relevant alternative business models. Because of this, engaging entrepreneurs in the process of social or environmental change may be a superior strategy for social activists and others hoping to effect lasting change.

Today, the green building sector is growing rapidly and it is conceivable that, within our lifetime, green building industries will revolutionize every aspect of the way buildings are constructed. If so, it will be due in no small part to entrepreneurial firms. Firms that create opportunities and take action where others do not will be the ones that help shape new industries in the green building sector and create markets that make the promise of sustainable building a reality.

References

Abernathy, William J., and James M. Utterback. 1978. Patterns of industrial innovation. *Technology Review* 80 (7): 40–47.

Aldrich, Howard E., and C. Marlene Fiol. 1994. Fools rush in? The institutional context of industry creation. *Academy of Management Review* 19 (4): 645–670.

Austin, James, Howard Stevenson, and Jane Wei-Skillern. 2006. Social and commercial entrepreneurship: Same, different, or both? *Entrepreneurship Theory and Practice* 30 (1): 1–22.

Coase, Ronald H. 1937. The nature of the firm. *Economica* 4:386–405.

Davidson, Donald. 2001. *Subjective, intersubjective, objective*. Oxford: Clarendon Press.

Deephouse, David L., and Mark C. Suchman. 2008. Legitimacy in organizational institutionalism. In *The Sage handbook of organizational institutionalism*, ed. Royston Greenwood, Christine Oliver, Kerstin Sahlin, and Roy Suddaby, 49–77. Thousand Oaks, CA: Sage.

Dew, Nicholas, S. Ramakrishna Velamuri, and Sankaran Venkataraman. 2004. Dispersed knowledge and an entrepreneurial theory of the firm. *Journal of Business Venturing* 19 (5): 659–679.

DiMaggio, Paul, and Walter W. Powell. 1983. The iron cage revisited: Institutional isomorphism and collective rationality in organizational fields. *American Sociological Review* 48 (2): 147–160.

Hannan, Michael T., and John Freeman. 1984. Structural inertia and organizational change. *American Sociological Review* 49 (2): 149–164.

Haveman, Heather A., Hayagreeva Rao, and Srikanth Paruchuri. 2007. The winds of change: The Progressive movement and the bureaucratization of thrift. *American Sociological Review* 72 (1) (February 1): 117–142.

Hayek, F. A. 1945. The use of knowledge in society. *American Economic Review* 35 (4): 519–530.

Hiatt, Shon R., Wesley D. Sine, and Pamela S. Tolbert. 2009. From Pabst to Pepsi: The deinstitutionalization of social practices and the creation of entrepreneurial opportunities. *Administrative Science Quarterly* 54 (4): 635–667.

Hoffman, Andrew J. 1997. *From heresy to dogma: An institutional history of corporate environmentalism*. San Francisco: New Lexington Press, Jossey-Bass Publishers.

Jensen, Michael C., and William H. Meckling. 1976. Theory of the firm: Managerial behavior, agency costs and ownership structure. *Journal of Financial Economics* 3 (4): 305–360.

Knight, Frank H. 1957/1921. *Risk, uncertainty and profit*. 8th ed. New York: Kelley & Millman.

Kraatz, Matthew S., and Emily S. Block. 2008. Organizational implications of institutional pluralism. In *The Sage handbook of organizational institutionalism*. ed. Royston Greenwood, Christine Oliver, Kerstin Sahlin, and Roy Suddaby, 243–275. Thousand Oaks, CA: Sage.

Lounsbury, Michael. 1998. Collective entrepreneurship: The mobilization of college and university recycling coordinators. *Journal of Organizational Change Management* 11:50–69.

Lounsbury, Michael. 2003. Social movements, field frames and industry emergence: A cultural-political perspective on US recycling. *Socio-Economic Review* 1 (January 1): 71–104.

Mair, Johanna, and Ignasi Martí. 2006. Social entrepreneurship research: A source of explanation, prediction, and delight. *Journal of World Business* 41 (1): 36–44.

Mair, Johanna, and Ignasi Martí. 2009. Entrepreneurship in and around institutional voids: A case study from Bangladesh. *Journal of Business Venturing* 24 (5): 419–435.

McMullen, Jeffery S., and Dean A. Shepherd. 2006. Entrepreneurial action and the role of uncertainty in the theory of the entrepreneur. *Academy of Management Review* 31 (1): 132–152.

Nelson, Richard R., and Sidney G. Winter. 1982. *An evolutionary theory of economic change*. Cambridge, MA: Harvard University Press.

Oliver, Christine. 1991. Strategic responses to institutional processes. *Academy of Management Review* 16 (1): 145–179.

Olson, Mancur. 1971. *The logic of collective action: Public goods and the theory of groups*, vol. 124. Cambridge, MA: Harvard University Press.

Scott, W. Richard. 1995. *Institutions and organizations*. Thousand Oaks, CA: Sage Publications, Inc.

Shane, Scott, and S. Venkataraman. 2000. The promise of entrepreneurship as a field of research. *Academy of Management Review* 25 (1): 217–226.

Sine, Wesley D., and Brandon Lee. 2009. Tilting at windmills? The environmental movement and the emergence of the US wind energy sector. *Administrative Science Quarterly* 54 (1): 123–155.

Stinchcombe, Arthur. 1965. Social structure and organizations. In *Handbook of organizations*, ed. James G. March, 142–193. Chicago: Rand McNally.

Suchman, Mark C. 1995. Managing legitimacy: Strategic and institutional approaches. *Academy of Management Review* 20 (3): 571–610.

Tushman, Michael L., and Philip Anderson. 1986. Technological discontinuities and organizational environments. *Administrative Science Quarterly* 31 (3): 439–465.

Urbano, David, Nuria Toledano, and Domingo Ribeiro Soriano. 2010. Analyzing social entrepreneurship from an institutional perspective: Evidence from Spain. *Journal of Social Entrepreneurship* 1 (1): 54–69.

USGBC. 2004. An introduction to the U.S. Green Building Council and the LEED certification program. https://www.usgbc.org/Docs/Resources/usgbc_intro.ppt (accessed December 29, 2011).

USGBC. 2011a. About USGBC. https://www.usgbc.org/ShowFile.aspx?DocumentID=4686 (accessed December 29, 2011).

USGBC. 2011b. About LEED. https://www.usgbc.org/ShowFile.aspx?DocumentID=4687 (accessed December 29, 2011).

Venkataraman, Sankaran. 1997. The distinctive domain of entrepreneurship research. In *Advances in entrepreneurship: Firm emergence and growth 3*, ed. Jerome Katz, 119–138. Greenwich, CT: JAI Press.

Weber, Klaus, Kathryn L. Heinze, and Michaela DeSoucey. 2008. Forage for thought: Mobilizing codes in the movement for grass-fed meat and dairy products. *Administrative Science Quarterly* 53 (3): 529–567.

Williamson, Oliver E. 1975. *Markets and hierarchies*. New York: Free Press.

Williamson, Oliver E. 1985. *The economic institutions of capitalism*. New York: Free Press.

York, Jeffrey G., and S. Venkataraman. 2010. The entrepreneur-environment nexus: Uncertainty, innovation, and allocation. *Journal of Business Venturing* 25 (5): 449–463.

Zimmerman, Monica A., and Gerald J. Zeitz. 2002. Beyond survival: Achieving new venture growth by building legitimacy. *Academy of Management Review* 27 (3): 414–431.

Zucker, Lynne G. 1987. Institutional theories of organization. *Annual Review of Sociology* 13:443–464.

7

Individual Projects as Portals for Mainstreaming Niche Innovations

Ellen van Bueren and Bertien Broekhans

Introduction

In the first decade of the twenty-first century, many innovative projects demonstrated that green construction can equal conventional ways of construction in terms of costs and quality (DBR 2010). Nevertheless, there are several organizational, institutional, and psychological barriers to sustainable construction that slow down and obstruct the mainstreaming of sustainable, construction (van Bueren and de Jong 2007; Hoffman and Henn 2008; Williams and Dair 2007, van Bueren 2009). The terms "sustainable construction" and "green construction" are used interchangeably in this chapter. Also, "building" and "construction" are used as synonyms, unless stated otherwise. Theories on strategic niche management help in understanding how these barriers prevent wide adoption of sustainable innovations.

Strategic niche management assumes that environmentally sustainable innovations, such as technologies and practices, radically differ from the existing socio-technical system or *regime*—the existing and dominant constellation of actors and their shared rules, assumptions, practices, infrastructures, and technologies of the mainstream construction sector. This regime, with other regimes, is embedded in a wider institutional landscape of formal rules, laws, and cultural values. The environmentally sustainable innovations can better evolve and mature in *niches* than in regimes. Niches are spaces dedicated to environmentally sustainable innovations and in which innovations are judged by a set of criteria that are augmented, whereas those in use by the regime are not. Eventually, the regime will adopt some of these innovations, but to do so, the regime and the innovation need mutual adjustment.

This chapter focuses on this "mainstreaming" process for sustainable innovations. By analyzing a particular construction project, we explore how a mainstream environment adopts niche innovations. This analysis will provide a more detailed knowledge of how the diffusion of niche innovations actually takes place and how projects function as a portal for mainstreaming.

The next section elaborates strategic niche management and the role of projects therein, followed by an introduction of the project studied in this chapter: the design and construction of a town hall in Leiderdorp, the Netherlands—a green project commissioned and delivered by mainstream actors. We analyze how both the environmentally sustainable construction niche and the conventional construction regime influenced this project. In the final sections, we conclude with how projects like the one studied help to understand the process of mainstreaming sustainable construction and how these processes can be improved.

From Green Niches to Regime Change: The Role of Projects

In this section we introduce the main theoretical concepts used in the analysis of the design process of the town hall. Strategic niche management theories are very effective in analyzing and explaining how incumbent *regimes* (e.g., construction, transport, agriculture) resist the adoption and diffusion of innovations developed within *niches* (e.g., sustainable building, electric cars, genetically modified crops) in the wider institutional/societal *landscape,* which forms the setting for regimes. The regime concept is a continually evolving hegemonic configuration of artifacts, actors, and institutions (Scrase and Smith 2009). It refers to dominant practices, activities, methods, and preferences that are often bound by formal and informal rules, or institutions, that make most societal actors such as experts, businesses, policymakers, and users blind and unreceptive to radically different technologies and practices. Instead, actors focus on improving, optimizing, regulating, and using existing designs.

As a regime, the conventional building and construction industry refers to both designs and the way they are produced, regulated, and used by most of us (Smith 2006, 2007; Brown and Vergragt 2008). There are all sorts of institutions within these regimes, ranging from formal ones like contracts and regulations, to informal ones like long-lasting relationships with suppliers or clients. The inertia these institutions create explains why 'business as usual' is preferred over the adoption of new innovations. A large number of studies emphasize that the building and construction industry is characterized by strong, stable institutions that are difficult to change (WBCSD 2008; UNEP SBCI 2009).

Environmentally sustainable construction, also known as green construction or green building, can be considered a *niche* within the established or incumbent regime of building design and construction. A niche is a place for innovation and experimentation which is sheltered from the pressure of the selection environment of the incumbent regime; niche innovations are judged by the rules in use in the niche and do not have to meet all of the regime requirements for adoption (Schot and Geels 2008). The development of all kinds of environmentally sustainable technologies as well as movements on eco-housing or sustainable building have been characterized as a niche by several authors (Smith 2006, 2007; Bergman, Whitmarsh, and Köhler 2008; Brown and Vergragt 2008; Seyfang 2010).

Both niche and regime are subject to macro societal (cultural, economic, and political) developments and dynamics that provide the context or landscape for action. These macro dynamics influence the willingness of regimes to change, as well as the possibilities for innovation adoption from niche to regime. Developments in this landscape influence regimes and generate stress and opportunities for regimes to change (Scrase and Smith 2009). For example, since the mid-2000s, climate change has become a major driver for the willingness of incumbent regimes to adopt niche innovations that reduce energy use and carbon emissions. This is also true in the building and construction field (Ahn and Pearce 2007; CB Richard Ellis 2009; DBR 2010; UNEP SBCI 2009). While the adoption of sustainable construction innovations by incumbent players develops very slowly (Ashuri and Durmus-Pedini 2010; WBCSD 2008), we can still observe the processes by which they occur.

Early theorization of strategic niche management in the 1990s focused on how sustainable innovations or practices developed in niches could then take over and replace established, incumbent regimes (Schot and Geels 2008). However, more recent work recognizes that in branching out, a niche interacts with the regime it intends to penetrate. Niche actors aim to transfer the innovations and the lessons learned in the niche to the actors in the regime. In this interaction, the regime exerts selection pressure upon the niche while the niche exerts pressure in return. This leads to a process of mutual adaptation by both the niche and the incumbent regime. Current theorizing acknowledges that under the influence of the wider contextual landscape of macro developments, mainstreaming is much more the result of a process of mutual adaptation of niche and regime toward each other.

Much of the recent scholarly work using a multi-level perspective of niche, regime, and landscape is oriented around setting a transition process in motion, and in particular sustainable transitions where experiments and innovation lead to the establishment of niches; and *successful* niche innovations can be transmitted to regimes. Strategic niche management then becomes the result of processes and interventions at multiple levels (Schot and Geels 2008). However, by emphasizing the importance of the development of long-term visions and reflexive capacity for the successful management of transitions, these texts offer little guidance to practitioners who try to contribute to this transition process in their daily work (Shove and Walker 2007). The texts also neglect industries characterized by nonhierarchical or project-based work—industries that cannot institute or guide "long-term visions" or "reflexive capacity" from above, such as building design and construction.

The actual processes by which incumbent regimes adopt and adapt niche innovations to their dominant rules and practices are an important area of study (Smith 2007). Mainstream regime projects differ from niches in which innovations were successful. In mainstream projects, actors confront tensions and conflicts between the niche and the regime in which the innovation is applied. Studying mainstream projects will contribute to more detailed knowledge of how niche innovations diffuse into such projects, which in turn can contribute to improving the design and management of mainstreaming niche innovations and practices.

The position of projects between niche and regime and how niche, regime, and landscape influence projects can be described as follows. Projects can be portals through which innovations are adapted and used and possibly transferred to the regime. In turn, projects also provide feedback to the niche on the innovations, whether and how they are adopted, and why they did or did not work out as intended. In this way, projects form a learning environment for both niche and regime.

Methodology

In this section we explain how we studied a case of sustainable construction in which a mainstream building actor decided to go green and develop "perhaps the most sustainable town hall" in the Netherlands. This ambition was stated by several people we have interviewed about the project in which they were deeply involved. We studied how actors in a real-time project deal with sustainable construction, and how niche innovations were applied in regime projects. The actions and interactions of niche and regime actors are influenced by macro developments at the landscape level, such as a growing environmental awareness and the economic crisis. These developments are beyond the direct influence of actors and usually take place slowly, on the order of decades, and can only be studied in hindsight over a long term. This case did not offer the opportunity to do so.

To study the Leiderdorp case, we interviewed the key stakeholders involved in the design process of a new town hall in this small town in the Netherlands: the property developer, several municipal project managers, a representative from the civil servants, the mechanical engineer, the structural engineer, the Regional Environmental Service, and the sustainable construction expert. We interviewed these actors at two moments in time: when the client—the municipality—approved (1) the preliminary design and (2) the final design. In between, we had several interviews with the project leader of the property developer. Afterward, toward finalizing construction, we interviewed with the property developer, the construction supervisor, and the municipality once more.

In the following sections, we analyze the project, based on the interviews, along the dimensions that have been used by Smith (2007) to distinguish niches from regimes. We reconstruct which influences of niche and regime are found in the project, and how this niche-regime interaction influenced the project's contributions to environmental sustainability—directly in terms of the building's construction and performance, and indirectly in terms of learning by actors on the adoption of niche innovations in a mainstream regime setting. In the final interviews with the property developer and the municipality we validated our findings; both actors confirmed that they recognized their project in our analysis. We conclude with identifying mechanisms that influenced the environmental sustainability of the project, and that can be used to further enhance mainstreaming.

A Competition to Design and Construct a New Town Hall

This section introduces the case of the design process of a new and sustainable public building—a town hall—for a medium-sized municipality in the Netherlands. The following case description draws from the interviews with key stakeholders involved in the project. In the following subsections we describe how the project started with a competition and how the sustainability requirements influenced the project.

Starting the Project

The project began with the town council's official approval to design and construct a new town hall. The existing town hall was outdated and in dire need of refurbishment and renovation. It sat next to a shopping mall, whose owner wanted to expand the mall on the plot of the old town hall. Thus the initiative for a new town hall was born. In January 2008, with a small majority of votes, the town council approved the plan for the new town hall, to be located on a piece of relatively cheap and undeveloped land between the town fringe and a busy highway.

Both the local government and the municipal civil service had little experience in property development. The responsible alderman (an elected politician) and the town council (of committed citizens for whom council membership is a side job) are the main players in the political arena. Because of their inexperience in building, the municipality hired project management expertise to guide the process of design and construction, assisted by two civil servants to manage the building project and the relocation of the municipal organization. The chief civil servant of the municipality invited several firms in this field to submit bids and selected the one that had also managed the construction of a town hall in a town nearby.

The externally hired project manager suggested a competition to award the design-build contract. The competition requirements stipulated that the building would reach environmental sustainability goals, following a regional policy guideline for sustainable building. This voluntary guideline combined a minimum level of sustainability assessed with the GPR tool[1], a Dutch assessment instrument for building design especially developed for local authorities, with additional requirements to public buildings. Complementary to nonvoluntary regulations, the regional policy guideline also set additional requirements for sustainable buildings, such as the use of certified timber, or raises the bar for already regulated issues, such as an energy efficiency requirement.

Required Design Specifications

The competition specified a number of requirements concerning floor space for offices, the council meeting room, a service desk for citizens, parking, and more. Flexibility was a key concern in the requirements for both the office space arrangement, and for the building itself. Another consideration: in the future it is possible that the town council and administration will merge with neighboring municipalities and the town will no longer need a municipal building of its own. The building also was to be cost efficient and meet environmental sustainability goals. The regional policy guideline advised local authorities to use the GPR tool to define a minimum level of sustainability. The GPR measures to which extent a building is sustainable on five dimensions: energy, environment, health, user value, and future value.[2]

At the time of the competition, the guideline demanded a minimum score of 8.0 on a scale from 1 to 10. Table 7.1 summarizes the sustainable construction demands based on national, regional, and local levels.

Table 7.1

Design requirements for sustainability from national and regional/local policies and project-specific requirements

National policy	Minimum energy performance requirement for public buildings as stated in the Dutch building code
Regional policy guideline	• 10% increased energy performance compared to the requirement for public buildings as laid down in the Dutch building code or the use of renewable energy and a well-insulated shell (heat resistance with closed façade parts Rc ≥ 3,5 m2.K/W, ground floor Rc ≥ 4m2.K/W, roof Rc ≥ 5 m2.K/W and HR++-glazing with U ≤ 1,2 W/m2.K. • Use of certified timber • Minimum GPR score of 8.0 for municipal buildings • Prevention of the use of heavy metals in outdoor construction to prevent emissions • Procurement of 100% renewable energy
Project-specific requirements	General requirement: The municipality should set the example and pay special care to sustainable building. Specified requirements: Climate-neutral building is more than a CO2-neutral building with its installations. Focal points are • Comfort • User costs • Reduced car use by offering bicycle facilities and public transport access and by offering digital, online services to the citizens • Minimized energy demand by creating a building with maximized insulation, use of passive solar energy, and use of excess heat produced by computers and appliances. The remaining energy demand for heating and cooling should be met with energy generated from renewable resources (generated by the municipality itself or procured), highly efficient artificial lighting and office appliances, sustainable procurement • Efficient waste management, separate collection of different fractions • Minimum GPR score of 8.0 for municipal buildings • 25% reduction CO2 emission • Flexible floor plan, building should be reusable for different purposes • 40–50% of the parking spaces should be created out of sight • Reduced water retention capacity (15%) should be compensated within the area

Course and Outcome of the Competition
The competition was made public on June 10, 2008. Four consortiums of design and construct firms competed; a jury judged their designs. The jury consisted of well-respected names in the field of architecture and property development, some representatives of local government and administration, and a university professor specialized in sustainable building. The jury had four groups of selection criteria. First, the competition requirements had to be met. In addition, scores weighted the criteria: 45 percent for the design, 35 percent for the costs, and 20 percent for the project management proposal.

On November 21, 2008, the jury announced the winner. The design-build contract was granted to the design with the lowest price and the highest sustainability score. The jury stated that the price had not been the sole decisive factor. The jury praised the high sustainability score, 8.8 instead of the requested 8.0, which was the result of smart, integrated design choices.[3] One of the most influential of these was the decision to combine construction and climate control, via concrete core activation, which reduces energy need and improves comfort, and the use of heat and cold storage through a ground heat exchanger. Although contributing to a high assessment score, these design and technology options had been tested and used in many niche projects in the Netherlands and abroad, but had not yet found their way into Dutch mainstream building regimes. In addition, the jury praised the design for its modest look and feel for a publicly financed building.

In June 2009, the property developer and the municipality signed the contract, by which the property developer promised to deliver the building presented in the winning design, including the environmental performance score of 8.8.

How Niche and Regime Influenced the Project

In this section we analyze the project along seven dimensions that distinguish niches from regimes (Smith 2007; Schot and Geels 2008). Niche and regime differ from each other in their guiding principles, the technology used, the structure of industry, the relations between users and markets, the focus of policy and regulations, the relevant knowledge, and their dominant culture. For these dimensions, we explore how the niche or the regime, or both, influenced the project. We rely on the work of Smith (2007) for characterizing the sustainable construction niche and the regime that dominates construction projects, comprising all actors in the value chain, including manufacturers, property developers, contractors, and architects. Many other studies support Smith's findings (Bergman, Whitmarsh, and Kohler 2008; Crabtree and Hess 2009; DBR 2010; Gluch, Gustafsson, and Thuvander 2009; Revell and Blackburn 2007) and are in line with observations of the characteristics of mainstream construction and of sustainable or green construction in other chapters of this book.

Guiding principles: Niche principles guided the initial design process, but were ousted by those of the regime afterward.

The first dimension concerns guiding principles that direct actor behavior in general. At the landscape level, such principles are democracy, individual freedom, and macroeconomic rules. Within this generally acknowledged setting, principles guiding behavior of actors in a niche will diverge from those guiding actors in the regime that is contested by the niche. Sustainable construction niches usually focus on maximizing ecological efficiency by minimizing resource use and closing loops, while in the mainstream construction regime, economic efficiency is the primary guiding principle.

In the project studied, guiding principles of both niche and regime alternately influenced the project. During the competition stage, the principles of the niche strongly influenced the design specifications and the process of selecting the winner. Of most prominent influence were the decisions by the municipality to require a minimum sustainability score (GPR), to make this score part of the assessment by the jury and to include it in the contract. However, the design also had the lowest price, which was in line with the guiding principle of the regime. Having scored best on both economic and ecological criteria, this design was easily selected by the jury as a winner.

The property developer put special effort into making the preliminary design as environmentally sustainable as possible. To achieve this, the developer formed a building team in which the developing and contracting division of the company collaborated. The developer also invited an architect, a mechanical engineer, and a structural engineer to become part of the team. Since none of the team members had significant experience with sustainable construction, the developer also hired a well-known sustainable building expert with detailed knowledge of and broad experience with the GPR method. The result was a building team that produced a very high score that even surprised the team members themselves.

During the post-competition stage of final design and construction, regime principles of economic efficiency dominantly guided the process, taking over from niche principles of ecological efficiency. After signing the contract, both the municipality and the property developer considered the delivery of the project "business as usual." As the one of the actors stated in an interview: "The Mercedes has been bought, now it just has to be constructed."

Technology: The use of niche technologies by mainstream actors was inevitable due to design requirements.

The second principle concerns technologies; the niche prefers products and technologies that minimize resource use, use renewable resources, and maximize user satisfaction. Whereas the niche stimulates the development and use of innovative, experimental technologies, the regime clearly prefers the use of tested and proven technologies, because mainstream actors are both familiar with these and trust that they will limit liability concerns.

Most of the project materials and technologies used had been widely tested and used in the sustainable building niche, but were new to some of the project participants and can still be regarded as niche technologies in the Netherlands. For example, the architect considered the heat and cold storage by means of concrete core activation highly innovative. The sustainable building expert who was added to the building team suggested this technology. While traveling in Switzerland the architect discovered that this technology was already quite common, and he used the remainder of his trip to study interior design solutions to optimize lighting and acoustics.

In this case, we see that even if the technologies were innovative to the regime, the regime preferred technologies that had been tried, tested, and proven within the niche. This situation formed the selection environment for the project. Keeping in mind that the municipality, its project manager, and the building team that won the competition—except for the sustainability expert—were inexperienced in the field of sustainability and sustainable construction, the team was hesitant to try "real" innovations—those that would be new to the niche as well as the regime. However, by including environmental sustainability in the design requirements, technologies that were innovative to the regime had to be used to achieve the requested GPR score.

After the competition, many actors were not very eager to consider additional ideas that could further enhance the environmental performance of the building, even though the property developer suggested some to the municipality. Several niche technologies were suggested in addition to those already laid down in the design: for example, solar panels were considered to be too expensive; a solar boiler lacked visibility and was therefore not wanted; and the alderman considered chairs from recycled materials inappropriate, stating, "I will not have my citizens sit on waste." So, by adapting niche technologies, the project achieved its high GPR score, but actors did not want to invest in additional niche technologies with an even better performance.

Industry structure: Principal-agent relationships reduced the benefits of innovative contracting after the initial design.

The third principle concerns the relationships between contracting parties in the industrial structure. The highly fragmented character of the construction value chain contributes to coordination problems among actors (WBCSD 2008). Price becomes a dominant coordination mechanism in mainstream construction, which leads to a focus on upfront capital costs and short-term return on investments—two criteria that favor mainstream products over sustainable products and technologies. In the sustainable construction niche, actors point to the need for collaboration, life-cycle costing, and innovation from a life-cycle point of view, which all could be enhanced by the longer-term involvement of actors in the process of building and construction. Innovative forms of contracting, such as the "design-build" contract used in our project, are therefore welcomed. In the Netherlands, such contracts are still relatively new to public authorities.

In mainstream construction regime, a concurrent shift was also being made toward more innovative methods of procurement and contracting (Priemus 2009). In the Netherlands, as in many other countries, traditional forms of contracting led to strategic behavior by project developers and contractors, leading to delays and budget overruns. The essence of the problem is that the principal has formal control over the agent, but to exercise this control, the principal has to rely on the information the agent provides. This gives the agent control over the principal. Design-build contracts used in the project studied in this chapter helped to reduce this principal-agent problem from happening.

In this project, the benefits of the choice for a design-build contract partly worked out as intended: the building team had three months to prepare the bid, and had to collaborate closely and rely on one another's knowledge and experience to optimize the design for functional and environmental performance and costs. The specialists involved, such as the mechanical engineer, acknowledged that the design-build format offered the opportunity to bring their knowledge and requirements into the project at an early stage, before irreversible design elements were added. The construction engineer stated: "It often happens that I get involved in projects too late." The contract form and tight schedule forced the designers and builders to collaborate in a team and to integrate their ideas in an early stage. There was no time for negotiations about the details of the bid and the division of costs, benefits and risks among them. All actors involved in the design team considered that they achieved a truly integrated design, which they attributed to this form of contracting.

However, once the jury selected a design and the contract was signed, principal-agent considerations started to dominate relationships among the actors involved. At this point in time, the relationships among the actors in the design team—which were informal during the preparation of the bid—were formalized in contracts, just as the relationship between the municipality and the property developer. Also, the communication between the municipality and the property developer had to take place via a project manager contracted by the municipality. This reduced the opportunities for informal communication and exchange of ideas among all principals and agents. Figure 7.1 gives an overview of these principal-agent relationships among the actors involved. Principle-agent relationships were present between the municipality and the property developer as a main contractor, but also within these "teams" and their hired experts. In the interviews we heard many examples of such relationships. To mention three:

1. The property developer saw that the municipality was struggling with the implications of the flexibility of the floor plan—which still demanded some structural choices—for its municipal organization, but did not dare to step in out of concerns that changes to the design would be made. The relationship between the property developer and the municipality was so indirect and formal that there was no opportunity to offer such informal help.

2. The developer considered the sustainable building expert's work complete, thereby depriving the design development and construction team of his expertise that was used to create the initial prize-winning design.

3. The contractor claimed that LED lighting could not be delivered in time and therefore traditional lighting was used. Because the contractor and developer were part of the same organization, there was no incentive to challenge this claim.

Thus, the design-build contract was desirable for both niche and regime actors, albeit for different reasons. During the preliminary design stage, the contract form contributed to the use of innovative technologies and resulted in a fairly priced design, but during the stages of finalizing the design and delivery the alleged advantages were no longer present.

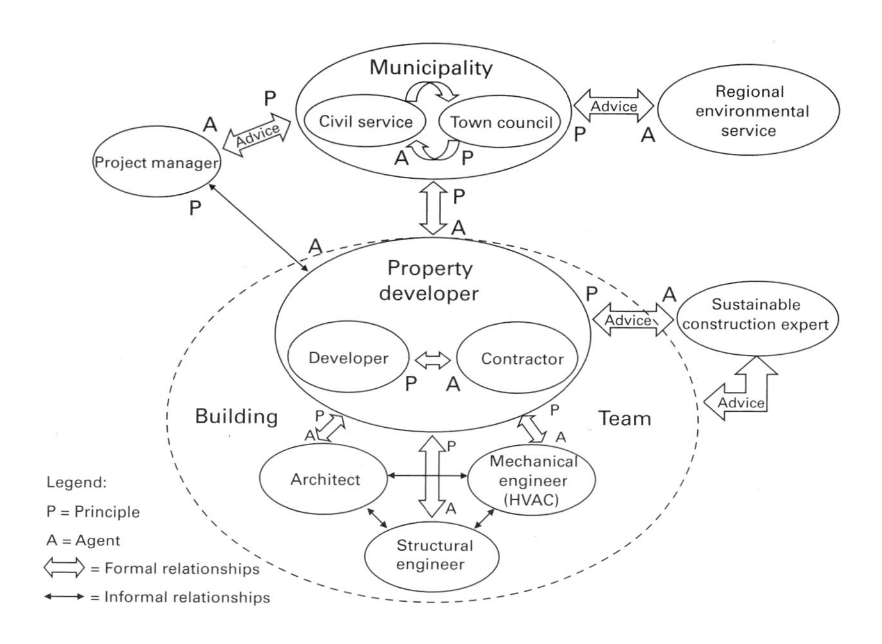

Figure 7.1

Relationships among the key actors involved in the Leiderdorp Town Hall Project, The Netherlands

Users and markets: The design process focused on the commissioner, not on the end-users.

A fourth dimension concerns the relationships among clients and markets. Within mainstream construction, client relations are especially focused on the commissioning actor, the (future) property owner, in this case the municipality. In the niche, there is more focus on the end-users, in this case the civil servants, citizens, and council members. From a regime perspective, the lack of attention to the needs of end-users can again be explained from the short-term involvement of construction actors, the use of proven and familiar technologies, and the need for quick wins. In contrast, actors in the sustainable construction niche are strongly focused on attending to the needs and wishes of the end-users and on the role of end-users in sustaining the environmental performance of a building.

In our project, the users only had a limited role in the design process. No special attention was given to user involvement. Neither the property developer nor the municipality considered this to be a problem, since the design was "sustainable" according to GPR ratings, and the contract would make sure that the promised sustainability would be delivered. However, toward the finalization of the design, the property developer and the municipality became aware of the importance of end-user involvement. For example, they realized that the use of waterless toilets and the paperless office concept needed some explanation and meetings were organized with the civil servants to discuss the implementation of these niche innovations.

Regarding user relations and markets, the regime thus seems to have dominated the project. Both the developer and the municipality seemed to consider a sustainable building to be a thing that can be designed and sold and bought at a market independently of the users' idiosyncratic preferences.

Policy and regulations: Sustainable building policies pushed ambition levels beyond the minimum required.

Policy and regulations form a fifth dimension to distinguish niches from regimes. Environmental concerns for the town hall were the result of a regional policy stimulating local authorities to commission buildings beyond minimum requirements. The regional environmental service department initiated the policy, and local authorities in the region agreed upon it. The policy was intended to push local authorities to raise sustainable building ambitions beyond the regulatory requirements, at least for public construction projects. Within the regime, actors—commissioners and developers/contractors—are known for limiting their ambitions to this minimum and attempt to achieve it with the least costs and efforts possible, whereas the niche actors continuously seek to maximize environmental sustainability. In addition, even when environmental ambitions are agreed upon, they tend not to be fully lived up to during project realization, because environmental values are traded for other values, such as cost reduction, increased returns, and quality, such as space and luxury.

In the town hall project, the inclusion of the promised GPR score in the contract made sure that both contract partners stayed focused on achieving this score. In the interviews, the municipality and the property developer confirmed this, but they also confessed that environmental sustainability concerns were not the only reason for sticking to the agreed score. Another and perhaps more urgent reason was that they did not want to disrupt existing contractual agreements. Any changes in the contract would need the consent of the town council. Political support for the new town hall was declining and it was not certain that the town council would approve of project changes. So, once the contract was signed, actors tried to deliver the building as promised—nothing less or more sustainable.

Knowledge: Niche knowledge was used instrumentally during initial design and was not related to construction.

The sixth dimension concerns knowledge. Actors involved in sustainable construction need knowledge and expertise in two fields: property development and sustainable construction. The municipality had neither of these. Municipalities of such modest size rarely commission construction projects, and such knowledge and expertise had to be hired. For knowledge of sustainable building, the municipality relied on the Regional Environmental Service, which upon request provides municipalities in the region with environmental knowledge and expertise.

The property developer and its building team partners were, of course, highly experienced in developing mainstream property, including public buildings, but did not have much knowledge of sustainable construction. This property developer in particular wanted to use this project to gain knowledge and experience in this field and, if successful, use the project as a showcase.

Both the municipality and the property developer thus needed to hire the relevant knowledge and expertise. They selected their partners on a number of grounds. The municipality hired a small process management consultancy firm "because we got on well during the interview" and the firm had managed the design and construction of an innovatively designed town hall of a municipality nearby. The property developer selected its team members on a variety of grounds: the contractor because it was part of the same company, the architect because he was a local, the mechanical engineer because he collaborated well in another project, the structural engineer because he was well regarded, and so on. When the building team was put together, the developer decided to add a sustainable construction expert to the team. The choice for the particular expert was a strategic one: the expert had a good reputation, one that was known by the jury members, he had a wide experience with the GPR tool and similar assessment tools, and the developer expected that he could learn from him.

In the niche, sustainable construction knowledge and experience are considered to be important factors for project success and one of the niche lessons is to select project partners on these grounds. The main reason for this is that niche innovations require an understanding of how the innovations contribute to sustainability. Without this understanding, it is difficult to adopt them as intended, and expected sustainability benefits will not be achieved fully. Surprisingly, the property developer and the municipality did not select their partners based on their knowledge of and experience with sustainable construction, except for the sustainable construction expert. When we confronted them with our surprise during the interviews, they explained that they did not consider this a problem. After all, attention for sustainability was firmly embedded in the process at several points in time: at the start, by means of the policy advising to use the GPR tool; in the preliminary design, by means of the sustainable construction expert; and in the formal agreements, by including the GPR score in the contract. The firm institutional embedding of attention for sustainability thus seems to have contributed to the sustainability of the project, despite the weak involvement of experts. Experts and knowledge were used instrumentally, when the mainstream actors needed niche knowledge to meet the design requirements. However, they did not value the continued contribution and involvement of niche experts in the process, whereas niche experiences show that the realization of sustainability ambitions as intended does require such continuing input. Although niche knowledge embodied by niche experts was recognized as valuable by the mainstream actors, they did not understand that niche technologies needed further expert involvement for successful implementation. Although the expert stressed the importance of his continued involvement, his argument was neglected.

Culture: Mainstream actors involved in the project approached the sustainability of the project pragmatically.

Culture is the seventh and final dimension to be distinguished. Actors within the regime usually adopt a realistic stance toward sustainable construction: when there would be a market demand, or when they would be forced by regulations, they will deliver. Within the niche, actors tend to consider sustainability as an intrinsic requirement for construction, just as sustainable development should be at the heart of all human activity. In the town hall project, the actors approached sustainability pragmatically. They tried to deliver a sustainable town hall within the regime constraints. In the project, niche innovations were adapted, but not much attention was paid to the implications of these technologies for the use and users of the building. The future will show what the actors will learn from the use of the building and how this may affect the position of sustainability requirements for design: considering them as add-ons or integrated in design and everyday use.

Table 7.2 summarizes the niche, regime, and the project for the seven dimensions we have discussed.

Table 7.2

Characterizing the project, niche, and regimes along seven dimensions

Dimension	Niche: Sustainable construction	Project: Design and construct of sustainable town hall	Regime: Mainstream construction industry
Guiding principles	Ecological efficiency	Balance between sustainability requirements and costs constraints	Economic efficiency
Technologies	Use of materials, products, and technologies that minimize resource use and maximize user satisfaction	Use of sustainable technologies, unconventional in the regime, but tried and tested in the niche	Tried and tested materials, products, and technologies
Industrial structure	Innovative contracting supports life-cycle costing and longer-term involvement of actors; this changes the cost-benefit analysis, enables sustainable investments	Design-build contract resulted in a sustainable design for a fair price. Delivery of promised sustainability against least costs	Profit from contracted price. Price based on upfront construction costs. Risk-averse behavior out of liability concerns. Legalistic relationships
User relations and markets	Active and highly involved end-users	Developer focuses on municipality as commissioner and not as end-user(s)	Users are passive and conservative consumers
Policy and regulations	Unconventional understanding of policies and regulations is legitimate to overcome regulative barriers	Keep to minimum contract requirements. When cost neutral, additional effort to increase sustainability	Keep to minimum requirements, lobby for less constraining requirements
Knowledge	Expert knowledge on sustainable building; experience important to make use of unique opportunities of location and project	Expert knowledge on sustainability used at the start of the process. Expert knowledge of property development is considered important throughout the process	Expert knowledge relevant to existing competencies and business practice
Culture	Idealistic: sustainable construction is intrinsic requirement	Pragmatic: sustainable building is delivered at reasonable costs	Realistic: markets and regulations should ask for sustainable building

Influences of Niche-Regime Interaction on Project Sustainability

The previous section described the process by which the sustainable construction niche and dominant regime influenced the town hall project. This section discusses these influences more specifically: how they contributed to the sustainability of the town hall both directly, in terms a sustainable design, and indirectly, in terms of learning about sustainability by the actors involved. This allows us to draw some final conclusions on niche-regime interaction in a particular project context and how such a project can support the mainstreaming of sustainable innovations in construction and possibly in other regimes as well.

In 2010, to the surprise of the property developer and the municipality, the building was nominated for a national green building award. This indicates that the actors succeeded in designing a building that addressed environmental sustainability concerns. In addition, a test of the final design by the Regional Environmental Service confirmed that the project achieved the promised GPR score. We can safely conclude that the niche inspired actors to use project requirements to enforce sustainability. This stimulated actors to use expertise and to collaborate toward an integrated and highly sustainable design. Some members of the building team acknowledged that they learned a lot from the sustainable construction expert, for example, on reducing material use, heat and cold storage, concrete core temperature control, the GPR tool, and the process of integrated design. This has raised these team members' curiosity and enthusiasm for sustainable construction, and they stated that they intend to expand this knowledge and use it in projects to come. They highly value the environmental sustainability of the project. The architect exclaimed with a big smile: "The civil servants will love it; during summer they will be surprised by the amazing indoor climate."

However, the surprise of the actors reveals something else as well. The full potential of the collaboration of the actors involved, as anticipated by the niche, was not fully realized during the project. Major discrepancies from the innovations as intended by the niche were

- the early exit of the sustainability expert;
- the limited attention paid to the building's life cycle and the end-users; and
- principal-agent relationships and the resulting strategic behavior.

This had consequences for the role that sustainability played in the later stages of the design process, and we will present some examples of how sustainability was no longer a factor in decision making.

An example of relatively uninformed decision making, and possibly of strong influence on the building's sustainability performance, is that of the heat and cold storage by means of concrete core activation. This suggestion by the sustainable construction expert strongly contributed to the design's high energy performance. After the contract was awarded, the property developer and the other actors considered involvement of this expert no longer necessary. The absence of his knowledge of the technologies used may explain why, toward the finalization of construction, they mistakenly opened the building in the midst of winter. As there was no heat yet stored from past use, the project required additional conventional heating in the first months. The result is that the municipal facility manager expected the energy bill for heating this new, highly energy efficient building to be higher than had ever been paid for heating and cooling of the old, poorly insulated building. Users expected and demanded higher temperature comfort in this new building, whereas they accepted the poor conditions in the old building and put on an extra sweater in winter and took the hot indoor climate in summer for granted.

These examples show how the environmental sustainability of the design requires an understanding of the technologies used, and how the initial design can be jeopardized or enhanced during the design and construction process. This is common knowledge in the sustainable building niche, and the risks of instrumental learning are a pitfall that is also identified in strategic niche management theory (Schot and Geels 2008). If mainstream actors only learn to implement certain technologies, but do not fully understand how these technologies work and weaken or strengthen each other, a lot of sustainability potential, synergy, and efficiency may be lost. However, the key actors in the project did not seem to be aware of such influences or did not consider these to be important, as long as contract agreements were met.

Conclusion: Projects as Portals for Mainstreaming Sustainable Construction

This chapter explored the use of sustainable niche innovations in the design and construction of a town hall, a project that was initiated and carried out by actors who usually operated in a regime context. In the project, we witnessed how sustainability innovations were used and modified under the influence of both a niche and a regime. The project was the arena where the inevitable process of mutual adaptation of the niche and regime took place. Projects, because of their restriction in time and place, are also places that offer opportunities for improving the conditions for mainstreaming sustainable innovations.

In our case study, we can identify three interrelated mechanisms that influenced the diffusion and adaptation of sustainable innovations. Identification of these mechanisms also presents opportunities to manage them and thus improve the conditions for mainstreaming of sustainable innovations in future projects.

The first mechanism concerns the attention to niche issues. This attention depends on its anchorage in different phases during the project. Our project shows how the sustainable construction niche gradually gained influence over the incumbent regime, the building industry, and the rules according to which it acts. The niche strongly conditioned the early design phase of the project: sustainability requirements were included in the design requirements and in the contract. However, once the contract was signed, the regime conditioned the later construction phase of the project: there were no incentives that pushed actors to consider environmental sustainability throughout the finalization of the design and construction. As a result, some decisions put the intended environmental sustainability goal at risk, and opportunities for further optimizing sustainability remained unnoticed or unconsidered.

The second mechanism is that early agreements about assessment of the sustainability of the design guide ongoing attention for sustainability and building performance. Decisions and agreements about the design's environmental sustainability built upon niche knowledge. This knowledge was available in user-friendly, readily available tools and metrics such as GPR; the latter offers niche lessons in a quantitative form that fits the regime context, and conditioned the environmental sustainability eventually to be achieved in the project of our case study. All participants in the town hall project were concerned about maintaining the agreed-upon score included in the contract, which would be assessed at the start and the finalization of the design. However, the instrumental aspects of the metric tool may contribute to perverse effects, such as nonchalance about sustainability issues later in the project, stress if the final design is not as sustainable as agreed, and strategic actions if designers choose to game the points instead of abiding by the spirit behind them. Both the property developer and the municipality realized that a lot of decisions were made after signing the contract, and they did not fully understand the consequences of those decisions for the project's environmental sustainability.

This brings us to the third mechanism: the effort and expertise of actors can reinforce the design and its performance. Niche lessons about the importance of the collaboration of actors throughout the design process, including experts and users, did not work as well as desired. Regime rules dominated the relationships after the contract was signed. The importance of collaboration, experts, and expertise of environmental sustainability was minimized.

These three mechanisms show that there are opportunities to more firmly protect niche innovations from regime influence by including sustainability not only in formal agreements and tools, but also in incentives, knowledge, and opportunities for attending to environmental concerns throughout the design and construction process. For example, ongoing benefits may be realized by requiring that a sustainable construction expert remains part of the building team, or by selecting the sustainability consortium based on knowledge and experience of all participants. Moreover, the continuous participation of experts will prevent instrumental use of expertise, and increase collaboration and learning among all participants.

Nevertheless, this case shows that it is possible for a regime-dominated project to create a highly ambitious niche building. Based on the case study, it is too early to state anything about changing niche and regime contexts and rules, but it does show that the coexistence of niche and regime is important to continue to challenge the regime to further adoption of the innovations developed in the niche.

Notes

1. GPR in Dutch stands for "Gemeentelijke Praktijk Richtlijn"; translated, this means practical guideline for municipalities. For more information, see http://gprsoftware.nl/english/ (accessed December 17, 2012).

2. Information on the GPR in English can be found at http://gprsoftware.nl/english/ (accessed December 17, 2012).

3. This was the average score. The design scores for the different GPR themes were: energy 9.0, materials 8.5, waste 9.0, water 8.9, and health 8.7.

References

Ahn, Yong Han, and Annie Pearce. 2007. Green construction: Contractor experiences, expectations and perceptions. *Journal of Green Building* 2 (3): 106–122.

Ashuri, Baabak, and Alev Durmus-Pedini. 2010. An overview of the benefits and risk factors of going green in existing buildings. *International Journal of Facility Management.* http://ijfm.net/index.php/ijfm/article/viewArticle/15 (accessed December 17, 2012).

Bergman, Noam, Lorraine Whitmarsh, and Jonathan Köhler. 2008. Transition to sustainable development in the UK housing sector: From case study to model implementation. University of East Anglia: Tyndall Centre Working Paper #120.

Brown, Halina, and Philip Vergragt. 2008. Bounded socio-technological experiments as agents of systemic change: The case of a zero-energy residential building. *Technological Forecasting and Social Change* 75:107–130.

CB Richard Ellis. 2009. *Who pays for green? The economics of sustainable buildings*. Los Angeles: EMEA research. http://portal.cbre.eu/portal/page/portal/research/publications/FPR_EMEA_ECONOMICS_OF_SUSTAINABLE_BUILDINGS_2009.pdf (accessed December 17, 2012).

Crabtree, Louise, and Dominique Hess. 2009. Sustainability uptake in housing in metropolitan Australia: An institutional problem, not a technological one. *Housing Studies* 24 (2): 203–224.

DBR. 2010. Green buildings—A niche becomes mainstream. Deutsche Bank Research, April 2010.

Gluch, Pernila, Mathias Gustafsson, and Liane Thuvander. 2009. An absorptive capacity model for green innovation and performance in the construction industry. *Construction Management and Economics* 27 (5): 451–464.

Hoffman, Andrew, and Rebecca Henn. 2008. Overcoming the social and psychological barriers to green building. *Organization & Environment* 21 (4): 390–419.

Priemus, Hugo. 2009. Do design & construct contracts for infrastructure projects stimulate innovation? The case of the Dutch high speed railway. *Transportation Planning and Technology* 32 (4): 335–353.

Revell, Andrea, and Robert Blackburn. 2007. The business case for sustainability? An examination of small firms in the UK's construction and restaurant sectors. *Business Strategy and the Environment* 16 (6): 404–420.

Schot, Johan, and Frank Geels. 2008. Strategic niche management and sustainable innovation journeys: Theory, findings, research agenda, and policy. *Technology Analysis and Strategic Management* 20 (5): 537–554.

Scrase, Ivan, and Adrian Smith. 2009. The (non-) politics of managing low carbon sociotechnical transitions. *Environmental Politics* 18 (5): 707–726.

Seyfang, Gill. 2010. Community action for sustainable housing: Building a low-carbon future. *Energy Policy* 38 (12): 7624–7633.

Shove, Elizabeth, and Gordon Walker. 2007. Caution! Transitions ahead: politics, practice, and sustainable transition management. *Environment & Planning A* 39:763–770.

Smith, Adrian. 2006. Governance lessons from green niches: The case of eco-housing. In *Framing the present, shaping the future: Contemporary governance of sustainable technologies*, ed. Joseph Murphy, 89–109. London: Earthscan.

Smith, Adrian. 2007. Translating sustainabilities between green niches and socio-technical regimes. *Technology Analysis and Strategic Management* 19 (4): 427–450.

UNEP SBCI. 2009. *Buildings and climate change: Summary for decision-makers.* Paris: United Nations Environmental Programme, Sustainable Buildings and Climate Initiative.

van Bueren, Ellen. 2009. *Greening governance: An evolutionary approach to policy making for a sustainable built environment.* Amsterdam: IOS Press.

van Bueren, Ellen, and Jitske de Jong. 2007. Establishing sustainability: Policy successes and failures. *Building Research and Information* 35 (5): 543–556.

WBCSD. 2008. *Efficiency in buildings: Business realities and opportunities.* World Business Council for Sustainable Development. Geneva and Washington, DC: WBCSD.

Williams, Katire, and Carol Dair. 2007. What is stopping building in England? Barriers experienced by stakeholders in delivering sustainable developments. *Sustainable Development* 15 (3): 135–147.

III

Operational, Organizational, and Cultural Change

8

Empowering the Inhabitant: Communications Technologies, Responsive Interfaces, and Living in Sustainable Buildings

Kathy Velikov, Lyn Bartram, Geoffrey Thün, Lauren Barhydt, Johnny Rodgers, and Robert Woodbury

Introduction

The deeply resonant statement of Winston Churchill, "We shape our buildings and thereafter they shape us," foregrounds the significance of the feedback loops that individuals have with their built environment and the active coevolution that they necessarily share, creating a context for considering the agency of both inhabitants and buildings as complex hybrids (McMaster and Wastell 2005). In the design of environmentally sustainable buildings, the concept of building/inhabitant/environment as a coevolving system can be mobilized by designers to promote social change, not only to increase the "intelligence" of building systems, but also to increase the intelligence of their inhabitants as well. Given that the actions of building inhabitants can account for significant variations in building energy use and their overall impact on the environment, both buildings and their systems can be designed to anticipate and even transform behavior toward more sustainable patterns of living and building use.

Statistics regarding energy use and greenhouse gas emissions due to building construction and operation are well documented, and provide foundational justification for the urgency associated with the green building movement. However, the design of more energy efficient buildings alone cannot address the challenge to reduce the negative environmental impacts produced by the built environment. Energy and resource consumption, in buildings and associated infrastructure, encompasses social, political, and personal dimensions that are as critical to achieving performance transformations as are technical innovations (Cole, Brown, and McKay 2010; Janda 2009; Stern and Aronson 1984).

In the residential sector, differences in individual behavior can produce large variations in energy consumption—in some cases as much as 300 percent—even while controlling for differences in housing typology, appliance efficiency, HVAC (heating, venting, and air conditioning) system composition, and family size (Janda 2009, 10; Socolow 1978). Some researchers estimate that approximately half of the energy used in the home can be attributed to the physical characteristics of a house and its equipment and that inhabitant behavior and activities determine the balance (Schipper et al. 1989, 277). Social scientists recognize that motivations behind consumption or conservation of energy are societal and cultural (Lutzenhiser 1993), and argue that achieving real and lasting energy reduction in the building sector requires deep social change (Janda 2009, 9–10) *in addition to* changes in building technology and construction.

In a recent study that monitored extended energy use patterns in a community of retrofitted Zero Energy Homes (i.e., houses with zero net annual energy consumption and carbon emissions), results showed that while the energy efficient and energy generating features of the buildings were effective in reducing the net energy consumption of these dwellings, the patterns of occupational energy use by inhabitants remained the same as the patterns of their neighbors living in conventional homes (Janda 2009, 10). Thus, it may be fair to say that choosing to live in an environmentally sustainable or high-performance home does not necessarily mean that one is living sustainably. Such data exposes a gap in design dialogue around environmentally sustainable buildings, which focuses heavily on the development of energy and resource optimization technologies and analyses but is far less focused on effectively supporting how people use these technologies and interact with them. Further, design of human interaction should not only be aimed toward the goals of optimizing performance, but should also aim to address questions of behavior and lifestyle as related to environmental sustainability.

Dwellings constitute one of the most valuable sites for confronting the problem of energy consumption in buildings, by advancing the design of inhabitant support interfaces that enable environmentally sustainable living practices. Residential buildings and their operation consume 57 percent of building energy in the United States (U.S. DOE 2009), representing more than one-fifth (22 percent) of the United States' total energy demand, and rising (Ehrhardt-Martinez, Donnelly, and Laitner 2010). Approximately 80 percent of residential energy use is attributed to single-family homes (ACEEE 2011). More immediately, this sector affects the entire population in a profoundly personal way. Environmental psychologists consider the home as the primary site of human occupation and the source of self-identification, one that is deeply rooted in social, cognitive, cultural, and behavioral development and formation (Tognoli 1987, 655–

660). The home, therefore, could become the perfect learning laboratory for driving sustainable change.

Aware Home and Smart Inhabitant

Perspectives regarding the human's role in a building differ across disciplines and each discipline also develops its own set of terms: architects refer to occupants or inhabitants, policymakers to residents, and interaction designers to users. However, we believe that all of these perspectives align, and we use the term "inhabitant" to bridge them all, as this term implies a more active relationship to the built environment (Cole et al. 2008). Interaction designers are keenly aware of the need to model how people use artifacts as an integral part of the design process (Bartram, Rodgers, and Woodbury 2011; Bartram and Woodbury 2011). This awareness only recently arose in architectural design dialogue, and an emerging area of research in the design of sustainable buildings is how building operation engages and involves building inhabitants (Cole, Brown, and McKay 2010; Leaman and Bordass 2001). Ray Cole stresses the need to model this concept of "inhabitant intelligence" into existing design concepts of intelligent buildings.

Buildings designed around inhabitant intelligence will provide flexible, adaptive task environments, refined control zones, and technologies that maximize inhabitants' access to adaptive opportunities (Cole and Brown 2009). Thus architects, engineers, and system designers face the challenge of reframing design strategies as a coevolution of human and building behavior to encourage as well as underpin environmentally sustainable use. This requires new models of design thinking beyond the typical "smart home" encompassing inhabitant behavior, motivational strategies, and exploration of how automation can impact inhabitants' daily rituals and perceptions of comfort. The general promise of the smart home, that intelligent operation could be offloaded to a computational component, is confounded by the human factor. The most daunting issues in smart home design are not technological capability but complexity and poor usability (Eckl and MacWilliams 2009; Cole and Brown 2009; Leaman and Bordass 2001; Baker 1996).

Environmental educator David Orr argues that buildings play a role as pedagogical tools that instruct inhabitants how to think about the connections between building and site, between interior and exterior environment, the origins of materials, and the value of resources such as energy. He posits that most buildings teach us that "disconnectedness is normal," energy is cheap, and that the process involved in producing and disposing of materials is irrelevant (Orr 2000). In this chapter, we argue for stimulating "inhabitant behavior" in service of environmental sustainability goals. By including the inhabitant, we redefine what the green home is beyond the performance of its technical components, as promoted by organizations such as the EPA and rating systems such as LEED.

Redefining the Green Home

An environmentally sustainable home must be considered more than a green building: it is also a living environment that has the capacity to encourage inhabitants to use fewer resources more effectively while fostering a compelling lived experience. Small, learned changes in behavior as to how we use our homes, such as turning off lights, reducing heat, uncovering or covering windows, or shortening showers, can result in substantial energy and water savings. However, transforming the way people use resources in a more fundamental way as part of their daily lives proves challenging. The combination of ubiquitous computing, interactive information systems, and computational intelligence offers an opportunity to enable inhabitants to dynamically interact with building technologies for feedback and control regarding performance and atmosphere, while empowering the inhabitant as an agent of behavioral change. The building, then, is more than just an artifact of spatial and material assemblage. Through its architectural and systems design, the building can produce a responsive environment that engages inhabitants in a dynamic, active, and integral role relative to its use, performance, and inhabitation. The North House research project is an experimental prototype constructed in 2009 that begins to develop shared knowledge of how designers could effectively integrate technologies and patterns of use into sustainable domestic design.

The North House

The North House (depicted in figure 8.1) is a prototype high-performance, energy-producing home which incorporates advanced technologies that not only work to manage building energy, resources, and comfort, but also make it possible for inhabitants to actively redefine their role within the operation of the home in order to achieve their environmental goals. Developed as a collaboration of architects, mechanical engineers, systems engineers, material scientists, building technologists, software engineers, interactivity designers, and industry partners, the 800-square-foot prototype explores both the physical design of environmentally adaptive

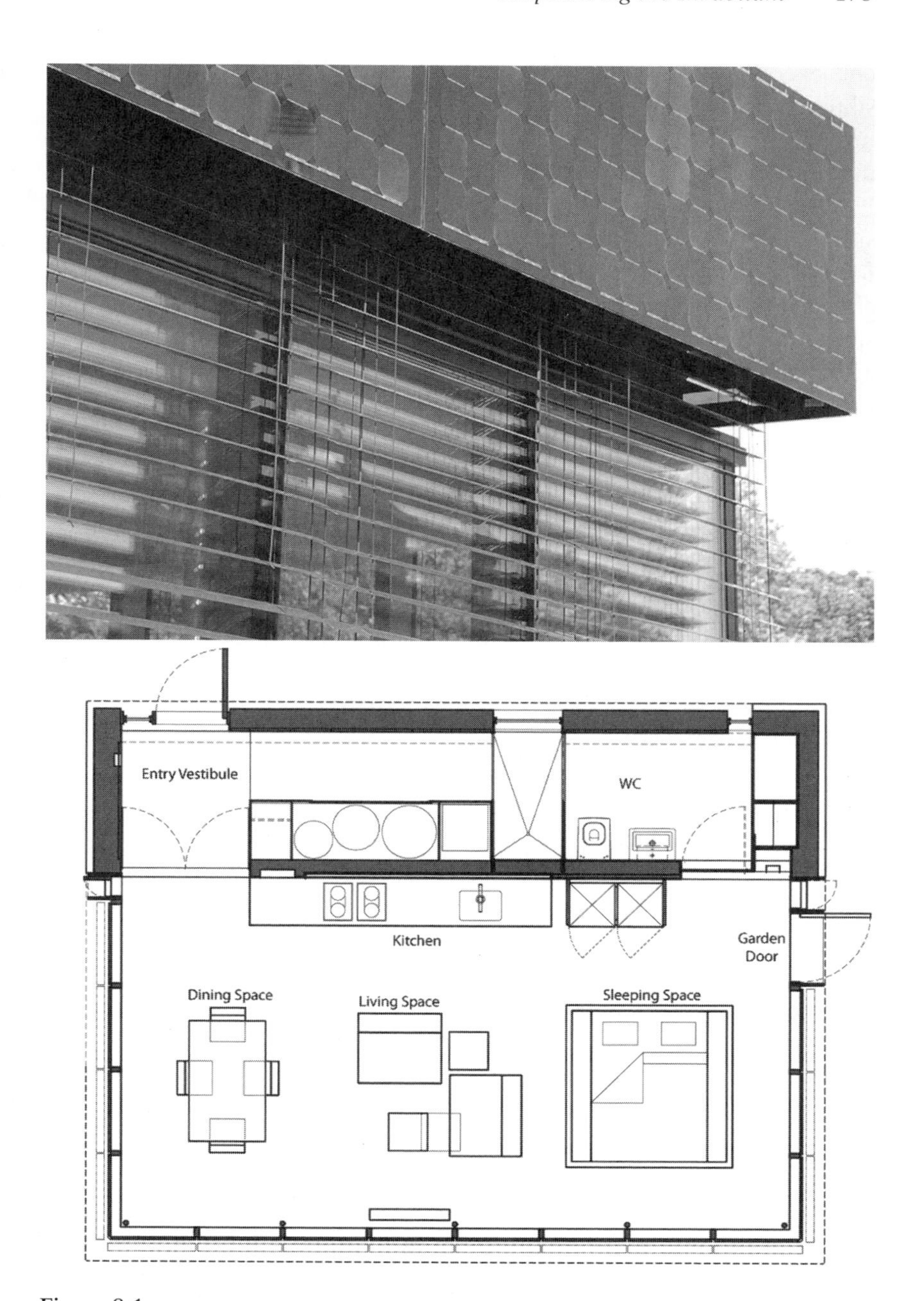

Figure 8.1

North House exterior facade of photovoltaic panels and dynamic shading (top) and North House plan (bottom)

and energy saving building technologies as well as the interactive possibilities of communication technologies, feedback technologies, and user interfaces toward increasing inhabitant intelligence and sustainable living practices in domestic green buildings. While the design of the North House incorporates a broad suite of design strategies aimed at promoting sustainable lifestyles, such as creative use of smaller living spaces, on-site food production, and prioritization of social spaces within the home, this chapter will focus on two innovations that incorporate feedback and response among the building, the inhabitant, and the environment. These are: (1) an envelope system that dynamically manages passive thermal gain and active energy production by adapting in real time to climatic conditions, diurnal cycles, and inhabitant inputs; and (2) an adaptive living interface system that provides the inhabitant with simple, intuitive controls, monitoring with meaningful feedback on the impacts of their behavior, as well as social motivation tools to foster sustainable patterns of living. North House proposes a whole building system that "learns" to perform more efficiently precisely because the inhabitant, who is part of the system, learns to occupy and operate it more intelligently.

Distributed Responsive System of Skins (DReSS)

The building envelope is the primary interface between building and environment. It performs not merely as a static barrier with openings, but also as a complex membrane that dynamically manages energy, material, and information exchanges, and operates "as part of a holistic building metabolism and morphology" (Wigginton and Harris 2002, 27). The current generation of high-performance envelopes consists of sophisticated assemblies that combine real-time environmental response, advanced materials, dynamic automation with embedded microprocessors, wireless sensors and actuators, and advanced design-for-manufacture techniques (Velikov and Thün 2012). The high-performance envelope developed for the North House, termed the "Distributed Responsive System of Skins" (DReSS), actively manages passive solar gain by combining a dynamically controlled exterior shading system (venetian blind type) with a highly insulated window system and phase change material embedded in the floor to provide thermal energy storage (Thün and Velikov 2012). The dynamic blinds of the DReSS respond to environmental conditions such as location, direction, and intensity of solar irradiation, exterior and interior temperature and humidity, wind velocity, and time of day to manage solar gain, heating the building almost entirely through passive means, and then defending against thermal gains from solar irradiation when temperatures rise (see figure 8.2).

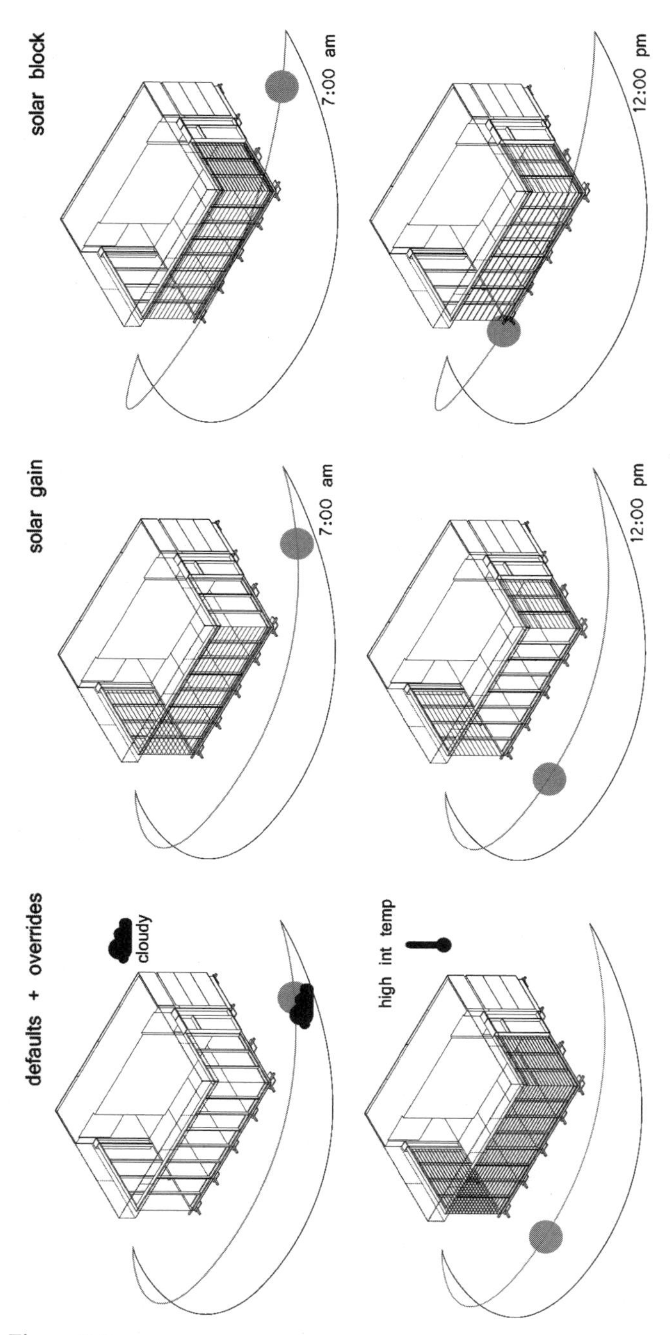

Figure 8.2

Window shading modes for different climatic conditions

Building applied photovoltaics (BAPVs) mounted horizontally on the roof, and building integrated photovoltaics (BIPVs) applied to the non-transparent areas of the envelope on the south, east, and west facades generate electricity at different times of the day and year, taking advantage of the variable sun angles characteristic of northern climates. A computerized central home automation system controls the DReSS for energy optimization. Building energy simulations predict that the DReSS assembly of high-performance glazing, active shading, and phase change materials could effectively eliminate cooling loads and reduce heating loads by 45 percent (Thün and Velikov 2012), while also providing ample daylight and reducing the need for artificial light in daytime hours (see figure 8.3).

Recognizing how important it is for individuals to have control over their own home, the automated system can be overridden, providing manual control of the blinds, temperature, and humidity based on specific needs and desires. In the feedback system of the North House, ALIS (described in detail in the following section) provides real-time exchange of information. For example, the inhabitant may open the window to receive fresh air, or open the blinds to allow sunlight to flood the living space. But if these actions impact the optimized energy performance of the building, both informational and ambient cues gently notify the inhabitant. As a result, inhabitants may choose to proceed with their desired actions, knowing that these will impact the home's energy efficiency, or may decide to make no overrides. Inhabitants can begin to map these configurations to their own activities, negotiating between personal inputs and automated responses to climatic conditions. The human-system relationship is anticipated to evolve as inhabitants learn to modify their own actions relative to information about energy impact they receive. This research agenda aims to eventually develop a computer algorithm that learns from inhabitants' repetitive actions, and begins to anticipate certain configurations, working toward the development of second-order cybernetic systems (Dubberly, Pangaro, and Haque 2009) within the home.

A combined environment- and user-responsive building envelope, such as the DReSS, offers advantages beyond the simple matter of energy conservation. By transforming windows from a thermal energy deficit to a positive energy resource, the building allows for a maximal amount of window area while providing inhabitants with the freedom of complete control over the benefits of daylight and views to the exterior. The importance of daylight and natural views on human physical and psychological health is supported by increasingly diverse research (Frumkin 2008; Loftness and Snyder 2008). In northern regions where daylight hours are

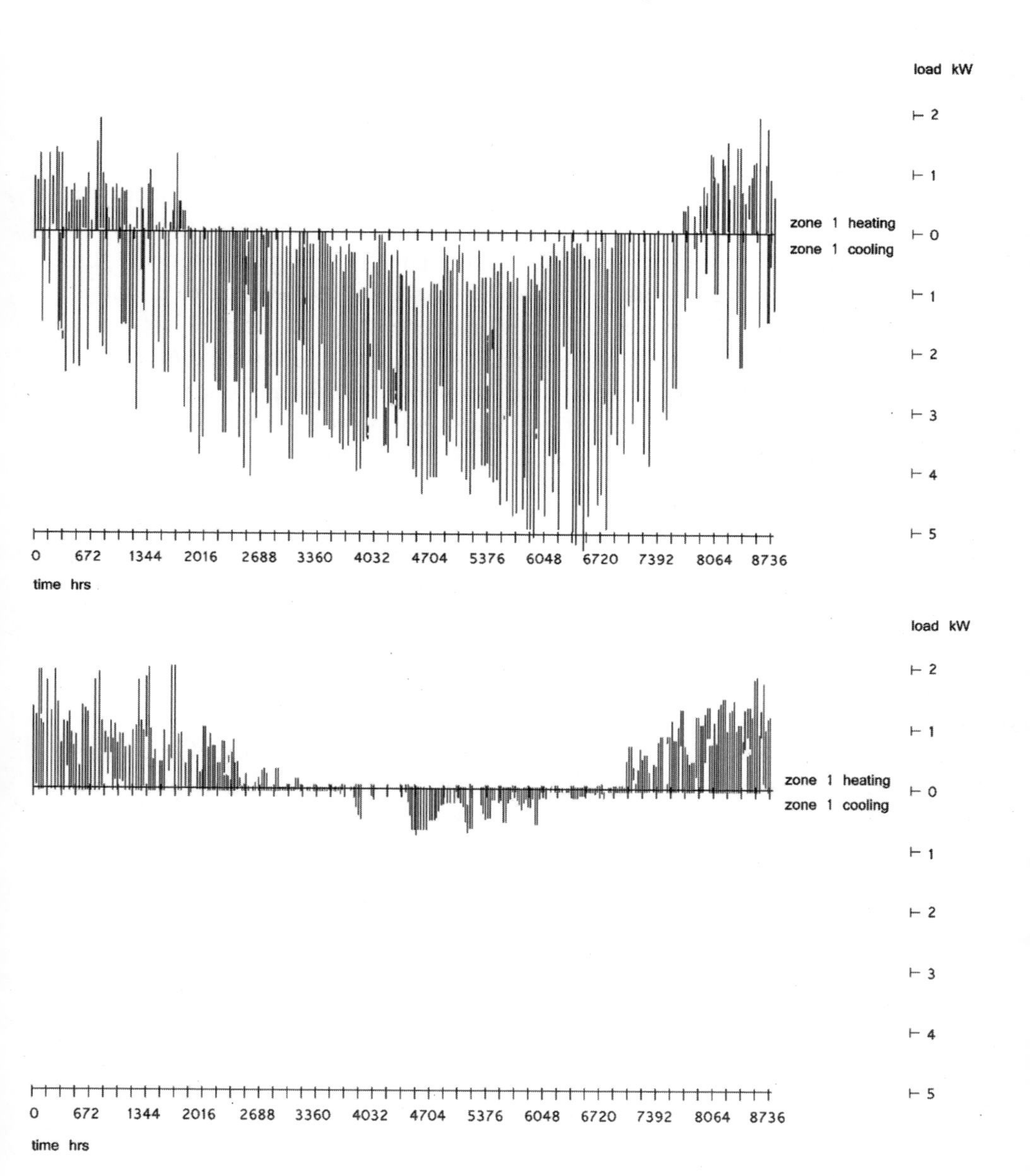

Figure 8.3

Modified ESP-r model of annual heating and cooling loads with no shading (top) and with dynamic shading (bottom)

significantly reduced in winter, these health and psychological benefits are all the more critical. (For more on biophilic design, see also chapters 14 and 16, this volume).

Taking advantage of new envelope technologies, the North House challenges the assumption that buildings for northern climates need to be thick-walled, small-windowed, and isolated from the exterior in order to maintain a consistent interior climate. The dynamic adjustments of the exterior blinds continually inform the inhabitant of the variable relationship between the interior and exterior, where heat and light are harvested from outside, and the living space is visually continuous with the exterior landscape, light, and seasons. Furthermore, the multiple configurations of the blinds serve as a constant expression of the changing relationship between exterior environment and the inhabitant within. While the North House engages in advanced building technologies, it is not intended to demonstrate that the answers to the challenge of environmental sustainability are purely technological. Instead, the design aims to develop a more nuanced and complex ecology of systems, almost passively educating human inhabitants to encourage evolving interactions.

The Adaptive Living Interface System (ALIS)

Beyond DReSS, the building control and inhabitant support system developed for the North House comprises four main components: (1) a control backbone that provides fine-grained measurement; (2) device control and automation logic that is based on a heavily customized commercial application (mControl™+Verhub); (3) a web services layer that manages data and commands between components; and (4) the Adaptive Living Interactive System (ALIS) that embodies the inhabitant's interaction with the home (shown in figure 8.4).

ALIS builds on a comprehensive information model incorporating control and device details; resource-specific production and consumption data in terms of standard units; pricing levels and standard usage equivalencies; personal and shared goals; and a hierarchical model of energy-control settings to enable "one-step" optimization. For inhabitants to sustain energy conscious practices, this infrastructure must integrate with their daily routines, habits, and rituals. Rather than deliver a whole new set of tools and devices to the user to manage the house, the ALIS functions integrate with familiar tools and systems. ALIS currently comprises an ecology of interfaces embedded as part of the architecture of the home and its devices: a set of variably configured user interfaces use web browsers on both embedded displays and home computers; a mobile application; and ambient feedback displays embedded in the house. Intuitive touch interfaces control the system, and easily comprehensible graphics present the information to inhabitants (Bartram, Rogers, and Muise 2010; see figure 8.5).

ALIS enables inhabitants to directly control and monitor lighting, shading, and interior climate control settings of the North House. In addition, inhabitants can configure energy-optimizing "modes" of operation as presets in ALIS controls: for example, turning off most lights and lowering the thermostat in a "Sleep" mode, or tuning settings and shutting down standby power in the "Away" mode. Inhabitants can activate these presets via one button from any ALIS control, or preschedule mode changes. Modes are entirely user configurable, and coexist with individual control settings for fine-grained control when desired. Finally, ALIS designers implemented a control hierarchy so that inhabitants can easily access master controls and the most commonly used switches, or selectively drill down to fine-grained controls on demand.

The ALIS computer interface offers the most detailed access to feedback about resource consumption. It provides the inhabitant with standard full access to the entire ALIS system. Inhabitants can configure modes, set goals, and schedule house operation from any computer connected to the Internet. ALIS also supports a variety of feedback displays and analytical tools. While they can display in any web-enabled device, they are optimized for the viewing experience of a "typical" computer screen.

The ALIS controls architecture is organized so that the top level comprises the Resource Dashboard that expresses resource use in variable terms: as standard units, financial figures, by usage ("Today's water use is equivalent to two baths"), and in relative terms ("25% less electricity than yesterday"). The research involved in the development of the Resource Dashboard explores appropriate contextual ways to present the information, as these vary not only by individual preference, but also by location and task. For example, in the garage the inhabitant may wish to see the power consumed by her electric car overlaid on the top-level dashboard view, and to see it in terms of "kilometers earned." Detailed information on resource production and use is available in both real-time and historical views, categorized in different ways (by type of device, by location in house, by time of use). Building energy-analysis tools provide detailed performance analysis and prediction. The Resource Dashboard integrates visualization components in the display of energy use data in multiple formats: as numerical financial data and graphs in the Overview screen, comparative and scalable performance data in the Resource Usage

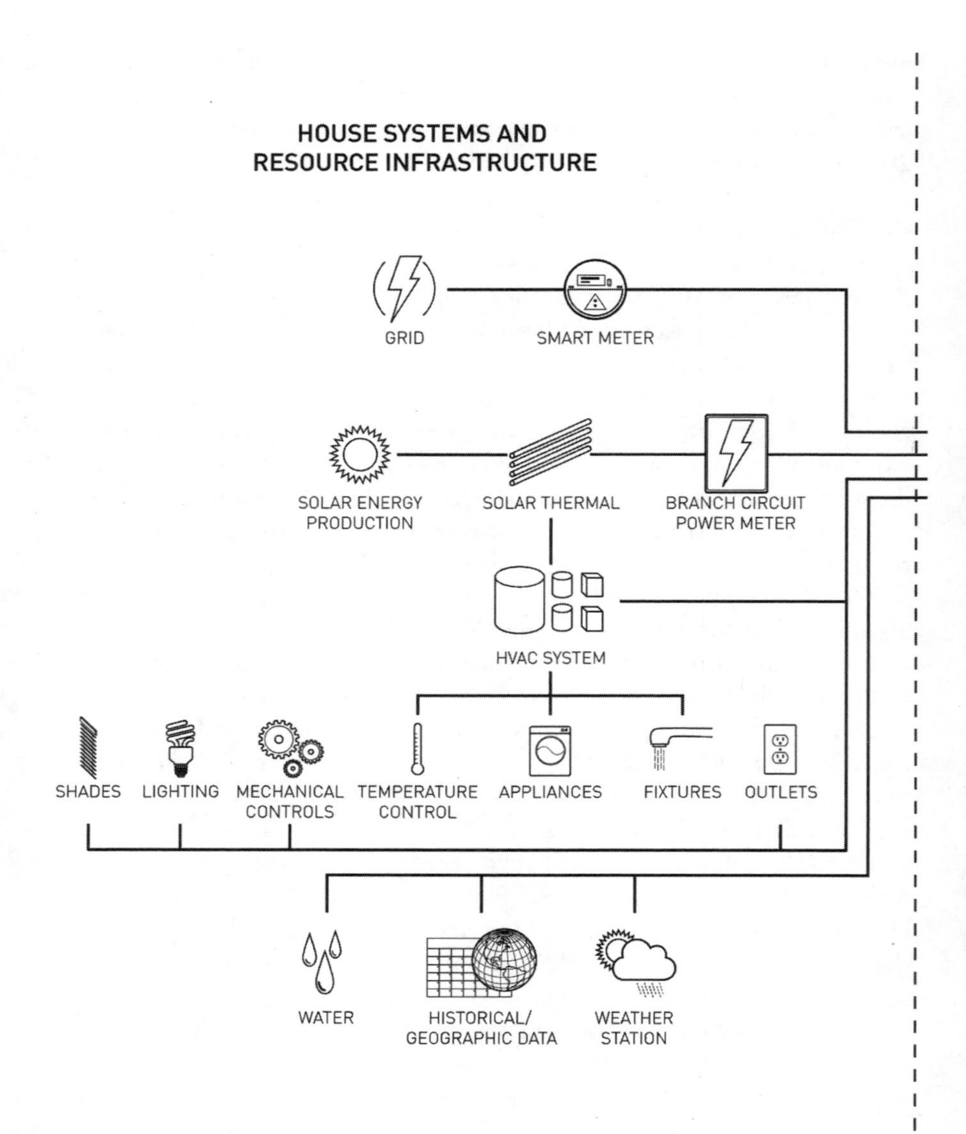

Figure 8.4

Schematic of home control systems and user interfaces

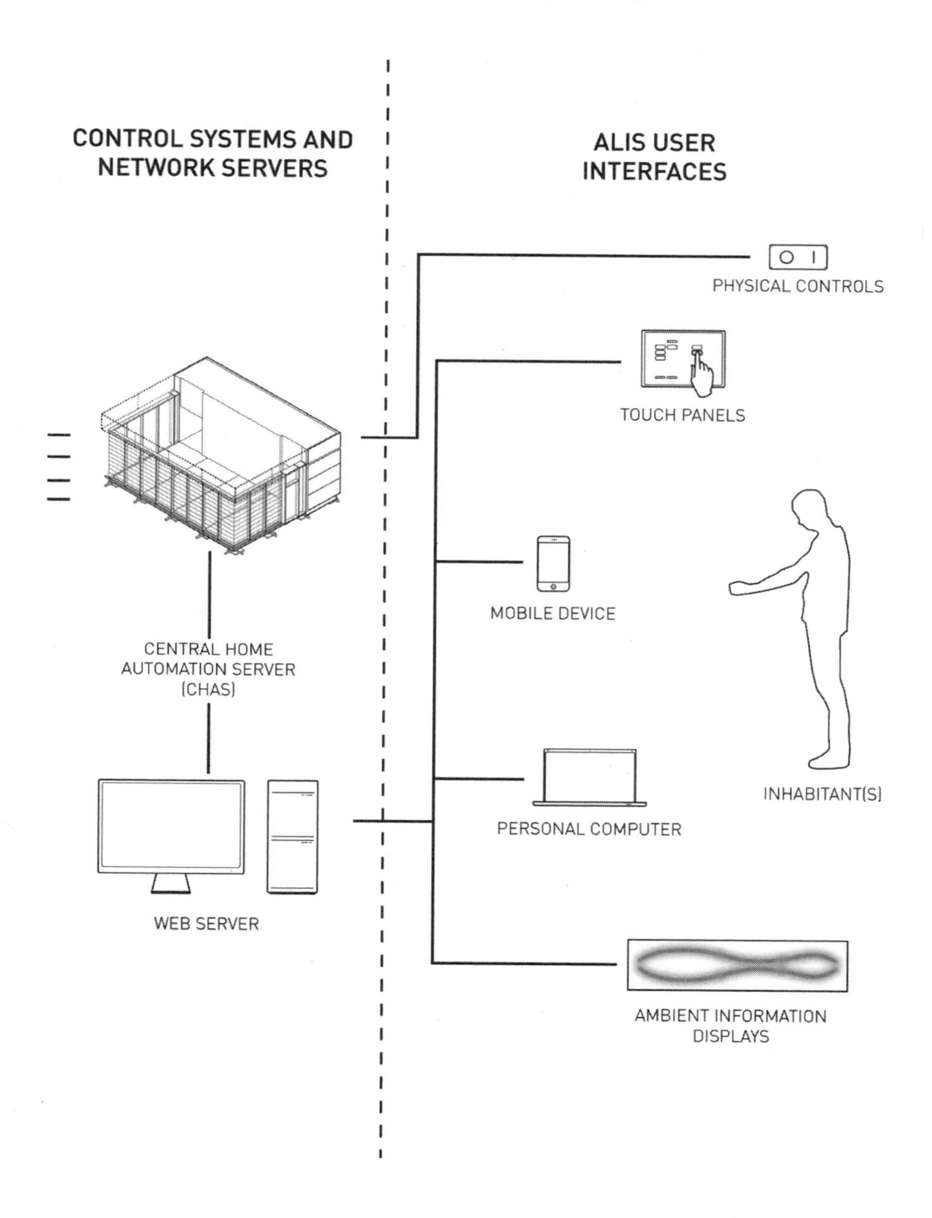
CONTROL SYSTEMS AND NETWORK SERVERS
ALIS USER INTERFACES
PHYSICAL CONTROLS
TOUCH PANELS
MOBILE DEVICE
CENTRAL HOME AUTOMATION SERVER (CHAS)
PERSONAL COMPUTER
INHABITANT(S)
WEB SERVER
AMBIENT INFORMATION DISPLAYS

Figure 8.5

Main control touchscreen for home control systems

tool, and simplified comparative performance data in the Neighborhood Network views. Inhabitants can set personal milestones and challenges that the system can measure—for example, "use 10% less energy than last month" (see figure 8.6, top).

Reception of Feedback and Direct Information

Presentation of energy use information in meaningful ways is especially important, as the general public remains largely undereducated about building energy use. Although the idea that energy conservation should be taught in schools is widely supported, and many people express a desire to reduce personal energy consumption, a recent study by the National Environmental Education and Training Foundation found that while a majority of Americans considered themselves to be knowledgeable about energy issues, only 12 percent of them could pass a basic energy quiz (NEETF and RoperASW 2002). Average families use neither scientific units (joules

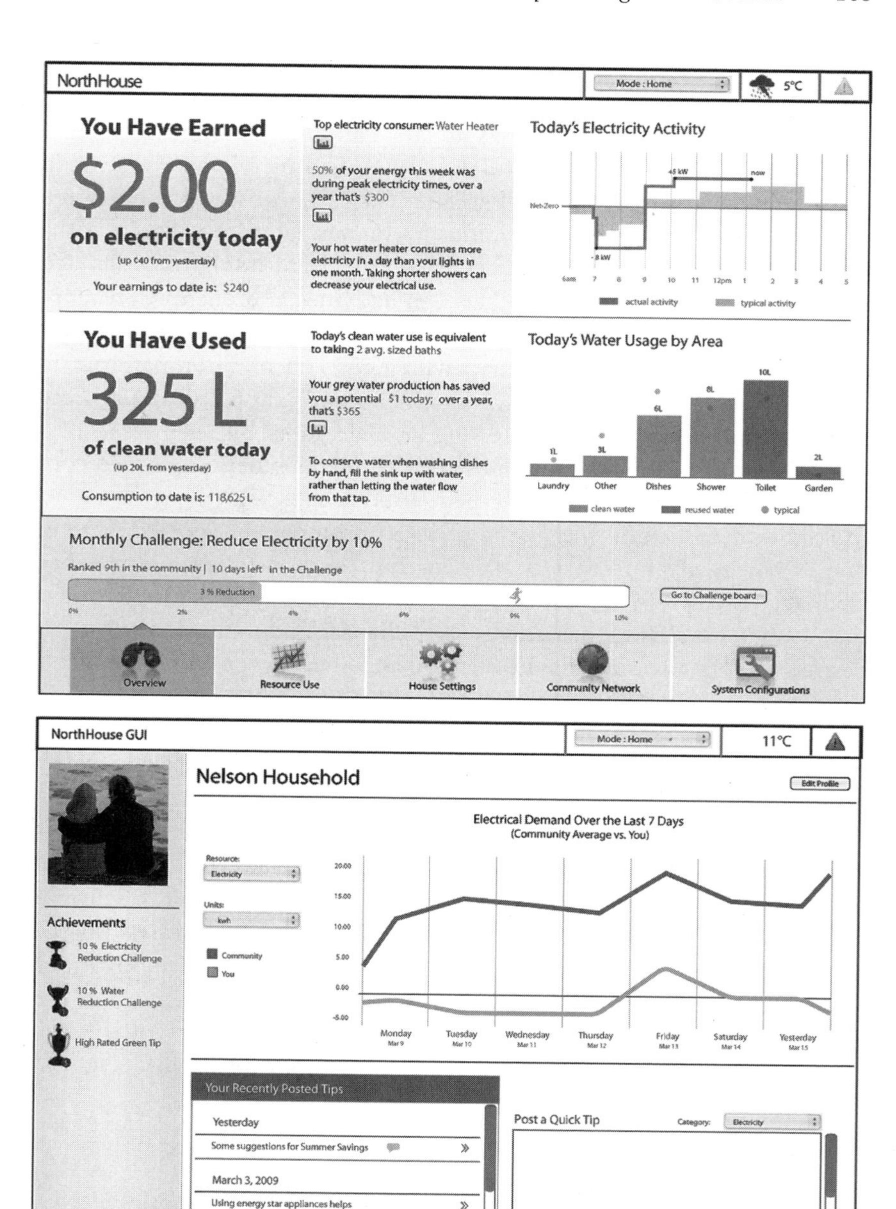

Figure 8.6

Resource Dashboard (top) and neighborhood network (bottom) sample screens

or BTUs) nor commercial units (kWhrs or CCFs) in thinking about energy consumption. Rather, they use "folk units," which are familiar, multipurpose, and easily visualized, such as dollars, gallons, and months. Furthermore, in the absence of meaningful ways of measuring relative energy load of specific devices, they attribute a value based on observable but often irrelevant criteria such as the quantity of human labor it replaces or the frequency with which they interact with a device, leading to misguided concepts about the relationship between activities and energy consumption (Kempton and Montgomery 1982). A popular metaphor to illustrate consumers' understanding of their own energy consumption patterns, in the absence of detailed feedback, is that of a family shopping at a grocery store with no price tags and receiving an aggregated bill at the end of a month's worth of purchases, from which family members must then construct a model for understanding their costs (ibid., 817).

Studies that include energy monitors to promote energy reduction in homes repeatedly find that consumers appreciate detailed performance feedback (Darby 2006), that they will enthusiastically consume information toward modifying their energy use behavior, and that many express a desire for more advanced data collection and analysis tools that could advance their understanding more thoroughly (Woodruff, Hasbrouck, and Augustin 2008). Sarah Darby presents a thorough review of research on the effectiveness of feedback on residential energy consumption (Darby 2006). She notes that in general, the effectiveness of feedback is highly dependant on (1) general context, the cultural attitudes and motivations of the population, as well as the characteristics of the technologies involved; (2) quality of feedback, the legibility and frequency of delivery; and (3) synergies with other factors or strategies, such as goal setting or receiving advice. Further studies explore the complexity of motivations behind resource management for individual homeowners, as well as the various types and contexts of feedback and information that inhabitants respond to (Chetty, Tran, and Grinter 2008; Woodruff, Hasbrouck, and Augustin 2008).

Leveraging and Social Interaction

ALIS also includes a community interface, called the Neighborhood Network, which encourages competition, comparison, and collaboration among community members. It takes advantage of the widespread use of social networking software to enable individuals to connect with a wider community of people pursuing similar goals around environmental sustainability, enabling them to share strategies, incentives, and successes (figure 8.6, bottom). It is available on both computers and mobile devices as part of the web application. Inhabitants can see a historical view of their energy consumption compared to a community average and set conservation goals, allowing them to compete with others in achieving those goals while sharing tips and comments. Those who follow through with their commitments receive awards (in the form of digital trophies) that are displayed to others within the community. These trophies not only apply to challenges, but also recognize the degree to which one has engaged with the community by posting conservation advice and replying to other members' feedback. ALIS uses existing social networking tools by way of a Facebook plug-in that connects to the server and automatically collects data. Automatically collected data as well as data from voluntary self-entered reporting is compiled for online publication. A challenge in sharing energy use information relates to data normalization: simple quantification between dissimilar houses does not support effective comparison. Instead ALIS relies on the more general metric of individual progress toward a user-defined set of goals.

Jennifer Mankoff and colleagues argue that social networking is an appropriate media for motivating the reduction of personal energy consumption (Mankoff et al. 2007). They point to social movement theory, which argues that the mobilization of political movements historically occurs through "informal networks, pre-existing institutional structures, and formal organizations" (Morris 2000, 445). Furthermore, sustained participation in a political movement directly relates to the degree to which it is "integrated with other life goals and activities of the participants." Networks serve important functions relative to social movements by "structurally connecting prospective participants to an opportunity to participate, socializing them to a protest issue, and shaping their decision to become involved" (Passy and Giugni 2000, 2001). Journalist Bill McKibben argues that the Internet transforms environmental activism from the realm of politicians and large organizations to something accessible to anyone with a "good idea and an internet connection," stating that "the fight against global warming requires all kinds of technology—solar panels and windmills, but also servers and routers" (McKibben 2007).

Spatial and Lifestyle Integration and Ambient Information

In order for feedback mechanisms to generate persistent behavioral changes, they must remain in place in the home, and integrate within daily life practices. Though some researchers argue that behavioral changes that develop over three months or more tend to persist for a year or more, shorter term tests with feedback mechanisms show that consumption patterns revert to normal once they are removed (Darby 2006, 15). Thus ALIS is not a single control panel or application, but an "information ecosystem" (Bartram, Rogers, and Muise 2010, 9), with embedded control and information points (both interactive and atmospheric) conveniently located throughout the home and incorporated into the everyday tasks and devices of inhabitants. For example, the central computer touchscreen is built in to the wall surface of the kitchen, making it easy and convenient to use and seamlessly integrating this technology within the domestic environment.

The mobile device provides feedback and control to the inhabitant of the home from her pocket—a simplified remote domestic system control. Mobile applications are an extension of the web application (figure 8.7), and offer a subset of the features available in the computer interface with each feature redesigned for use on a mobile device. For example, the controls available from the mobile device emphasize ease of use through logical groupings. These "master" controls allow the inhabitant to adjust the lights for a whole room, or shades for a whole house facade, with a single control. More fine-grained control of individual fixtures is still available, but a hierarchy of control makes the most used items easily accessible. Graphing displays are simplified; a playful "Spinner" interface allows inhabitants to select pairs of performance variables through a slot-machine style interaction to compare and graph on demand. The mobile application provides alerts about house status, energy consumption, and production thresholds, and allows inhabitants to keep in touch with their neighborhood network, offering monitoring of challenge progress and community chat. Perhaps most important, mobile access enables contact with and control over the systems of the house while the inhabitant is away, enhancing both the concept of, and reinforcing the engagement with, this networked domestic ecology.

Yet not all feedback provided to inhabitants needs to be direct, quantified, and detailed. In an age of information overload, there is increasing interest in the cognitive value of "calm technology," that is, technology and information that inhabits the periphery of human attention, and that provides attunement to conditions without requiring attentive focus (Weiser and Brown 1996). The Ambient Canvas is an atmospheric indicator embedded within the kitchen backsplash surface of North House that provides sensible feedback on the balance of produced versus consumed

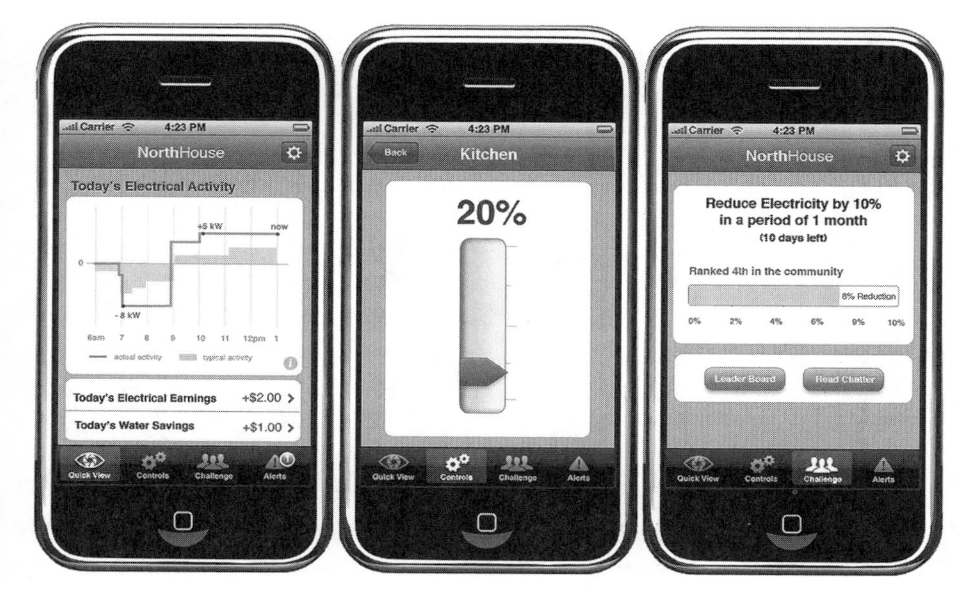

Figure 8.7

ALIS mobile application

energy, and how the inhabitant is progressing toward a self-specified goal (for example, reducing energy by 10 percent from the prior week). Unlike typical graphical displays that may use numbers or charts to convey information, the Ambient Canvas combines LED lights and filters of various materials to produce light effects on the kitchen backsplash (shown in figure 8.8). This subtle feedback on performance and energy efficiency does not require active attention on the part of the inhabitant, and integrates into the home cohesively. This approach promotes awareness of resource use to assist and influence sustainable in-the-moment decision making through intuitive and sensorial domains.

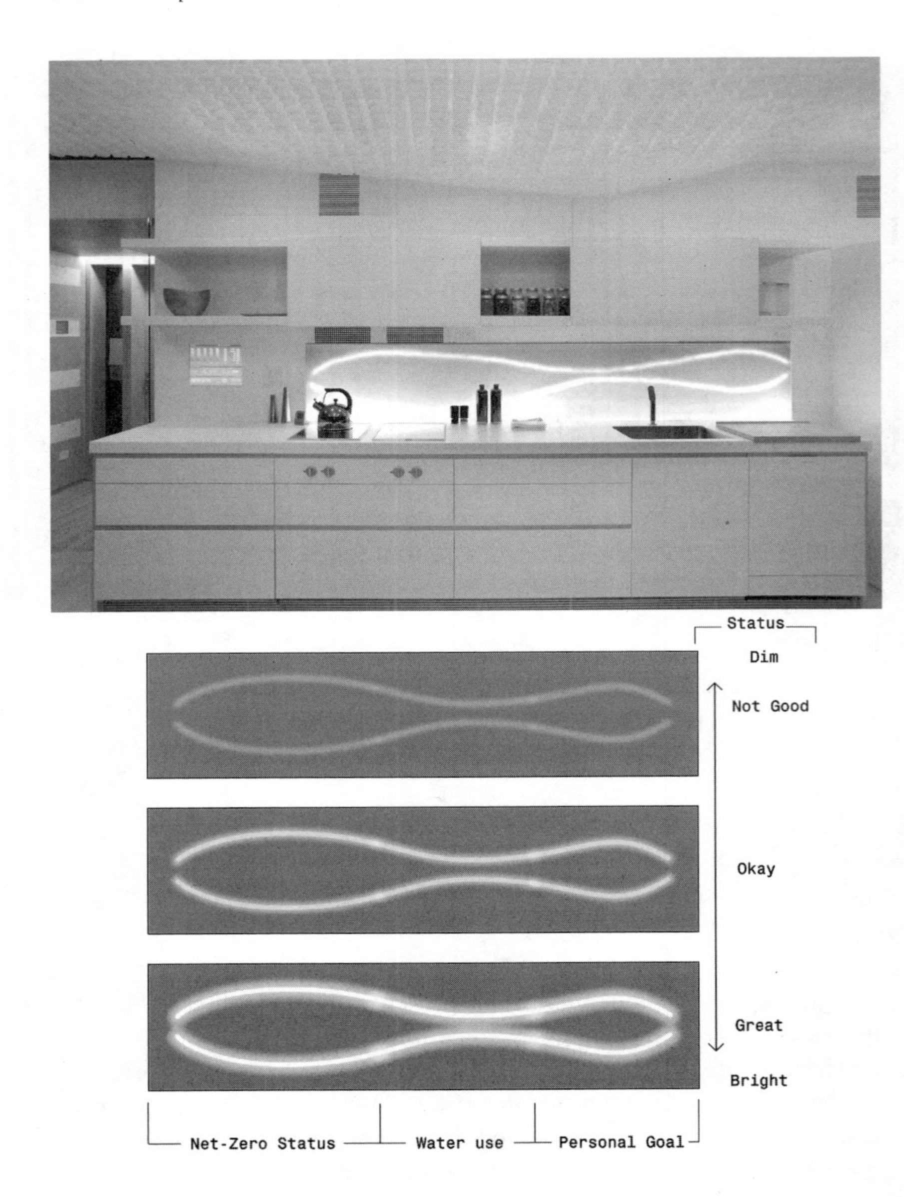

Figure 8.8

Ambient display behind kitchen backsplash that provides peripheral feedback on resource use performance.

Conclusion: Prospects for Active Design Research

Understanding how inhabitants use energy in the home is an emerging area of active design research and this awareness is still nascent in architectural design discourse. Current computational design tools help architects and engineers to predicatively simulate and evaluate building and technology performance. However, simulation models are limited because they are not yet able to take into account the prediction of how inhabitants behave and impact performance. Designing for societal uptake of technology-based transformations requires a user-centered approach. Architects, engineers, and system designers face the challenge of reframing design strategies as a coevolution of human and building performance that will encourage as well as underpin sustainable use. This requires new models of design thinking beyond the typical "smart home," encompassing inhabitant behavior, motivational strategies, and exploration of how technology can impact inhabitants' daily rituals and sense of comfort.

We propose that developing more informative models of inhabitant behavior requires an understanding of three critical areas of building use: context, the complexity of technological intervention, and appropriate interaction with building performance and automation. Contextual factors may include human activities in the space, demographic variables (age, gender, mobility, and income of the inhabitants), time of day, or current presence in the building, to name a few. They influence how people use their home and its resources in different situations. Technology deployment introduces new patterns of use: we need to understand what impacts it may have on human interaction with the living environment and on the inhabitant's comfort and acceptance. Finally, if automation or assistance is to be implemented, we need to consider its operational behavior: are we using the right conditions to affect automated or intelligent decisions about controlling the residents' environment?

Moreover, we argue that this approach demands new processes of design as we move from the concept of a building as an artifact, through that of a building as an assemblage of systems, toward the model of an integrated system comprising spatial organization, responsive materials, technology, environment, and humans participating as an interconnected domestic ecology. A critical part of the process of systems design is the iterative development of prototypes and successive refinement of working models that allow designers to explore and test how well the ideas, artifacts, or systems work in synergistic and dynamic relation, and relative to human engagement. While a building is a singular proposition, a research prototype is capable of evolving for development and evaluation through iterative cycles of improvement. This kind of progressive development cannot be simulated in a lab or studio: it requires constructed testbeds, capable of facilitating inhabited observation and evaluation. Finally, and perhaps most important, the aims of this work cannot be achieved without constant and meaningful collaboration from the outset with multiple disciplinary experts, including architects, engineers, interaction designers, and computer programmers, as well as sociologists, environmental psychologists, and artists.

Acknowledgments

We thank the the U.S. Department of Energy, the National Renewable Energy Laboratory, the Ontario Power Authority, British Columbia Hydro, Natural Resources Canada, the University of Waterloo, Ryerson University, Simon Fraser University, and the *rare* Charitable Research Reserve for their financial support of the North House project. We also acknowledge the numerous graduate students who collaborated on the two-year project, and also the generous contributions of the professional industry partners involved. Graduate research was partially funded by the MITACS ACCELERATE graduate internship program in the provinces of Ontario and British Columbia. A detailed team credit list can be found at http://www.rvtr.com/files/TEAM_NORTH_CREDITS.pdf. All images are used by permission of the authors.

References

ACEEE. 2011. *Residential sector: Homes and appliances*. American Council for an Energy Efficient Economy. http://www.aceee.org/sector/residential (accessed August 29, 2011).

Baker, Nick. 1996. The irritable occupant: Recent developments in thermal comfort theory. *Architectural Research Quarterly* 2:84–90.

Bartram, Lyn, Johnny Rodgers, and Kevin Muise. 2010. Chasing the negawatt: Visualization for sustainable living. *IEEE Computer Graphics and Applications* 30 (3): 6–12.

Bartram, Lyn, Johnny Rodgers, and Robert Woodbury. 2011. Smart homes or smart occupants? Supporting aware living in the home. In *Proceedings of IFIP INTERACT*, ed. Pedro Campos, Nicholas Graham, Joaquim Jorge, Nuno Nunes, Philippe Palanque, and Marco Winckler, 52–65. Lisbon: Springer.

Bartram, Lyn, and Robert Woodbury. 2011. Smart homes or smart occupants? Reframing computational design models for the green home. In *Proceedings of the AAAI Spring Symposium*, 2–9. Palo Alto, CA: Association for the Advancement of Artificial Intelligence.

Chetty, Marshini, David Tran, and Rebecca E. Grinter. 2008. Getting to green: Understanding resource consumption in the home. *Proceedings of UbiComp* 08:242–251.

Cole, Raymond J., and Zosia Brown. 2009. Reconciling human and automated intelligence in the provision of occupant comfort. *Intelligent Buildings International* 1:39–55.

Cole, Raymond J., Zosia Brown, and Sherry McKay. 2010. Building human agency: A timely manifesto. *Building Research and Information* 38 (3): 339–350.

Cole, Raymond J., John Robinson, Zosia Brown, and Meg O'Shea. 2008. Recontextualizing the notion of comfort. *Building Research and Information* 36 (4): 323–336.

Darby, Sarah. 2006. The effectiveness of feedback on energy consumption. Environmental Change Institute, University of Oxford. http://www.eci.ox.ac.uk/research/energy/downloads/smart-metering-report.pdf (accessed May 2, 2010).

Dubberly, Hugh, Paul Pangaro, and Usman Haque. 2009. What is interaction? Are there different types? *Interaction* 16 (1): 1–13.

Eckl, Roland, and Asa MacWilliams. 2009. Smart home challenges and approaches to solve them: A practical industrial perspective. In *Intelligent interactive assistance and mobile multimedia computing*, ed. Djamshid Tavangarian, Thomas Kirste, Dirk Timmermann, Ulrike Lucke, and Daniel Versick, 119–130. New York: Springer.

Ehrhardt-Martinez, Karen, Kat A. Donnelly, and John A. Laitner. 2010. Advanced metering initiatives and residential feedback programs: A meta-review for household electricity-saving opportunities. Report to the American Council for an Energy-Efficient Economy. http://aceee.org/research-report/e105 (accessed August 29, 2011).

Frumkin, Howard. 2008. Nature contact and human health: Building the evidence base. In *Biophilic design: The theory, science and practice of bringing buildings to life*, ed. Stephen R. Kellert, Judith H. Heerwagen, and Martin L. Mador, 107–118. Hoboken, NJ: Wiley.

Janda, Kathryn B. 2009. Buildings don't use energy: People do. In *Architecture, Energy and the Occupant's Perspective—Proceedings of the 26th conference on Passive and Low Energy in Architecture (PLEA)*, ed. Claude Demers and André Potvin, 9–14. Quebec, Canada: Les Presses de l'Université Laval.

Kempton, Willett, and Laura Montgomery. 1982. Folk quantification of energy. *Energy* 7 (10): 817–827.

Leaman, Adrian, and Bill Bordass. 2001. Assessing building performance in use 4: The Probe occupant surveys and their implications. *Building Research and Information* 29 (2): 129–143.

Loftness, Vivian, and Megan Snyder. 2008. Where windows become doors. In *Biophilic design: The theory, science and practice of bringing buildings to life*, ed. Stephen R. Kellert, Judith H. Heerwagen, and Martin L. Mador, 119–132. Hoboken, NJ: Wiley.

Lutzenhiser, Loren. Social and behavioral aspects of energy use. *Annual Review of Energy* 18:247–289.

Mankoff, Jennifer, Deanna Matthews, Susan R. Fussell, and Michael Johnson. 2007. Leveraging social networks to motivate individuals to reduce their ecological footprints. *HICSS*: 87–97. http://www.cs.cmu.edu/~assist/publications/07MankoffHICSS.pdf (accessed May 2, 2010).

McKibben, Bill. 2007. The power of the click. *Los Angeles Times*, October 16. http://articles.latimes.com/2007/oct/16/news/OE-MCKIBBEN16 (accessed April 18, 2010).

McMaster, Tom, and David Wastell. 2005. The agency of hybrids: Overcoming the symmetrophobic block. *Scandinavian Journal of Information Systems* 17 (1): 175–182.

Morris, Aldon. 2000. Reflections on social movement theory: Criticisms and proposals. *Contemporary Sociology* 29 (3): 445–454.

NEETF and RoperASW. 2002. Americans' low "energy IQ:" A risk to our energy future. The National Environmental Education & Training Foundation, Washington, DC. http://www.csu.edu/cerc/researchreports/documents/AmericansLowEnergyIQ2002.pdf (accessed April 18, 2010).

Orr, David. 2000. Architecture as pedagogy. *Journal of the Mississippi Academy of Sciences* 45 (4): 210.

Passy, Florence, and Marco Giugni. 2000. Life-spheres, networks, and sustained participation in social movements: A phenomenological approach to political commitment. *Sociological Forum* 15 (1): 117–144.

Passy, Florence, and Marco Giugni. 2001. Social networks and individual perceptions: Explaining differential participation in social movements. *Sociological Forum* 16 (1): 123–153.

Schipper, Lee, Sarita Bartlett, Dianne Hawk, and Edward Vine. 1989. Linking lifestyles and energy use: A matter of time? *Annual Review of Energy* 14:273–320.

Socolow, Robert. 1978. *Saving energy in the home: Princeton's experiments at Twin Rivers*. Cambridge, MA: Ballinger.

Stern, Paul C., and Elliot Aronson, eds. 1984. *Energy use: The human dimension*. New York: W. H. Freeman.

Thün, Geoffrey, and Kathy Velikov. 2012. North House: Climate responsive envelope and control system. In *Design and construction of high performance homes*, ed. Franca Trubiano, 265–281. London: Routledge.

Tognoli, Jerome. 1987. Residential environments. In *Handbook of environmental psychology*, ed. Daniel Stoklos and Irwin Altman, 655–690. New York: Wiley.

U.S. DOE. 2009. *Buildings energy data book*. United States Department of Energy. http://buildingsdatabook.eren.doe.gov/Default.aspx (accessed August 29, 2011).

Velikov, Kathy, and Geoffrey Thün. 2012. Responsive building envelopes: Characteristics and evolving paradigms. In *Design and construction of high performance homes*, ed. Franca Trubiano, 75–92. London: Routledge.

Wigginton, Michael, and Jude Harris. 2002. *Intelligent skins*. Oxford: Elsevier Architectural Press.

Weiser, Mark, and John Seely Brown. 1996. The Coming Age of Calm Technology. Xerox PARC, October 5. http://www.johnseelybrown.com/calmtech.pdf (accessed October 10, 2011).

Woodruff, Alison, Jay Hasbrouck, and Sally Augustin. 2008. A bright green perspective on sustainable choices. In *Proceedings of ACM 2008 Conference on Human Factors in Computing (CHI '08)*, 313–322. New York: ACM.

9

Building Up to Organizational Sustainability: How the Greening of Places Transforms Organizations

Christine Mondor, David Deal, and Stephen Hockley

Introduction

The rapid market uptake of green building has transformed the design and construction industries, yet little attention has been given to how a green building project transforms the organizations for which the project was built. As architects and green building consultants, we began with a simple observation that seemed widespread throughout our many projects—organizations that begin sustainability efforts with tactical green building goals, such as saving energy or achieving a LEED (Leadership in Energy and Environmental Design) certification, frequently widen their scope to include broader sustainability goals within the organization. We questioned whether place-based sustainability initiatives are cause or effect of organizational transformation. Why do certain organizations dive deeper into organizational sustainability than others? What organizational practices and infrastructure makes them more likely to succeed, and in what ways does the organization change?

We argue that green building projects are often a gateway or a *gravitational assist* that gives momentum to a broader, organization-wide definition and commitment to sustainability. Gravitational assist describes the transformation of organizational culture to include concern for holistic sustainability through place-based initiatives. Similar to the centrifugal force of a planet's gravitational pull when it is used to reorient a spacecraft on a long journey, the knowledge gained through a green building project can provide the momentum to accelerate a deeper commitment to sustainability within the organizational culture.

This chapter establishes a definition for "organizational sustainability" and then explores the relationship between green building projects and broader environmental and social commitments. We describe a systemic approach to sustainability through three types of actions—those concerning *people*, *process* and *place*. We document how place-based initiatives inform an organization's systemic approach to sustainability, and illustrate characteristics of transformational projects and organizations that are primed for change. Finally, we examine how tactical efforts shape organizational values and summarize the most effective means to create systemic sustainability transformation that lasts beyond the green building project.

This study focuses on three nonprofit organizations: the Greater Pittsburgh Community Food Bank, the Pittsburgh Opera, and Phipps Conservatory and Botanical Gardens. In each case, buildings and grounds are critical to fulfilling the organization's mission, and each organization expanded its focus from an initial green building project to organization-wide efforts toward sustainability. We selected these organizations for comparison because each has a public mission and shares similar staffing and governance structures. Through interviews with the organizations' executive directors, we found that each nonprofit believes that its green building project(s) were the catalyst to advancing discussions on sustainability within the organization (Christopher Hahn, pers. comm., Feb. 2010; Richard Piacentini, pers. comm., Feb. 2010; Joyce Rothermel, pers. comm., Feb. 2010). These commonalities make it possible to compare actions and motivations and to observe characteristics of the gravitational assist phenomenon.

A Definition for Organizational Sustainability

"Organizational sustainability" is a term used to describe the adoption and application of sustainability principles across all levels and efforts of an organization, including operations and governance. Organizational sustainability is different in meaning from simply "sustaining an organization." Rather, it addresses issues and context beyond the traditional spatial, temporal, and monetary systemic boundaries of the organization. For example, it can include consideration of environmental and social cost implications or benefits of new investments, manufacturing processes, or product design. A comprehensive approach to organizational sustainability is accomplished through ongoing assessment cycles at all levels, where sustainability metrics are measured over time and influence the organization's mission, goals, and initiatives. Organizational sustainability has the potential to become the strategic lens through which all practice is viewed.

However, the way an organization implements sustainability practices depends on how the group defines and values sustainability; there are some commonly shared definitions which serve to guide such organizations. The term "sustainable development" was most famously articulated in the Brundtland Commission Report (WCED 1987, 43) stating "sustainable development is development that meets the needs of the present without compromising the ability of future generations to meet their own needs." One of the most common frameworks for measuring the success of sustainable development is the "triple bottom line" of social equity, economic performance, and environmental stewardship (Elkington 1998, xiii). The triple bottom line has become a simple yet significant way for organizations to expand their reporting frameworks beyond economic indicators and to account for more difficult-to-measure social and ecological goals. This broadly accepted understanding informs the definitions of sustainability adopted by this chapter's three organizations.

The "triple bottom line" definition can be seen clearly in large, multinational corporations like HSBC, an international financial institution, whose 2008 sustainability report (HSBC Corporate Sustainability 2008) included a letter from Chairman Stephen Green that cites HSBC's "responsibility to manage our business across the world for the long term by making a real contribution to social and economic development and by protecting the environment in which we operate." As its overarching target, the corporation committed to carbon neutrality in 2005 and this commitment influences all levels of decision making from staffing, to investment policy, to public giving, to the operations of its data centers, and the building of environmentally friendly facilities. The motivations for such a commitment are numerous and include international regulations as well as HSBC's own need to limit risk on worldwide investments. Like many other large corporations, HSBC has significant resources to position itself as a leader in sustainability. The company has a team of internal managers dedicated to integrating sustainability issues into practice and makes this process public through its corporate sustainability report.

However, the motivating forces and mechanisms to pursue organizational sustainability are less clear for smaller organizations that may have more limited resources. Sustainability reporting frameworks such as the Global Reporting Initiative (GRI) are structured for larger organizations with resources to perform ongoing and inclusive processes that measure and monitor complex performance indicators. Large and small companies are often frustrated with the proliferation of sustainability "certifications" and struggle to understand the value to their organizations or garner the resources that are required to achieve them.[1] These factors have tended to confuse the market and inhibit action, especially for smaller organizations that lack expertise or resources to define a value proposition for sustainability.

In Small Organizations, Sustainability Grows from Place

In our work, we observe that smaller organizations (such as nonprofits or privately held companies of under two hundred people) often encounter issues of sustainability at the building project scale before they experience a larger strategic shift. Small and medium-sized organizations often rely on their physical facilities for their operations. Building projects represent significant capital investment and the design, construction, and certification of a green building can provide a strong framework and common language for measuring the success of an investment. In addition, building-related forces such as an "enlightened" donor, a funder request, or potential marketing opportunities can have impact across the entire organization. In this realm, place-based projects influence organizations toward strategic sustainability initiatives and can help explain this gravitational assist phenomenon in smaller organizations.

First, the U.S. Green Building Council's LEED certification program has dominated the green building industry since the late 1990s. LEED has achieved a high level of recognition from professionals and the general public and has become synonymous with environmental leadership. The LEED system of green building standards distilled simple core concepts that can be applied to building projects regardless of scale.[2] Its key principles raised industry norms for design and construction, as is evidenced by the incorporation of LEED-based criteria into building codes in California and Maryland, and elsewhere across the country (Guevarra 2008; Eco-Structure staff 2011). Even if a project does not submit for certification, LEED may have influenced the project team by providing a simple framework to discuss sustainability.

Second, smaller organizations are often facility dependent and can be highly influenced by the quality of their offices, warehouses, or public venues. Indeed, all three organizations selected as case studies in this chapter rely on their facilities to support their core operations. Their stories illustrate the dynamic relationship between strategic and tactical approaches to green building and suggest some of the factors that might increase the likelihood of green building projects enabling gravitational assist in smaller organizations.

For example, when the Greater Pittsburgh Community Food Bank built its new headquarters, the core concepts of sustainability seemed to be aligned with its core mission focused on improving social equity. The Food Bank closes supply chain loops by using its facility to gather, process, and redistribute donated and purchased food. In 1999, the Food Bank moved from a poorly suited facility that needed accessibility and safety improvements into a new 100,000-square-foot warehouse and offices. Given the large capital investment, the Food Bank's funders required that the building use the nascent LEED rating system, and the new headquarters became one of the first ten LEED-certified buildings in the country. As a pilot project, LEED 1.0 provided guidelines for workplace quality, but many elements of the system had yet to be defined. The greatest increase in efficiency came not from LEED-mandated building performance, but from the consolidation and modernization of the Food Bank's logistic systems in the new facility. According to Executive Director Joyce Rothermel, the building was seen as another tool to meet the Food Bank's goal of "more food to more people," but strategic sustainability concepts that were seeded during construction took time to incubate within the organization. So while the building served as a gateway to strategic sustainability initiatives, the Food Bank's shift in focus did not occur immediately.

Similar to the Food Bank's move into a new facility, the Pittsburgh Opera moved into new headquarters in 2009 in Pittsburgh's Strip District neighborhood. In contrast to the Food Bank, the Pittsburgh Opera quickly pursued a comprehensive approach to a greener headquarters and operations following the belief that a high-quality indoor environment had a strong correlation to better performances on stage and across the organization. According to Executive Director Christopher Hahn, sustainability might never have become relevant to the Pittsburgh Opera if the organization had not acquired a building. The historic factory building had been the birthplace of Westinghouse Air Brake in 1869 and was a source of pride as the first facility that the Opera owned and operated. Like the Food Bank, funders mandated the use of the LEED system. However, the directive came too late to use LEED for New Construction (LEED NC) so LEED for Existing Buildings, Operations, and Maintenance (LEED EBOM) was adopted. Executive Director Hahn describes the Opera's prior view of sustainability as limited because the organization handled few materials in its operations and did not own any facilities. When the Opera acquired its new practice space and offices, the organization faced new challenges related to building operation costs but also had increased control of the environmental quality, flexible use of space, and associated building improvement opportunities. The building became a place to host the community, to showcase energy efficiency, and to implement sustainable practices in purchasing and waste management, none of which had been priorities in the previously rented offices.

In contrast with the Pittsburgh Food Bank and Opera, the Phipps Conservatory and Botanical Gardens has always had the environment and issues of sustainability at the core of its mission, even if initially this was not made explicit. The Phipps historic glass greenhouses and gardens are the main attraction for the organization, and so facility stewardship greatly affects its success. The early educational mission of Phipps was to provide horticultural instruction and pleasure to the public. Ironically, the Phipps Conservatory's collection of plants was not necessarily in service to *ecological* goals that we might recognize today, as many of the early plants were exotic species gathered from the 1893 World's Columbian Exposition in Chicago.

Emphasis on sustainability developed much more recently through the design and construction of a Welcome Center (2005) and the Tropical Forest Conservatory (2006). As its mission aligned with the emerging concepts of sustainability, Phipps was able to leverage a capital campaign and subsequent phased development projects to construct some of the most innovative greenhouses in the country. Executive Director Richard Piacentini describes how the requirement to certify the first of four facility improvement projects, the Welcome Center, opened conversations on sustainability. The organization's commitment deepened with the construction of production greenhouses and the Tropical Forest Conservatory, which is believed to be the most energy-efficient conservatory structure in the world.

With each investment in facilities, Piacentini describes the increasing commitment to greening Phipps's facilities as also raising the bar of sustainability performance across the organization's mission and operations. The construction efforts have provided opportunities for Phipps to use the buildings as "living laboratories" for ongoing sustainable practices as its staff constantly looks for opportunities to save water or energy. Indeed, Phipps's mission statement now has sustainability as a core value—"to advance sustainability and promote human and environmental well-being through action and research." Currently, Phipps is constructing the last of the planned expansion projects, the Center for Sustainable Landscapes (CSL). Phipps's design and construction goals are targeted at the highest level of LEED certification, Platinum level, and the project has adopted two emerging rating systems—the Living Building Challenge and the SITES rating system—that require much higher levels of performance than LEED. Phipps has also brought its historic facilities into the green building circle by adopting LEED EBOM to provide structure and improvements across operations.

Gravitational Assist: Vectors of Action and Motivating Forces

To understand how gravitational assist can be leveraged, we need to better understand how organizational sustainability is operationalized. Organizational sustainability can be described in three *vectors of action*, people, process, and place, and the depth of commitment can be influenced by *motivating forces*, including internal forces, external forces, and regulatory or quantitative limitations.

Organizational Sustainability: Vectors of Action

Vectors of action categorize the initiatives that an organization pursues, even when an organization does not have a stated definition of sustainability (such as the triple bottom line). We observed in this study and across our client cohort that smaller organizations rarely begin with an abstract definition of "sustainability," even when their mission is seemingly aligned with sustainability principles. It is more likely that the definition is built up from numerous initiatives and shaped accordingly. A "top down" approach would include broadly understood goals and governance structure to allow each unit within a company to contribute to the corporate goals and to evaluate their effectiveness. However, this type of meta organization rarely happens first and many organizations find an entry point into sustainability through one of three vectors for action cited earlier: people, process, and place, as illustrated in figure 9.1.

people	process	place
Labor	Materials	Site disturbance
Health and safety	Energy	Vegetation
Productivity	Emissions/waste	Stormwater
Diversity	Transportation	Water efficiency
Education	Econ. performance	Energy efficiency
Human rights	Investment policy	Carbon reductions
Community building	Purchasing policy	Renewable energy
Public policy	Green cleaning	Heat island reduction
		Materials life cycle
		Clean air
		Daylight and views

Figure 9.1

People, process, and place vectors of organizational sustainability

People

People is a category that acknowledges that individuals and the persons that comprise an organization are central to any sustainability initiative. Their knowledge, beliefs, and actions are key to determining achievable goals and implementation strategies. Actions and intentions can often be as effective as material or technological approaches to sustainability.

Process

Process is a category that defines what we do, how we accomplish it, and the material and energy flows that are needed. Processes must be viewed over time and with different lenses to understand how they can be efficient and effective on multiple levels.

Place

The category *place* has a dual role. Places influence our experience and our understanding of sustainability and they can enable us to implement sustainable practices. Places are often tactile demonstrations of environmental commitment and, at the fundamental level, contain the materials and resources that are at the core of many environmental issues.

Organizations are usually structured in service or production units such as "human resources," "procurement," or "facilities," and the initiatives of these units can often be aligned with the people, process, and place categories. These three categories together allow for a broader view of the directions that an organization can take and provide an opportunity to visualize the effect of gravitational assist, as illustrated in figure 9.2.

When frameworks such as GRI, the LEED rating system, or the Sustainability Tracking and Assessment Rating System (STARS) developed by the Association for the Advancement of Sustainability in Higher Education's (AASHE), are examined with respect to people, process, and place, patterns emerge with most systems emphasizing one category over another. Understanding this emphasis is important for organizations when they are selecting the frameworks most applicable to their sustainability goals and organizational mission.

Organizational Sustainability: Motivating Forces

Organizations have distinct influences that cause them to move toward sustainability commitments. In our work there are three motivating forces that describe the majority of engagements. An organization is more likely to move to action when there is more than one factor exerting influence or when the type and magnitude of the factor is especially relevant to the organization's core.

FIRST-ORDER KNOWLEDGE
function of the enterprise

SECOND-ORDER KNOWLEDGE
enterprise/environment interface

GRAVITATIONAL ASSIST
caused by place-based initiative

THIRD-ORDER KNOWLEDGE
change in enterprise/environment interface
dynamics of systems
expanded knowledge resulting from place-based initiative

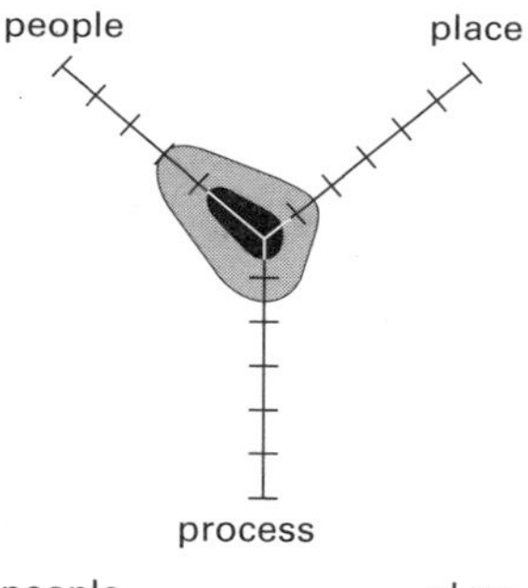

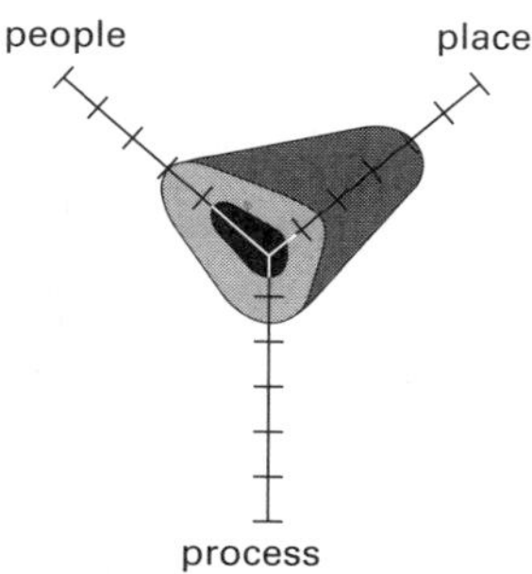

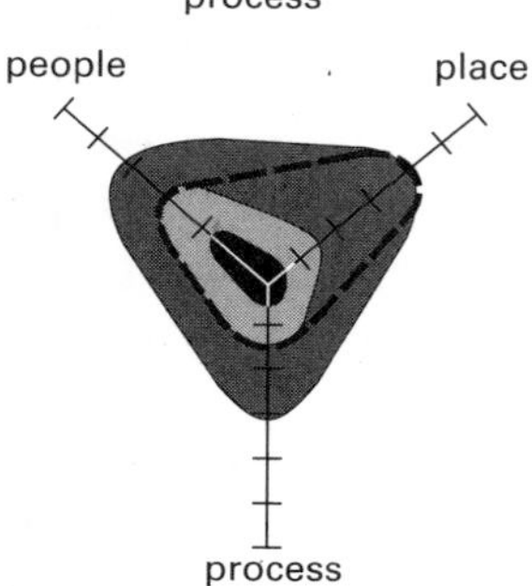

Figure 9.2

The term "gravitational assist" describes the acceleration of organizational sustainability through an increase of actions along the people, process, and place vectors.

Internal or Cultural Forces

Sustainability is seen as integral to the mission or vision of the company or is essential to its practices. Initiatives are typically proactive and can be formalized through mission, vision, and policy or may be culturally accepted in a more informal manner.

Examples: (1) A company had significant intellectual property in the innovation of ultra-efficient mechanical systems and benefits from a market emphasis on sustainability. (2) A high-tech company is known for its creative employee culture that values progressive environmental practices.

External or Competitive Forces

Pressures to adopt sustainability measures come from market demand, market expectations, or specific competitors' initiatives. The implementation is usually reactive or imitative of the market leader.

Examples: (1) A small paint company with limited research and development funds incorporates low VOC products into its line because there is increased demand and competitors are quickly developing products to fill the need. (2) A research company's employee recruitment is competitive and workplace quality is a differentiating factor.

Regulations and Quantities

Action is taken as a result of regulations, incentives, or policies. Organizations are incentivized or required to be in compliance.

Examples: (1) Tax credits incentivize companies to install solar panels and inspire a company to consider adopting sustainability initiatives in other areas. (2) A government contract requires that suppliers evidence basic reporting of sustainability commitment and practices.

Across various projects, we anecdotally observe that organizations pursuing sustainable practices because of internal or cultural forces (values) undertake efforts that have depth and longevity. To understand the potential effectiveness of these forces, we look to Donella Meadows's landmark analysis of places to intervene in a system to leverage change (Meadows 1999, 19). Originally created in response to the North American Free Trade Agreement (NAFTA), Meadows articulated a nine-point, later twelve-point, continuum of levers that can with small shifts produce big changes within a system. The list has been applied to many different system types from ecosystems to economic systems, and is relevant in this study to understand the forces creating sustainable change within an organization. Meadows describes each lever with examples from different systems, but we have observed that these levers can be demonstrated within a singular system and propose assembling the nine levers into three categories:

1. Most effective: *Internal or cultural forces* reflect the goals and paradigm of the system or its reason for being. For example, the purpose or value of a municipal water system is to provide water for a community. The goal of the system affects its design, structure, and flows so this category offers significant potential for creating change in a system.

2. Moderately effective: *External or competitive forces* relate to system structure, information flows, and system rules. In the municipal water system example, referenced above, the system's design and structure enable functionality.

3. Least effective: *Regulations and quantities* address the flow and magnitude of material in the system. In the municipal water system, these represent the behavioral or operational controls that regulate the flow of materials and resources.

Meadows maintains that the most effective way to change a system is to change the goals or the paradigm in which the system is designed. Decisions of design or operation of the system result from the core values for which the system was created. In this way, place-based initiatives have transformative power to change an organization's operational systems when the projects challenge and change organizational values. The case studies in this chapter describe various degrees of organizational transformation as a result of the differential penetration of sustainability into each nonprofit's core values. The more the core values were affected, the greater the longevity and depth of initiatives.

Duckles (chapter 12, this volume) distills five common themes that organizations use to describe their green buildings and organizational narratives. These themes—profit-driven green, practical green, deep green, green innovators, and hidden green—relate to our gradient of motivations and can be observed in the way that the organizations in our case studies present their sustainability commitment and green building projects. In our chapter's cohort and in other projects, we have observed that profit-driven and practical narratives often signify an organization influenced by regulatory or quantitative limitations or external forces. Similarly, deep green and innovating green narratives seem to be common in internally or culturally motivated organizations. The Food Bank could be characterized as being influenced by quantitative limitations and the corresponding narratives have been practical. Phipps Conservatory and the Pittsburgh Opera have talked about internal and cultural forces and their narratives (especially Phipps) are highly supportive of Duckles's deep green and innovating green categories.

Prior to embarking on green building projects, the Greater Pittsburgh Community Food Bank, the Pittsburgh Opera, and Phipps Conservatory and Botanical Gardens each had initiatives that could have supported sustainability goals. However, none had a mission-level focus on sustainability and none of the organizations had a comprehensive framework to evaluate their efforts. As a result of their green building projects, each organization refined its mission to reflect its approach to sustainability and each has developed specific methods of decision making and evaluation of sustainability issues.

Early in the design and construction process, the Food Bank identified an internal or cultural affinity toward sustainability. Because this was not formalized into its mission or policies, it had little long-term influence on the organization's direction at that time. Without the cultural transformation, motivation for green building never reached the internal or cultural level and instead resulted from the regulations and incentives of funder pressure. During and after the project, the Food Bank staff often described the mission with the memorable phrase "more food to more people." Efficiency, at least as it supported the distribution of more pounds of food, was the main path to creating equity in the community. The organization's goal seemed straightforward and had an easy-to-visualize metric: the ubiquitous can of food. The clarity of this mission proved to be a double-edged sword—initiatives that did not contribute directly to this metric were difficult to justify.

For example, at the time of building construction, the Food Bank participated in many process- and people-focused actions. The Food Bank ran a farm, gleaned unpicked food from other farm fields, gathered excess food from events or restaurants, and participated in other programs that redirected material from the waste stream. Other proposed projects had returns that were not directly related to the food metric and were jettisoned during the design and construction, including a methane digester, a production greenhouse, and vermicomposting. The economic bottom line orientation of the Food Bank may have allowed clarity, but it presented difficulty in creating a definition of sustainability that might have allowed for broader adoption of sustainability initiatives across all three categories of action of people, process, and even place.

In addition, the LEED 1.0 framework was evolving simultaneously with the project delivery and the pilot version offered less guidance regarding the value of broader sustainability initiatives. Much later versions of LEED NC and frameworks like LEED EBOM and the Living Building Challenge reset the goals of green building by giving metrics to strategies initially viewed as externalities, such as energy production, green purchasing, and minimizing waste during operations. By incorporating these metrics in a project's goals, the project team is able to question larger organizational goals. The Food Bank project team members, however, spent much of their energy finding materials and developing documentation methods to meet the requirements associated with a newly developed LEED rating system; this hampered consideration of broader initiatives.

The Pittsburgh Opera's embrace of sustainability transformed internal culture and created a formal administration system to perpetuate the changes. Prior to their move to their historic headquarters, the Opera's mission statement had no reference to sustainability and the Opera had few initiatives that could be described as green. In fact, Christopher Hahn describes the difficulty of starting a recycling program in their old facility because behavior patterns were difficult to uproot without the "massive change" of the move to the new building. After the Opera accepted the mandate to be certified in the LEED EBOM framework, the board of directors and the staff of the Opera began to question how sustainability might be reflected in their core documents, with the board adding it as a core value in 2010. To define this direction, Hahn describes the Pittsburgh Opera organization's ongoing and mutually supportive relationship with the community. Regional embeddedness became essential to the Opera's success and then it considered sustainability leadership important to broader community improvement. This, combined with the operations and maintenance goals of the LEED EBOM framework, has allowed the organization to incorporate actions that reach beyond place-based initiatives to include purchasing, waste minimization, and even costume sharing and recycling.

Phipps Conservatory differs from the Food Bank and the Opera because its approach to sustainability was shaped fundamentally by the *iteration* of its green building projects. This decade-long commitment changed internal culture and formalized the organization's commitment through policy and leadership. Richard Piacentini describes how place-based projects raised the bar for all operations. Phipps aligned core activities such as education and programming with sustainability, created a sustainable horticulture program, and hosted many conferences on sustainability topics. Piacentini describes how the building projects shifted the organization's long-term vision to become a "leader in sustainability," which meant that no area of operations could go unexamined. It was Phipps's place-based initiatives that sparked the deep commitment to people-based and process-based sustainability actions like sourcing, purchasing, waste reduction, community outreach, education, and encouragement of behavior modification.

Sharing Knowledge to Create Gravitational Assist

Our inquiry proposes that an organization's capacity to share knowledge, or its knowledge management practices, directly impacts the effectiveness of gravitational assist. A green building project is both a material project and a knowledge-intensive process. Green building frameworks require thinking at multiple scales and project teams often find that the recommended strategies are beyond the immediate project-based concerns and engage broader issues. In this study, we observed that early knowledge development on sustainability focuses on project-based motivations, such as how to meet energy saving regulatory requirements or how to capture LEED credits or points. At the next level, it can also encompass thinking about the value of the effort relative to external dynamics, such as a focus on benchmarking and competitive market position. When deep knowledge about sustainability is created, the process affects the cultural "hardwiring" of an organization and can result in high-level strategic commitments that give broad structure for current and future actions, such as adopting a triple bottom line pledge to sustainability.

Kathia Laszlo (2001) adapted a model of business knowledge from earlier work by Ervin and Christopher Laszlo (Laszlo and Laszlo 1997) that helps in understanding the different types of knowledge. The model describes business knowledge of three kinds by observing the evolution of business practices from the industrial revolution to today. *Business knowledge of the first kind* is concerned primarily with the function of the enterprise and was typical for companies prior to the 1950s. *Business knowledge of the second kind* was prevalent from the 1950s to the 1990s and involved an expanding of knowledge to include the enterprise/environment interface. Last, K. Laszlo asserts that the contemporary model, or *business knowledge of the third kind*, accounts for prior orders of knowledge but emphasizes the situational understanding of system dynamics and includes the contours of change in the enterprise/environment interface.

K. Laszlo's third knowledge type, dynamic change in the system, emphasized the evolution of the enterprise/environment context and is descriptive of the type of knowledge needed to create resilient organizational sustainability. Nattrass and Altomare (1999, 187–200) describe "evolutionary corporations" as transitioning from the industrial model of "organization-as-machine" to learning organizations or "organization-as-brain." These organizations are conscious of their capacity to create environmental change at an evolutionary scale and need to be self-conscious of their learning processes. Green building projects seem to be especially effective ways to enable "evolutionary" organizations through K. Laszlo's third kind of knowledge; the development of a green building can create organizational learning structures and knowledge management systems. Consider that a green building project: has a defined scope and duration; requires an internal governance structure and requires collaboration across an organization; can involve "thick" or profound examination of an organization's mission and goals; is a tangible project where sustainability principles can be shared and physically demonstrated; and creates "massive change" or immersive displacement of older, less sustainable patterns.

These characteristics of a building project help to set the stage for adoption of further sustainability initiatives by articulating explicit environmental values and indicators of success, creating organizational infrastructure to examine sustainability strategies, and demonstrating the resource connectedness to the larger context.

It is worth noting here the special role that frameworks serve in helping to shape the development of organizational knowledge. Frameworks provide a delimited understanding of sustainability that often serves as a microcosm of larger organizational sustainability issues. This represents a narrowing of the knowledge funnel and enables an organization to constrain the seemingly infinite possibilities for exploration of sustainability (Pasteur, Pettit, and van Schagen 2006).

This structure of an organizational learning process can also be informed by the emerging field of transition management. Since 2001, experts in the Netherlands have been defining the theory and best practices for directing sustainable development and long-term societal change. According to Loorbach (2007), transition management emphasizes normative processes and governance, both of which are highly dependent on the sharing and construction of knowledge. His call for sustainable development to be a "guiding notion that enables science and society to search for long-term collective goals and ambitions, to experiment in the short term, and to regularly assess progress," reflects the ongoing evaluative processes experienced by the organizations studied in this chapter, first at the project scale (green building) and then later at the broader organizational level.

In addition, the green building projects share structural principles that are similar to transition thinking, thus suggesting that the focused project might help prepare a group of learners for larger questions of sustainability. Some of the principles that Loorbach articulates for transition thinking are shared with green building projects and can be seen in the organizational evolution of the Food Bank, Pittsburgh Opera, and Phipps Conservatory. Both green building projects and transition thinking: consider different domains, levels of scale, and system states; consider the long term (more than twenty-five years) when evaluating short-term actions; use an integrated multi-actor process; practice backcasting and forecasting to plan for uncertainties; and enable social learning through learning by doing and doing by learning.

Transition management principles may help to explain why the Food Bank has the least formalized infrastructure supporting organizational sustainability, even though it has had the longest germination period. As an early adopter of LEED, the project was the first green building effort for the Food Bank as well as the design and construction team; we propose that organizations with multiple iterations have more opportunity to develop deeper commitment and expertise. In addition, current standards such as the *Whole Systems Integrated Process Guide* (ANSI 2007) and the *Integrative Design Guide to Green Building* (7group and Reed 2009) were not yet available. The guides advocate for an *integrated design process*, a robust method that is inclusive of a broad array of stakeholders and the design and construction team members. The inclusiveness is intended to encourage multidimensional thinking and consideration of larger time scales than would normally be considered in smaller teams. The process is highly normative, with charrettes and workshops helping to construct knowledge and articulate values that form the basis of decision making throughout the process. The lack of such a process combined with other factors such as the novelty of the LEED framework and the lack of a mandate from the board or staff created less opportunity to influence the Food Bank as a learning organization.

In contrast, Phipps Conservatory demonstrates how a structured decision-making process, internal leadership, and multiple iterations can provide gravitational assist that transforms organizational culture. The Center for Sustainable Landscapes, the final building project in the campus master plan, used an integrated design process to structure the delivery of the design and construction. The charrette workshops were attended by diverse stakeholders including board members, staff members, funders, supporting nonprofits, contractors, consultants, educational institutions, and others who might not otherwise be included in a conventional streamlined process. Each stakeholder had to represent his area of expertise, while taking on the concerns of others in the discussion. The inclusion of so many people, the multiscale character of the discussion, and activities that required collaboration and consensus meant that organizational knowledge was constructed organically and the concerns relating to the project spread to other operational areas.

Executive Director Richard Piacentini describes the organizational decision-making process as one driven by broad goals for sustainability, yet implemented at the tactical level. He describes how with each decision, the bar for performance is raised. "Why wouldn't we do it [the sustainable] way?" is the question that drives decisions, often asked by Piacentini. He shares how "the low hanging fruit was picked a long time ago" and how the types of issues they encounter often do not have proven best practices to easily adopt. There is a great need for ongoing learning development in the organization to "identify, test and implement" strategies. By asking "How can we do this better?" Phipps benefits from incremental learning whose results are aggregated and evaluated for possible mandates of leadership. Piacentini describes how small steps lead to large actions. During construction of the Tropical Forest Conservatory, every strategy was incrementally questioned. Almost as an afterthought, the team members realized they had created the most energy efficient conservatory in the world. This aggregation of smaller actions became a mandate for future expanded leadership and allowed Piacentini to leverage support from funders and his board of directors.

The Pittsburgh Opera experienced transformational change, yet it did not benefit from multiple projects, nor did the Opera use an integrated design process. In this case, the penetration into deep practice resulted from the breadth of the rating system chosen and the relatively flat organizational structure of the nonprofit. The commitment to achieving LEED EBOM certification meant that most of the staff members were directly affected by the need to gather and evaluate information, creating a forum for discussion that eventually was reflected in a new mission statement that incorporated sustainability. Executive Director Christopher Hahn attributes the rapid uptake of sustainability issues to the relatively young age of his staff members, most of whom have encountered environmental values in their upbringing and education. Hahn's own childhood experiences in South Africa were formative to his leadership and enabled him to step into the role as sustainability advocate both inside and outside the organization.

Internal and external stakeholders gave voice to perspectives from multiple domains, scales, and systems during the learning process at the

Pittsburgh Opera. For example, prior to owning a facility, one of the few areas of environmental concern was the sourcing of costumes. The Opera adopted sustainability initiatives that served the community and expanded sustainability across all areas of practice as a result of constructing a green headquarters. Staff members began to consider how their daily roles and responsibilities could accelerate sustainable practices; Opera staff began purchasing local and organic food service, office supplies with recycled content, and even created a green event standard for those renting the space for parties and conferences. The Opera formed an internal Green Team and created a governance role of Sustainability Coordinator to ensure that communication occurred and that newly established sustainability goals were being met. Despite their deliberate creation of infrastructure, knowledge management continues to provide challenges. Hahn describes the organization as "information hungry, but not information rich" and identifies this as the next threshold in their decision-making processes.

Organizational Sustainability Principles for Action

Organizational sustainability is an ongoing and iterative commitment to apply sustainability principles across the workings of an organization, and the market will continue to see companies, nonprofits, and even entire disciplines build their environmental knowledge in response to internal and external pressures. The places where we work will continue to be an important entry point to sustainability for many organizations because built environment projects affect the individual's understanding, operational procedures, and organizational knowledge. Buildings and landscapes are ubiquitous in our workplaces, and our efforts to design, build, and occupy these facilities offer excellent opportunities to create or sharpen environmental knowledge. Place-based initiatives can provide a gravitational assist that gives momentum to a broader, organization-wide definition and commitment to sustainability. To accomplish this, place-based sustainability projects commonly: have a defined scope and duration; require internal governance structures and collaboration; offer logical opportunities to examine mission and goals; can dislodge older patterns making way for a newer and more sustainable direction; are tangible demonstrations of sustainability principles; and involve transferable sustainability principles that can be applied in other areas of operation. As a result of building projects, organizations are often better prepared to adopt environmental values and key performance indicators, to plan and execute sustainability initiatives, and to articulate their connectedness within their community and context. This examination of place-based initiatives within three nonprofit organizations demonstrates that gravitational assist can be leveraged to create greater value for the organization by engaging the following principles for action:

Framework Selection
Leverage appropriate systems or frameworks. Broadly accepted frameworks for organizational sustainability and place-based sustainability, such as STARS or LEED, help to structure efforts and to institutionalize organizational knowledge by limiting the breadth of relevant issues, defining indicators of success, and providing pathways for execution.

Gravitational Assist
Strategize early about how a sustainability framework can leverage organizational knowledge to other areas of operations. The principle of gravitational assist describes the migration of values and knowledge from one area of sustainability to another. While any single framework might be seen as restrictive in achieving robust organizational sustainability, when viewed in the context of organizational knowledge development sustainability frameworks can be seen as transformational in launching an organization into cycles of sustainability improvement.

Beyond Certification
Select a sustainability framework or certification standard that is most relevant and deliberately rigorous to increase the depth of organizational knowledge. As the proliferation of third-party ratings, certifications, and labels continues, frameworks that are well defined and function at multiple scales are more likely to catalyze actions across organizational operations. The case studies demonstrate that frameworks with underdeveloped informational resources or ill-defined processes (such as LEED 1.0 Pilot at the Food Bank) can make it less likely that future action will spread across an organization. In comparison, those frameworks which are more established, have a broader scope, and have significant reference and support resources (such as LEED EBOM 2009 at Phipps and the Opera) can help facilitate action by allowing the team to focus on the concepts and not simply the logistics.

Iteration
Create cycles of improvement to deepen the commitment to organizational sustainability. Success takes practice and ongoing projects can use

frameworks multiple times to improve results. Choose frameworks that require ongoing measurement or verification to improve performance and require more formalized collaboration, reporting, and reporting methods.

Leadership

Create organization-wide leverage through leadership. While project-based sustainability initiatives such as the LEED certification of a building can effectively introduce sustainability to an organization, it will be difficult to leverage benefits across the organization without support from leadership. Administrative or board leadership allows for the creation of organizational infrastructure to share knowledge and can directly increase the likelihood of the lessons learned benefiting other operational areas.

Buildings are large investments for organizations and their legacy continues long after the design and construction teams have left the site. We hope that concept of gravitational assist and these five principles for action will improve our understanding of the transformative nature of our buildings and landscapes, and bring additional value to organizations that commit to sustainability.

Notes

1. The green building sector experienced a period of differentiation where competing or duplicative systems were either accepted or not accepted within the market. As a result, certain systems have become dominant and others have assumed supporting roles or disappeared altogether. This is not likely to occur with such rapidity in the broader organizational sustainability sector, as market forces encouraging corporate or organizational sustainability are more complex and the metrics for comparison are not as aligned as those for buildings.

2. Although LEED concepts are applicable across a range of project scales, we often hear complaints from building owners that the actual execution of a rating submission requires expertise and additional resources that favor larger organizations and projects.

References

7group, and Bill Reed. 2009. *The integrative design guide to green building: Redefining the practice of sustainability*. Hoboken, NJ: John Wiley & Sons, Inc.

ANSI. 2007. *Whole systems integrated process guide*. American National Standards Institute. Washington, DC: WSIP.

Eco-Structure. 2011. Maryland adopts international green construction code. *Eco-Structure*, May 12. Washington, DC: Hanley Wood.

Elkington, John. 1998. *Cannibals with forks: The triple bottom line of the 21st century business*. Gabriola Island, BC: New Society Publishers.

Guevarra, Leslie. 2008. California adopts green building code for all new construction. *GreenBiz.com*. July 17. www.greenbiz.com/news/2008/07/17/california-adopts-green-building-code-all-new-construction (accessed December 21, 2012).

HSBC Corporate Sustainability. 2008. *HSBC Holdings plc sustainability report*. London: Group Corporate Sustainability.

Laszlo, Ervin, and Christopher Laszlo. 1997. *The insight edge: An introduction to the theory and practice of evolutionary management*. Westport, CT: Quorum Books.

Laszlo, Kathia C. 2001. The evolution of business: Learning, innovation and sustainability in the 21st century. Paper presented at the annual conference of the International Society for the Systems Sciences (ISSS), Asilomar, CA.

Loorbach, Derk. 2007. Governance for sustainability. *Sustainability: Science, Practice, & Policy* 3 (2): 1–4.

Meadows, Donella H. 1999. *Leverage points: Places to intervene in a system*. Hartland, VT: The Sustainability Institute.

Nattrass, Brian, and Mary Altomare. 1999. *The natural step for business: Wealth, ecology, and the evolutionary corporation*. Gabriola Island, BC: New Society Publishers.

Pasteur, Kath, Jethro Pettit, and Boudy van Schagen. 2006. Knowledge management and organisational learning for development. Paper prepared for the Knowledge Management for Development Workshop. Brighton, UK.

WCED. 1987. *Our common future*. World Commission on Environment and Development. New York: Oxford University Press.

10

Green School Building Success: Innovation through a Flat Team Approach

Michelle A. Meyer, Jennifer E. Cross, Zinta S. Byrne, Bill Franzen, and Stuart Reeve

Introduction

Green schools cost less to operate, freeing up resources to truly improve students' education. Their carefully planned acoustics and abundant daylight make it easier and more comfortable for students to learn. Their clean indoor air cuts down sick days and gives our children a head start for a healthy, prosperous future. And their innovative design provides a wealth of hands-on learning opportunities.

—U.S. Green Building Council 2010

With over 130,000 schools nationwide and 20 percent of the population of school age, green schools have large-scale environmental consequences and also provide numerous benefits supporting the educational mission of school districts. Political support from the Conference of Mayors, the National School Boards Association, and Congress bring attention to green school initiatives, with up to 65 percent growth annually expected in the education sector, the fastest growth in green building adoption of any sector (Yudelson 2008). But the transition to green buildings can be difficult, as schools face obstacles similar to other organizations when implementing these green projects. Such obstacles include internal and external resistance based on individual values, beliefs, and motivations, as well as organizational structures and norms that inhibit green innovation (for discussion of green building obstacles, see Al-Homound 2000; Doppelt 2003; Hoffman and Henn 2008). Public organizations also face bureaucratic stagnation, conflict over fiscal responsibility, and the perceived political risk associated with green construction (Johnson 2000; Pearce, Dubose, and Bosch 2007). Because school districts require voter approval for capital expenses, even when green initiatives can save public money, political opinions and concern about affecting children's educational outcomes—whether substantiated or imagined—can stall green building adoption (Fernandez and Rainey 2006; Turner Construction 2005). While green schools show promising growth potential, generating support for green building can still be particularly challenging for school districts.

Yet one district in the Rocky Mountain West region of the United States turned initial resistance to green building into broad-based support, becoming an innovator and national leader in green building before green schools had national political support or LEED certification was widely used. In this chapter, we describe how the green building success of this district was a *byproduct* of a new focus on relationships and staff participation in the design, construction, and maintenance of schools to emphasize the importance of symbolic and interactional organizational processes on green school adoption.

Unfortunately, academic literature on organizations and the environment has often focused on environmental impacts, information dissemination, individual environmental attitudes, or economic rationality rather than organizational processes (Lutzenhiser 1993; Shwom 2009). We argue that understanding resistance and developing genuine employee support for organizational change is especially crucial for green building, because ideological disagreements on environmental protection and climate change can plague green building projects with politically charged debate (Hulme 2009). Our case study illustrates how reorganization of typically hierarchical organizational command structure into participatory flat teams affects the symbolic and interactional patterns within an organization. Organizational research has shown that flat teams, with responsibility and participation shared, support environmental initiatives by facilitating the communication of a green building vision to all staff and generating overall awareness and excitement about the green building initiatives. Also by changing interaction patterns, diversifying the decision-making power, and promoting communication and feedback loops, flat teams create space for learning about green design options (Al-Homound 2000; Cebon 1992; Doppelt 2003; Fernandez and Rainey 2006; Kotter 1995). Our results support and extend these findings.

Particularly interesting to those considering green building adoption, our case study describes processes that can counteract the perception that political views on the environment will prevent green building success. The success of this district was strengthened by the inclusion in a flat team of people with diverse and sometimes hostile opinions about environmental protection and climate change. For example, some of our interviewees called themselves "climate skeptics" and refused to be involved in projects focused solely on environmental protection or emission reduction. The team approach we discuss created a space for expression of these various viewpoints, without attacking individuals' environmental attitudes or ideological stance on these politically divisive topics. In this chapter, we detail how the creation of the "Green Team" in this school district's Operations Department provided the framework for three interrelated symbolic and interactional processes important to successful green building implementation, even in the face of anti-environmental sentiment: (1) collaborative development of a unified symbolic vision; (2) increased staff participation in design processes and decisions; and (3) increased building literacy. We begin our story with a description of the school district and its success, and then follow with discussion of each process.

The Case Study of Green School Building Success

The school district serves more than twenty thousand students in over fifty buildings. One elementary and one high school are certified LEED Gold and Silver, respectively, and two-thirds of the school buildings received the Energy Star label in 2010—a label that fewer than 4 percent of all schools in the United States have received. The district has won numerous state and federal sustainability and energy awards, and sustained energy conservation displays its remarkable success. The new design for a green elementary school *with* air conditioning cost $100 per square foot to build in 2001, and operates at $0.43 per square foot for gas and electric services. In comparison, at that time traditionally built elementary schools in the region *without* air conditioning cost on average $110 per square foot to build and $0.57 per square foot for gas and electric. The new design also uses about two-thirds less water than other schools in the district![1]

This success story began in 1998, when green building and high performance design were nascent concepts in the mainstream construction industry. District officials began planning for a $175 million capital bond to build six schools. During previous bond cycles for school construction and renovation, administration ignored green and energy efficiency suggestions proposed by the district's energy manager and other concerned staff. The administration and school board typically chose designs based on upfront costs without consideration of life-cycle costs, energy use, or other green benefits—common obstacles to green building adoption (Hoffman and Henn 2008; Turner Construction 2005). Operations Department employees continue to describe these older schools as "failures" and "wastes" of money and resources. One employee even remarked, "I would burn them all down if I could." Thus, transitioning to green building was a dramatic change in design and construction processes—and we set out to understand how this dramatic change occurred.

Our case study was designed to create an in-depth description of the change process used by this exceptionally successful green building adopter (Creswell 2007; Lofland et al. 2006). In 2009, we conducted semi-structured interviews with sixteen internal school district staff members and nine external supporters who participated in the green building processes. We interviewed the district's Operations Department leadership and various department managers—for example, mechanical, facilities, security, electrical, HVAC (heating, venting, and air conditioning)—as well as administrators, board members, and teachers. A majority of these interviewees were staff employed by the district before the green building initiative (the average length of employment was twenty years, with a range of one to thirty-five years). Non-school district interviewees included architects, employees of local utility companies, and engineering consultants.

The argument presented here grew from a grounded theory approach to qualitative data analysis. We coded and recoded the data in stages: first identifying a variety of concepts and topics in the data, then examining relationships between the concepts to identify any general themes, and finally identifying a core concept or process and recoding the data in relationship to this idea (Charmaz 2001; Strauss and Corbin 1990). From this approach, we learned how changing the interactional patterns within the district and with external partners such as architects and construction contractors led to the district becoming a national leader in green building.

The Green Team and Green Building

When tasked with preparing for the 2000 bond, a group of Operations Department managers discussed doing "something different." After learning of integrated design and new green products at an American Institute of Architects (AIA) meeting in 1999, one manager sent out a memo recommending the creation of what would become known as the "Green Team" to research green products and practices that the district could include in the new school prototype. The initial Green Team included fifteen to twenty school employees such as managers within the Operations Department (facilities, construction and project development, maintenance, operations, and energy managers), as well as representatives from custodial, electrical, outdoor maintenance, plumbing, HVAC, food services, transportation, and security. As described by the manager sending the initial memo, the goal was to represent all "different departments that take care of the buildings. [Because] when the architects are gone and the janitors are gone, these guys inherit that world." To increase the likelihood of successful organization initiatives, research shows that leadership teams should include representatives from many levels and departments of the organization (Kotter 1995). In terms of green building projects, these teams should include those directly involved in implementing changes, namely facilities and maintenance staff (Doppelt 2003; Etzion 2007; Pellegrini-Masini and Leishman 2011). The Green Team met both of these standards—a flat team representing different departments and levels of responsibility with specific inclusion of operations and maintenance staff.

Thus, in 1999 the Green Team began as a research group, with each member assigned one hour per week of Internet research on items related to green design and energy and water savings. They divided the building by architectural divisions, with team members choosing a division that interested them and fit their expertise. Members reported their findings at monthly meetings where they discussed the pros and cons of each item or procedure, to determine "how green we thought this really was and how practical it was from a facilities standpoint." These discussions commonly led to a desire for more information and a contact to the manufacturer for a demonstration, or to sending a team member to another organization to see the product in use. These research-focused meetings continued for a little over a year into early 2000, at which point the district hosted an architectural design competition. Architectural firms developed prototypes that included the green ideas chosen by the Green Team. The winner was selected in September 2000, based on cost, integration of the green design elements, and visual appeal.

Two months later, voters passed the bond and the district contracted with the winner to begin the project. The Green Team expanded to include the architects and contractors to ensure the implementation of their green design choices. Administration, teachers, and external support organization representatives (i.e., local utility engineers) also joined the Green Team. Today, the Green Team continues to meet periodically to evaluate sustainability plans and departmental goals in reducing resource consumption and eliminating waste. With another successful bond passed in 2010, the Green Team is planning green renovations to their older schools so they meet the green standards of the buildings designed in 2000.

On the surface, the creation of a Green Team appears straightforward. However, a closer look reveals complex, ongoing processes to successfully generate support and overcome individual- and organizational-level resistance to adopting new building design. We will now describe three processes that took place within the Green Team. These processes buttressed each other in a perfect storm of vision, support, and knowledge to bring about a building design and construction revolution in this district.

The Green Team Develops an Inclusive Vision

First, the flat team approach allowed for the collaborative development of a green building vision—"high performance buildings"—that all individuals could support without changing their personal values or agreeing to an environmental agenda. After nearly a year of team meetings, the superintendent published a memo about the district's collective vision for green building: "We believe that by working together in an integrated approach, we can build higher-performance schools that provide a superior learning environment while reducing life cycle costs through conservation of energy and natural resources." This vision statement highlights the numerous reasons to support green building design, also called sustainable or high performance design in the construction industry, including consuming fewer environmental resources, creating more comfortable and healthier buildings, and reducing lifetime maintenance and utility costs (Johnson 2000). Although often used interchangeably, the terms "green," "sustainable," and "high performance" each elicit different reactions making word choice important. Developing one unified vision involved collaborative discussions within the new team, where these differences were expressed and the vision was broadened from "green" to "high performance," creating a new symbolism for green building; something that even nonenvironmentalists could embrace.

Problems with a Green Building Vision

Creating an appropriate vision was necessary for encouraging full participation and internal support for the construction changes, as well as addressing perceived and real external resistance to the district's dramatic changes. Early in the process, Operations Department leadership began by describing the desired design and construction goal as "green building," which emphasized the environmental impact of buildings. Nearly 70 percent of the interviewees held personal beliefs that supported this environmental perspective of green building including a desire to conserve resources, preserve natural areas, increase sustainability, and reduce carbon emissions and pollution.

However, this environmental focus generated resistance from district staff and some administrators, and was believed to be opposed by the voting public. At the time, gasoline and electricity were affordable; many assumed that green building would only benefit environmental extremists, while being too expensive. Also, some members of the school board—who had final approval over building prototypes—disagreed ideologically with anything called "green." Operations Department leadership noticed that an appropriate vision was needed to address this resistance, as noted in the following quote from a Green Team initiator:

> There were a lot of political issues. . . . I remember we had problems with the lexicon. We said "green buildings" and folks thought that this meant we had something to do with Greenpeace and saving whales . . . , back then, there were green parties and the radical elements that used the word green, and we were just talking about doing something that again was a better building. . . . So we had to start being careful about how we used language, and that's where terms like "high performance buildings" came in. . . . If we said "green" [people] thought "you're some radical environmentalist."

Successful green building requires that environmental language be translated to fit with organizational norms and language, as well as with various stakeholders' views (for further discussion of language see chapter 12, this volume). Support for one unified vision set the stage for the entire green building initiative, as noticed by one Operations leader: "Being able to bring everybody along and have a unified goal so you don't have somebody out there in left field trying to sabotage what you're doing." Collaborative vision development reduced the need to change employees' attitudes toward the environment, deny their political views, or coerce participation.

The discussion about language choice in the district's team addressed this issue and led to one unified vision of high performance design that spoke to many, sometimes conflicting, individual beliefs and attitudes as well as to the core mission of the school district in educating children. One Operations employee described the resulting final vision: "We want a building that is light on the environment, using recycled materials, doesn't smell like a new car when you walk in, is non-toxic to people, has delightful interiors, is day-lit, is energy-efficient, is acoustically well-designed, is a place where people want to be, uses less water—all this great stuff. . . . Actually, saving money is a consequence, is an outcome of designing this high performance building. And isn't that wonderful?" The following three nonenvironmental perspectives discussed in team meetings became incorporated into the high performance vision.

Financial Concerns

Green buildings can reduce operation and maintenance costs over the long term, but the original "green building" wording did not highlight these benefits. Many interviewees described their perspective using financial reasoning: "Money gets wasted a lot of times. We see buildings that are just a waste, and it's sad, because it doesn't have to be." A budget audit in the mid-1990s also brought attention to costs indirectly related to education. As one administrator noted, "I was being directed to contribute to budget savings . . . and one of them was the energy budget. So this energy conservation piece was a budgetary issue."

Highlighting, in the vision, the financial benefits of green building was important because prior to the transition, district administration touted fiscal responsibility to the taxpayer by getting the largest building for the least upfront cost. One administrator acknowledged that the biggest criticism from both inside and outside the district was the assumed expensiveness of green design—another common green building obstacle.

Acknowledging the importance of costs, the Green Team focused on the life-cycle costs *while* remaining on budget: "You have a budget, and you're not increasing your budget to do this [high performance building]." Showing the cost benefit of green design required detailed planning by the team: "At the end of the day, there is a financial savings for being more efficient. And like we said, there might be some initial costs, but you do the planning, calculate through those numbers and in five years, if you're ahead, that's a no-brainer when you buy that equipment." For example, one department manager and Green Team member described how including financial concerns in the vision led to continued success in reducing operation costs: "There's a lot more to it [than just energy savings]. If you look at my total budget over the last 10 years and you look at the square footage . . . and my budget for repair and maintenance it's almost the same. . . . We've doubled square footage, and my budget has not doubled. It's been pretty much flat." He explained that reducing the run-time of equipment can substantially prolong the life of the equipment and reduce the number of repair tickets, which together lower maintenance costs. One district administrator acknowledged the importance of this long-term accounting of building costs by noting that the district still uses one nearly one hundred year-old school building and could use these new schools just as long: "If we get this [high performance building] right, this building will serve these kids for 75 to 100 years."

Educational Concerns

Green buildings improve occupant comfort and productivity and a more conducive learning environment than traditional school designs. These improvements that were included in the district's high performance building vision were not emphasized with green building language. Daylighting, passive/active design,[2] comfortable temperatures, lighting and acoustics appropriate for lectures, and better air quality motivated many interviewees to support the high performance buildings vision. For example: "Now, a student and teacher in a high performance building really doesn't say every day, 'Oh we're saving energy, we're saving dollars.' What students and people suddenly understand is that, 'I like being here. I'll stay here longer. This is a good place to be. I'm not getting as sick as much as I used to. I don't have as much stress as I used to have.'"

Others emphasized the ability of green buildings to create actual teaching opportunities, something many interviewees felt a green building vision could not articulate clearly. A former board member and parent discussed the new see-through walls to teach about plumbing and electricity and the worm composters for biology lessons. School administration denied these items in previous bonds citing perceived high costs but the new high performance building vision emphasized the role of infrastructure in educational opportunities. Other educational aspects include students conducting LEED building tours and teachers incorporating building and energy use data into curriculums. Because these motivations were expressed in the team process, the educational benefit of green design became part of the vision and was used to convince voters that the design changes directly address the educational mission of the district.

Professional Concerns

Before the flat team approach was implemented, operations and maintenance employees were "second-class citizens," only called when something went wrong. Internal district staff in particular mentioned a lack of professionalization among operations staff as a central problem with previous construction projects. The new vision of high performance buildings incorporated the need for facilities employees input in the design process and allowed Operations Department leadership to encourage facilities and operations staff to do their best work, as described in the following quote from an Operations Department leader. "There should be a balance with the facility operating at a high level, which also meant that the educational opportunities [for facilities staff] were there at a high level. So the culture when I came in was . . . people looked at them as not being craftsmen or tradesmen. I wanted to break that down and bring some esprit de corps into the groups."

Focusing on high performance buildings created "the opportunity to involve all these staff members and recognize them for the qualities that they bring, the skill, the professional level that they bring, the experience, to the table when you design a school. So now we had this opportunity to really cross-connect with the professionals, the engineers, and the architects," stated one interviewee. The goal of empowering staff became part of the high performance vision as a way to incorporate the expertise of all staff members to build better buildings, which many felt did not fit within a green building vision.

The discussion within a flat team led to the collaborative development of this vision of high performance buildings, which encouraged support and overcame resistance to the transition from conventional construction by incorporating different individual perspectives on what is important in school design. The green building vision did not adequately symbolize all the goals of the new design changes and thus limited support for the initiative until the language was evaluated and high performance buildings were decided upon. As the district worked toward this unified goal, the flat team approach created further space for genuine participation and maintained the focus on dialogue in the development and construction of the new buildings.

The Green Team Increases Participation

The second process that occurred within the flat team approach was genuine participation in decisions about building design that maintained the buy-in and support started with the collaborative vision. The flat team approach encouraged participation and authentic dialogue among diverse individuals, held team members accountable to all aspects of the high performance building vision during product selection, and eliminated the need for specific organizational rewards or disincentives to encourage participation.

Participation and Authentic Dialogue

As organizational researchers have noted, genuine participation and communication are necessary for successful organizational change (Fernandez and Rainey 2006; Kotter 1995). Green building projects often require more collaboration and "authentic dialogue" between all the parties, from building owner to architect to construction company, to reach efficiency and resource use goals (see also chapter 3, this volume). The wide base of participation in the Green Team meant the district did not rely on only those with pro-environmental values to generate change, which research has shown can limit the success of environmental initiatives. Before the Green Team, only two district staff members specifically focused on environmental issues (the energy manager and the environmental coordinator). These staff members could only handle small-scale projects, such as recycling or setting up an energy tracking system. They had been unable in previous bond cycles to introduce green design elements into school construction, but the use of a team spread the workload and the feeling of ownership in the process through the organization.

To ensure that every team member felt valued and expressed their opinions, Operations Department leadership focused on relationship building and became "cheerleaders" for the *process* of participation rather than the outcome of constructing a green building:

> So everybody's job is valuable. Everybody's job is important. Part of that is everybody's input. There's ideas that need to come to the table. When you empower people, and some people, it's like, "Great, but I'm good right where I'm at." "You know what? Thank you. I honor that as well." But there's other people who are like, "My opinions, you need my opinion because." It doesn't guarantee we use that specific idea, but it's part of the collaborative discussion.

Noting that in bureaucracies the typical method of change is top-down, one Operations Department leader stated the importance of genuine participation and dialogue to this district's green building success:

> I wasn't going to go say, "We are building high performance buildings whether you like it or not." People would sabotage that. It would not work. You have to say, "Here's some interesting things. Go find out about them. I trust you enough. You're the experts. Go out and find out about them, and you tell me what we should be doing.". . . That's a big part of this job, that relational facet, without that, this collaborative effort doesn't happen, and without that, the empowerment doesn't occur.

If the collaboration process had been superficial or used to get approval for predetermined plans, team members, including this manager, believed that broad-based support would be missing and resistance would have grown: "So many times you just go 'words, words, words' I'm kind of paying attention to you but really not. I definitely feel that this group, our operations group, pays attention to people and what they say." The weekly research assignments allowed team members to lead discussion on their findings, followed by voting on the implementation of the product. In team meetings and in hallway discussions, Operations Department leadership encouraged dialogue by directly asking staff to voice concerns and ideas. The flat Green Team provided an avenue for proof of products fiscal, environmental, and educational effectiveness and for complaints—a safe space to bring in work orders as evidence or to just say, "I told you so," and then learn from those mistakes. As one Operations Department manager remarked, "We started really embracing the different departments and getting that buy-in and it's not a mandate, it's 'let's look at this, look at the data, show it to us.' Very collaborative."

High Performance Accountability

The flat team approach prevented the high performance vision from becoming a facade for a solely environmental agenda. In other words, it prevented the reduction of the complex high performance building vision back into the simpler green building vision. The continued team participation held people accountable to all four aspects of the high performance vision during product selection without the use of one specific green building champion, as one team member discussed: "The Green Team's role [was] to help us bring projects to the table and help us make sure that we were holding each other accountable. . . . I think there are champions that can come in and help seed that, but I really do think that the buy-in of the group had made change more acceptable."

As one district leader described, some team members would become excited about the environmental benefits of certain products and overlook other concerns. "We need that, somebody looking over our shoulder and going, 'Why are you doing this?' You can either answer that intelligently or take a step back and be honest and go, 'Good question.'" The flat team approach equalized opportunity to question others' choices, and the ensuing debate led to winners and losers on some issues as consensus resulted in some products chosen primarily for an environmental (or financial or educational or professional) reason. Because the flat team approach encouraged equal participation, employees continued supporting the high performance vision and compromised on specific products rather than leave the process entirely. As stated by one mid-level Operations Department manager, "So there's several individuals—they're a little resistant, but then all of sudden, 'It's OK.' . . . Now in the background, if people fully agree with it or not, some may say, 'I'll maybe compromise.' But I think in the end, staff are better aware of why we're doing what we're doing and on board with at least the fact that this is collaborative."

Motivation without Specific Rewards

Our interviewees never discussed explicit rewards or disincentives to participate in the transition to green building. Instead, participation in building decisions within the Green Team was seen as a reward in itself because of previous experiences when their knowledge was ignored and inefficient buildings were built as a result. As one district employee described, the workload increased, but few complained: "It definitely did affect workload in the sense that you had to take time out of your everyday routines. [New assignments] had to get done, [such as] code issues, tests, procedures that you've got to continue working to go to these meetings, to do the research, to take a trip to take a look at something. But nobody complained about that, because they were involved. They felt like they were being heard."

Many organizational scholars highlight incentives or disincentives to encourage participation in change processes including environmental initiatives and green building adoption (Al-Homound 2000; Cebon 1992; Doppelt 2003). The flat team approach in this school district motivated participation among a diverse group of individuals without specific rewards or compulsion, even as workloads increased. Participation, employees noted, made them feel lucky to be in a district that supported input from all the departments, as stated by one district employee: "Just from talking to people I know that work in other school districts, they're surprised and jealous that we get to do what we do . . . I feel fortunate. I like waving it in front of these other guys when I see them. 'Look what I do.' It's been good." The participation in a team atmosphere generated ownership, personal responsibility, motivation, and commitment, and overcame resistance to "unleash the potential of people to work toward sustainability" (Doppelt 2003, 80).

The Green Team Increases Building Literacy

The third process that occurred within the flat team approach was information sharing (team participation), which increased the building literacy of individual team members and the district as a whole. The Green Team approach created an environment of information sharing that affected knowledge acquisition and the success of green building in two ways. First, the Green Team approach was used to incorporate the expertise of Operations Department tradespeople (electrical, HVAC, plumbing, etc.) into the design of the new school prototypes. Second, Internet research and the subsequent debate over design choices during team meetings improved the environmental product knowledge of all individual staff members, thereby improving the district's overall environmental literacy.

Incorporating Existing Expertise

As previously discussed, Operations Department leadership felt that involving the people who would eventually take care of the buildings, at the beginning of the design phase, was important in avoiding the mistakes made in previous administration-led building projects. Incorporating the expertise of the tradespeople supported the green building success of the district, as described by one department manager:

> So back in the '90s, with the [previous] bond issue, none of the maintenance guys were allowed onsite, were allowed any input into building new buildings . . . that's the worst thing—you know, people get the good and the bad, and that's definitely the bad. So when we hit the bond in 2000, all the maintenance people had an opportunity to put input into the buildings, into the design of the buildings. They each had their own expertise from their trade specialties, and we also challenged them to go out on the Internet, if possible go visit manufacturing plants, other schools, go out and find out what . . . the best products [are].

The diversity of team members' expertise and organizational roles helped the Green Team incorporate and exchange existing knowledge of various products and practices. Because of the inclusion of teachers, administrators, and janitors along with tradespeople and architects, the green building prototype was viewed holistically and knowledge flowed among members of the educational, administrative, and operational arenas (for more on the value of collaboration, see also chapter 3, this volume). The frequent discussions about their findings increased communication and generated feedback loops about cost and resource use, information important to incorporating green organization practices (Al-Homound 2000; Doppelt 2003). For example, Green Team members learned about the impact of daylighting on electronic presentations from teachers, waterless urinals on cleaning requirements from janitors, and bus route access on commuting by bike from administrators.

Inclusion of existing knowledge particularly affects construction projects. The typical linear approach to construction limits interaction between building users and the construction staff after design selection, as each different trade completes one part of the project and then passes the project to the next trade. Research shows that this linear approach to construction projects makes it difficult for individuals to accept new knowledge, assimilate new with existing knowledge, and apply new knowledge to real projects (Bresnen et al. 2003). The flat team approach created the framework for knowledge sharing because Green Team members participated in meetings with architects, engineers, and contractors to ensure that their product selections and their expertise about building operation were incorporated. The district used an extended integrated design approach, in which all participants in building construction (architects, lighting designers, landscape artists, construction managers, etc.) met with the Green Team, as described by an Operations Department leader in the following quote: "There's a sense that our architect knows our plumbers on a first-name basis, because they were interacting. It isn't: they were on a pedestal. We were looking at everybody being on the same level and providing those pieces of information that they bring to the discussion as being important and honoring those people." The exchange of information between tradespeople and school staff helped ensure that the new buildings met the standards the district desired.

Encouraging New "Green" Expertise

Beyond incorporating existing expertise into building construction, the Green Team encouraged assimilation of new "green" knowledge. Researchers note that the novel language used for green products or practices is often unfamiliar and requires extensive research and training, which can make green building seem daunting for an organization (Hoffman and Henn 2008). In 1999 when the Operations Department began contemplating green building, LEED was not fully developed and green products and practices were generally limited. In addition, no local architect, designer, or builder had expertise in these principles. The Green Team had to generate its own knowledge through members conducting Internet searches and from other organizations across the country that used green design. As the manifest purpose of the Green Team, team members researched a variety of novel products with purported green benefits, then discussed each product with the whole team. The team processes' effect on knowledge development is seen in the following quote from a staff member: "You're able to talk through, 'Why should we make this change rather than stay with the old tried-and-true?' Or 'This technology looks very good and could save us quite a bit initially and does X for the environment, but what's its track history? What's its life cycle? Is it gonna cost us more down the road for replacement parts and repairs?'"

The lack of knowledge about environmental options often causes organizations to view potentially cost-effective green options too "risky" to implement (Pellegrini-Masini and Leishman 2011). The literature shows that the habitual nature of organizational operations creates a fear of change and adds inertia to green building adoption, especially in the public sector (Fernandez and Rainey 2006; Hoffman and Henn 2008; Pearce, Dubose, and Bosch 2007). The Green Team's discussion of each design option improved individual knowledge of specific building components, especially those who saw green products as financially or operationally risky. The trust developed in the Green Team encouraged employees with ideas to voice them in a safe context—important to knowledge sharing and acquisition about new green products (Milliken, Morrison, and Hewlin 2003). In the following quote, a team member describes how the team process generated knowledge and overcame individual feelings of riskiness about green design, which he referred to as "gut-level decisions":

So there's that freedom to explore what's best and make wise decisions as opposed to gut-level decisions . . . I keep bringing it back to our team, because from our energy manager to our directors, to our department heads, we'd all get together and show and tell, if you would, where we'd sit down and talk about what they'd found out, what they'd learned, or bringing the representative back, the manufacturer's people in, where they could talk about this product and have everybody else be able to ask a few questions, not just take it at face value, and then reach a consensus.

To illustrate, one team member believed that geothermal technology was risky and unproven, and argued adamantly to continue with the traditional choice. After team discussion showed that many team members felt the technology was a good solution, this reluctant member visited another school district using the technology: "We visited, I think it was their facilities building that has geothermal, and then I also visited a school . . . I may have been against the idea, or reluctant . . . but at least when I was saying, 'If you're gonna do that, put these machines where we can get to them and work on them to maintain them.'" When he returned, though still uneasy about the technology, he compromised as long as the equipment was accessible for future maintenance. Because this team member was able to learn more, see the technology in action, and identify necessary features of building design and equipment installation, his concerns over what he felt was risky green technology were quelled and he agreed to the installation of geothermal.

Summary: Interactional Processes and Green Building Success

As our case study shows, a flat team approach can increase successful green building by affecting the interactional and symbolic influences within an organization in three ways: developing a collaborative vision, generating genuine participation and authentic dialogue, and increasing building literacy. Awareness about the high performance vision and the Green Team now extend throughout the organization with all questions about the buildings directed to the team. Also, the district has incorporated environmental sustainability into general operations through annual departmental reporting and goal setting, and every new staff member learns about the high performance buildings and sustainability in the district through the Operations Department "Buildings 101" course.

Often organizational research on creating green changes, including adopting green building, focuses on generating awareness or appropriate knowledge; coercing participation through rewards, compulsion, or disincentives (Al-Homound 2000; Cebon 1992; Doppelt 2003; Hoffman and Henn 2008); or building on a real or manufactured crisis (Fernandez and Rainey 2006; Kotter 1995). Although green building knowledge acquired through team research was obviously central to specific product adoption, interviewees emphasized the Green Team dynamics, processes, and interactions as much more significant than just knowledge generation. These processes centered on the incorporation of multiple views and listening to resistant voices, even those hostile toward environmental and climate change issues. Even the interviewees who called themselves "climate skeptics" expressed excitement about the new building designs and, more important, gratitude for having a voice in the process. The Green Team allowed a space for expression of these conflicting views and created the opportunity for a collaborative development of a vision for green building adoption that *all* members could support. Our case study results indicate that "high performance buildings" may not be the best vision for some organizations considering green building. Each organization will have its own types of resistance and support that can be leveraged if collaborative processes are implemented to create and support a vision specific to an organization's internal culture and external context. Listening to all members of an organization and valuing participation over an end outcome (e.g., carbon emission reduction or increased recyclable content) create the support necessary to reduce individual resistance.

The multiple viewpoints expressed were central to the learning that occurred within the team. The learning opportunities presented by forming this team immediately helped those who discounted green products and it made those who supported the green label accountable to the organization's goal of providing "better schools." Learning within Green Team discussions also helped the district overcome beliefs that "green" costs too much, that energy or maintenance costs are secondary, or that green equates with radical environmentalism—beliefs that commonly prevent green building adoption (Hoffman and Henn 2008). A flat team environment encourages learning and expression of viewpoints that, in turn, lead to further learning and successful transition of an organization's entire infrastructure.

Conclusion

School districts have unique opportunities to be leaders in green building. First, they generally own and operate their own buildings, with over 80 percent of schools responsible for building operations (U.S. Green Building Council 2010). Other construction projects, such as housing developments, detach the costs from the benefits of green building by separating those making initial decisions (a developer) from those faced with the long-term consequences (a homeowner). As individuals in our district pointed out, the schools constructed today may still be used in fifty years by the same community—which is a big incentive to think about construction choices today. In addition, many school districts employ a variety of building experts in the maintenance and operations departments. These individuals have valuable insight into the long-term consequences of design decisions that administrators and school board members, as well as architects and engineers may be less qualified to assess or simply overlook.

Yet, the ability of school districts as owner-operators to initiate and even consider green building requires a new way of thinking about building construction. The manifest mission of the team in this case was to increase knowledge of green options. Because this district has a robust Operations Department, its leaders had the personnel available to take on this task. In smaller districts or organizations that lack an operations department whose job it is to maintain the buildings, generating a group focused on learning about green ideas may be more difficult (Pellegrini-Masini and Leishman 2011). But our results may not be limited to teams within operations departments. Using the ideas presented herein, a group of any individuals connected to an organization could move forward with green design, or at least begin the discussion of green options.

"Some school districts say they care about energy conservation and being green, but from electricians to plumbers to kitchen staff to custodians to principals to teachers, I mean district-wide, they have buy-in and expect everybody to participate in that. I think that's the big thing that is different." As noted by the most recently hired facilities employee, this district is "different" and wildly successful in green building initiatives because of the interpersonal processes fostered through the Green Team. All organizations, not just public organizations, can become successful in green building if they use a team-based approach so that leadership (1) listens to employees—especially those who seem to be resistant—to understand what motivates them and incorporate these motivations into the green building vision; (2) encourages genuine participation in the building design process; and (3) promotes literacy throughout the organization on environmental and building-related topics that can then be translated to the general public or consumers for continued support of green initiatives.

Notes

1. For specifics on green building and energy conservation products and practices implemented in the district, see Poudre School District 2011 and Schelly et al. 2010.

2. Passive/active design describes the separation of classrooms and libraries (more passive spaces) from playgrounds, lunchrooms, and activity rooms (more active spaces) to reduce student distraction.

References

Al-Homound, Mohammad S. 2000. Total productive energy management. *Energy Engineering* 97 (5): 21–38.

Bresnen, Mike, Linda Edelman, Sue Newell, Harry Scarbrough, and Jacky Swan. 2003. Social practices and the management of knowledge in project environments. *International Journal of Project Management* 21:157–166.

Cebon, Peter B. 1992. Organizational behavior, technical prediction and conservation practice. *Energy Policy* 20 (9): 802–814.

Charmaz, Kathy. 2001. Grounded theory. In *Contemporary field research: Perspectives and formulations*, ed. Robert M. Emerson, 335–352. Prospect Heights, IL: Waveland Press.

Creswell, John W. 2007. *Qualitative inquiry & research design: Choosing among five approaches*. Thousand Oaks, CA: Sage.

Doppelt, Bob. 2003. *Leading change toward sustainability: A change management guide for business, government and civil society*. Sheffield, UK: Greenleaf.

Etzion, Dror. 2007. Research on organizations and the natural environment, 1992–present: A review. *Journal of Management* 33 (4): 637–664.

Fernandez, Sergio, and Hal G. Rainey. 2006. Managing successful organizational change in the public sector. *Public Administration Review* 66:168–176.

Hoffman, Andrew J., and Rebecca Henn. 2008. Overcoming the social and psychological barriers to green building. *Organization & Environment* 21 (4): 390–419.

Hulme, Mike. 2009. *Why we disagree about climate change: Understanding controversy, inaction, and opportunity*. Cambridge, UK: Cambridge University Press.

Johnson, Scott D. 2000. The economic case for high performance buildings. *Corporate Environmental Strategy* 7 (4): 350–361.

Kotter, John P. 1995. Leading change: Why transformation efforts fail. *Harvard Business Review* 73:59–67.

Lofland, John, David Snow, Leon Anderson, and Lyn H. Lofland. 2006. *Analyzing social settings: A guide to qualitative observation and analysis*. 4th ed. Belmont, CA: Wadsworth/Thomson Learning.

Lutzenhiser, Loren. 1993. Social and behavioral aspects of energy use. *Annual Review of Energy and the Environment* 18 (1): 247–289.

Milliken, Frances J., Elizabeth W. Morrison, and Patricia F. Hewlin. 2003. An exploratory study of employee silence: Issues that employees don't communicate upward and why. *Journal of Management Studies* 40 (6): 1453–1476.

Pearce, Annie R., Jennifer R. Dubose, and Sheila J. Bosch. 2007. Green building policy options for the public sector. *Journal of Green Building* 2 (1): 156–174.

Pellegrini-Masini, Giuseppe, and Chris Leishman. 2011. The role of corporate reputation and employees' values in the uptake of energy efficiency in office buildings. *Energy Policy* 39 (9): 5409–5419.

Poudre School District. 2011. Sustainability. http://www.psdschools.org/about-us/district-operations/sustainability (accessed November 1, 2011).

Schelly, Chelsea, Jennifer E. Cross, William S. Franzen, Pete Hall, and Stu Reeve. 2011. Reducing energy consumption and creating a conservation culture in organizations: A case study of one public school district. *Environment and Behavior* 43 (3): 316–343.

Shwom, Rachael. 2009. Strengthening sociological perspectives on organizations and the environment. *Organization & Environment* 22 (3): 271–292.

Strauss, Anselm, and Juliet Corbin. 1990. *Basics of qualitative research*. Newbury Park, CA: Sage.

Turner Construction. 2005. *Survey of green building plus green building in K-12 and higher education*. New York: Turner Green Buildings.

U.S. Green Building Council. 2010. Green school buildings. http://www.greenschoolbuildings.org/Homepage.aspx (accessed April 1, 2010.

Yudelson, Jerry. 2008. *Marketing green building services: Strategies for success*. Burlington, MA: Architectural Press.

11

Generativity: Reconceptualizing the Benefits of Green Buildings

Ronald Fry and Garima Sharma

> I would say the sub-contractors really got enthusiastic about the way this project had to be done and brought in their own ideas. What we found unexpectedly was that they brought in more than just their skill sets, not only their hands but brought their brains about how we could do it better. At the end of day they brought their hearts too since they felt this project will be distinctive. If you ask people to come in, do the work and go home they don't get as much satisfaction as they derived from this project. It was a revelation to us—we hadn't thought about engaging people in this way. They came in early and shared ideas in a roundtable. The collaboration was excellent.
>
> —Project manager for building and remodeling

What are the benefits of green buildings? How are these benefits generated? Who receives these benefits? Despite the interest of practitioners and scholars in the advantages of green buildings, most work to date describes these in operational and tactical terms such as energy efficiency or productivity gains. There is growing interest in moving to strategic or value-added measures for a "far reaching relationship between the buildings and strategic performance" (Heerwagen 2002, 35). But this shift does little to explain the path from green buildings to value addition. Moreover, this view maintains the narrow focus that the benefits accrue for the organization occupying the building and ignores the transformative consequences of green buildings for multiple stakeholders.

In this chapter, we propose generativity as a possible benefit. We describe it as the organization's potential to produce enduring, expansive and transformative consequences (Carlsen and Dutton 2011). In doing so, we integrate perspectives from (1) positive organizational scholarship (POS): a framework that focuses on organizational dynamics that create positive states of functioning, and (2) appreciative inquiry (AI): a methodology that focuses on the search for the positive core of organizations, to suggest that green building can generate "microfoundations" of higher order capability-building and capacity-enhancing dynamics (Dutton and Glynn 2008) within organizations. In line with positive organizational scholarship's claim that the factors that bring about a problem state may not be the same as those that cause a positively deviant state (ibid.), we purposefully focus on the life-giving dynamics created by the practices of successful construction and operation of green buildings. Specifically, we (1) elucidate how green buildings generate transformative benefits such as enhanced organizational capacity for environmental responsibility, (2) describe the expression of generativity at multiple levels within an organization, and (3) highlight the role of organizational and team contexts in explaining the relationships of green building practices to generative outcomes.

Our focus on generativity addresses the inadequacy in the existing conversation on the benefits of green buildings in two ways. First, by focusing on capacity building we provide a multilevel perspective of the benefits accruing to individuals, organization, environment, and other stakeholders. In other words, the outcomes generated by the practices of green construction, such as knowledge and energy at the individual level, are converted into organizational-level outcomes such as enhanced *processes of organizing* for the growth of the organization and its stakeholders, and expanded *meaning* of individual and organizational accountability toward the environment such that these outcomes are diffused beyond the organization's boundaries. Thus, instead of understanding the benefits as the financial gain for the organization and reduced impact for the environment, a generative focus on capacity emphasizes deeper level shifts in meanings and organizing such that the benefits continually expand upward and outward (Fredrickson and Dutton 2008).

Second, our focus on team and organizational context as necessary for connecting green building practices to individual and organizational-level outcomes has implications for practice. Organizations engaging in green building projects can create such facilitating contexts through managerial action (Ghoshal and Bartlett 1994) to achieve the desired outcomes. Thus, unlike most existing discussions on benefits that focus on "why," highlighting the team and organizational contexts begins to suggest the "how" of generating the desired outcomes.

Green Buildings and Business Benefits

The term "green building" has been used in many ways. Most describe it as the less resource-intensive and efficient design, building, operation, and remodeling of facilities. Paumgartten defines green buildings as "any building that meets the high standards set forth by USGBC's LEED green building rating system, the preeminent metric system by which new buildings are judged to be environmentally conscious"(2003, 126). Hoffman and Henn define green building as "strategies, techniques, and construction products that are less resource-intensive or pollution-producing than regular construction" (2008, 392). Some studies use the term "sustainable building" to describe building practices that strive for economic, social, and environmental performance in an all-encompassing way. Kibert calls this "sustainable construction" and suggests that green building is a subset of sustainable construction since it represents just the structure and does not address the "social and economic issues of the habitat, as well as the community context of buildings" (2003, 492). In essence, the terms "LEED certified," "sustainable," and "green" are often used interchangeably. Moreover, some definitions focus on outcomes generated from the process of construction, some on outcomes from the operation of the building, and some on both. We define green buildings as those (certified or not) that are built for the efficient use of resources and healthier internal environments (Kats 2003), and use materials as well as follow techniques of construction that are less resource intensive.

Specifically, we adopt a practice perspective to suggest that green building practices have transformative and expansive—generative—consequences. A focus on practices as "recurrent, situated activities" (Lilius et al. 2011, 876) reframes the benefits of green buildings in two ways. First, it allows inclusion of the entire spectrum of benefits—those emerging from the process of construction and those from occupying and operating the building. For example, practices related to the process of construction, if accompanied by clarity of goals and passion behind those goals, can generate a sense of aliveness and learning in those executing the project. This is related to outcomes such as better health and creativity. Similarly, practices during building operation such as automatic shut-off of lighting can generate awareness in incumbents of the potential for environmental responsibility from simple acts. This can cascade to other individual acts of environmental stewardship, such as employee volunteerism.

Second, a practice perspective allows expansion of benefits from those accruing to the focal organization (e.g., a business occupying the building) to benefits for multiple stakeholders. For example, diverse stakeholders in the team executing a green building project build knowledge resources within and among the team members by experimenting with and reconfiguring existing practices. These knowledge resources are taken back to individual organizations of different stakeholders. Thus, by focusing on practices we can consider a wider spectrum of benefits and multiple stakeholders who may be accruing those benefits.

Similar to the multitude of terms used to describe green buildings, there exist many typologies of the benefits of green buildings. At a broad level, for the builder and organization occupying the building, the benefits can be categorized into what Kibert (2005) call "hard" and "soft" cost benefits. Others have used similar terms such as "tangible" and "intangible" benefits (McNamara et al. 2011). Wilson (2005) provided the following categories of "hard cost" or tangible benefits: (1) first-cost savings given by streamlined permitting and approvals, reduced infrastructure costs, reduced material use, savings in construction waste disposal, savings from downsizing mechanical equipment, and tax credits and other incentives; (2) reduced operating costs given by lower energy costs, lower water costs, greater durability and fewer repairs, reduced cleaning and maintenance, reduced costs of churn, lower insurance costs, and reduced waste generation within the building. The categories related to soft cost or intangible benefits are (3) other economic benefits given by increased property value, more rapid lease-out, more rapid sales of homes and condominiums, easier employee recruiting, reduced employee turnover, reduced liability risk, staying ahead of regulations, positive public image, and new business opportunities; (4) health and productivity benefits given by improved health, enhanced comfort, reduced absenteeism, improved worker productivity, improved ability to recruit, improved learning, faster recovery from illness, and increased retail sales. Additionally, categories that describe benefits to other stakeholders include (5) community benefits such as reduced demand on municipal services; (6) environmental benefits such as reduced global warming impacts; and (7) social benefits such as support of sustainable local economies.

Acknowledging the relevance of these myriad benefit categories, we suggest that an alternate way to understand this is not through static accomplishments but through generative moments that are embodied, affective, aesthetic, and relational (Carlsen and Dutton 2011, 20). In other words, the everyday practices of green building construction and operation can lead to vitality, awakening, , and growing (ibid) for individuals, the organization, and its stakeholders such that the benefit includes but are broader than the oft-studied outcomes such as health, productivity, engagement, employee retention, and reduced environmental impact. We call this benefit "generativity" and elaborate on it as follows.

What Is Generativity?

Carlsen and Dutton (2011, 15) describe generativity as "strips of experience that bring the feeling of energy and aliveness to people and have the potential to produce more enduring, expansive and transformative consequences." Generativity has been an emerging theme in the field of positive organizational scholarship (Dutton and Glynn 2008) and the practice of appreciative inquiry (Bushe 2005a). Studies aligning with the POS perspective focus on generativity as an individual or collective's state of growth, creativity, resilience, or broadly a condition of a healthy, near optimal state (Dutton, Glynn, and Spreitzer 2006). Additionally, the work in this area focuses on the organizational dynamics or processes that unfold to generate such positive states of functioning. Similarly, AI describes the search for the positive core of organizations. The generativity in AI comes from the focus on positive deviance—those success factors, best practices, strengths, and other life-giving forces that cause the human system to be at its very best (Barrett and Fry 2005). Shared recognition and understanding of this positive core can be transformative by shifting the identity or qualitative state of being of the system (Bushe 2005a) as well as by creating new models and (lay) theories to perceive the old with a new lens (Cooperrider and Srivastava 1987). The empirically established benefits of generativity are many such as creativity and increased psychological and physical well-being at an individual level (Fredrickson 2004); mutual support and performance at the team level (Losada and Heaphy 2004); and momentum for change and long-term sustainability (Bushe 2005a) at the organizational level.

Two themes emerge from the existing literature on generativity from the perspectives of POS and AI. First, it can be understood as an enhanced *state* of the system reflected in shifts such as a new lens (Bushe 2005a) or positive meaning (Spreitzer and Sonenshein 2003). This new view or meaning signals a transformation in the underlying *processes* or dynamics to achieve that state (Schnackenberg, Singh, and Hill 2011); what Dutton, Glynn, and Spreitzer call "capacity-creating dynamics" (2006, 641). For example, Fredrickson and Losada describe the benefits of positive affect as the broadening of thought-action repertoire, as well as the building up of enduring resources such as "social connections, coping strategies, and environmental knowledge" (2005, 679). Thus, organizational practices which focus on the positive core and actively discover the life-giving properties of the system (Cooperrider and Whitney 2005) generate additional organizational capacity through enhanced collective meaning and processes of positive organizing.

The second theme is the multilevel expression of generativity. Dutton and Glynn (2008, 693) describe the research on generativity at individual, group, organizational, and inter-organizational levels. Barrett and Fry (2002, 3), in a similar vein, describe how the change process in AI and its consequences begin with the individual and diffuse to teams, organizations, and networks of organizations. Thus, the capacity-building dynamics of generativity have microfoundations in individuals and their interactions (Salvato and Rerup 2011; Lilius et al. 2011) and hold consequences for the organization, its members, and its networks.

This multilevel aspect of generativity also appears in the research on adult development. Here, generativity has been described as a concern with civic responsibility and leaving a legacy for future generations (Erikson 1950, 266–267, cited by Singer et al. 2002); concern for preserving what is good and making things better; and reflected in themes such as creating (product, project, new ideas), maintaining (product, tradition, project), offering (giving of self or self's products), and purposive positive interaction with the next generation (McAdams and de St. Aubin 1992). As an idea, generativity thus has its roots in the concern and creation for posterity and is therefore aligned with the study and practice of constructing green.

By focusing on generativity we join a shift in the current literature to consider broader benefits of green construction (see chapter 9, this volume). The remainder of this chapter will delineate the path from practices to the capacity for environmental responsibility.[1] We do this by linking green building practices to individual thriving (Spreitzer et al. 2005). We describe how the organization and team's context can foster this sense of thriving. The positive affect and individual resources (such as knowledge, energy, and relationships) from thriving translate into expanded organizational capacity for environmental responsibility in two ways. They expand the collective meaning and self-definition of the organization's responsibility toward the environment and other stakeholders (Brickson 2005) and they enable processes for positive organizing that diffuse the individual resources and actions. As a result, the organization can not only replicate the success of the focal green building project but also generate other actions from the expanded capacity for environmental responsibility. In this way, the benefits of green buildings magnify from organizational benefits of cost saving, morale, and productivity to generative benefits for the organization and its stakeholders.

We illustrate the proposed dynamics of generativity with examples of eight successful green building projects. These examples have been culled from interviews conducted with the CEOs, heads of sustainability, or individuals associated with the green building project in each organization. These organizations were identified from an existing database of synopses of business innovations from the World Inquiry project (World Inquiry's Innovation Bank 2010), an initiative of the Fowler Center for Sustainable Value at Case Western Reserve University. The World Inquiry is a global effort to identify, amplify, and promulgate examples of innovations that highlight the idea of business prospering in mutual benefit with the earth's ecosystems and the world's societies (Fry 2008). The eight organizations came from diverse industries ranging from manufacturing to service, and sizes from a large multinational to small family-owned organizations. We use quotes from these interviews to bring to life the proposed relationship between green building practices and generativity.

Generative Benefits of Green Building Practices

The proposed path from green building practices to organizational capacity building is depicted in figure 11.1. We adopt the perspective that generativity is deeply processual and begins with individual moments "characterized by feeling alive through deep inspiration, connectedness, bursts of insight, and expansion of thought" (Carlsen and Dutton 2011, 19–20). In figure 11.1 we call this a "state of individual thriving." Moreover, we view generativity as being embedded (Dutton and Glynn 2008); that is, such moments are facilitated by specific characteristics of team and organizational contexts. The arrows in figure 11.1 describe the characteristics of these contexts. Finally, individual moments accumulate to generate new collective *processes* and expanded *meaning(s)* of responsibility toward environment. We call this the "organizational capacity for environmental responsibility."

Green Building Practices and Individual Thriving

The term "thriving" refers to a positive affective state characterized by vitality or aliveness, and learning. Spreitzer et al. describe these qualities as follows: "Vitality refers to the positive feeling of having energy available, reflecting feelings of aliveness. Learning refers to the sense that one is acquiring, and can apply, knowledge and skills" (2005, 538). We posit that green building practices are associated with thriving and explained by the mechanisms discussed in the sections that follow.

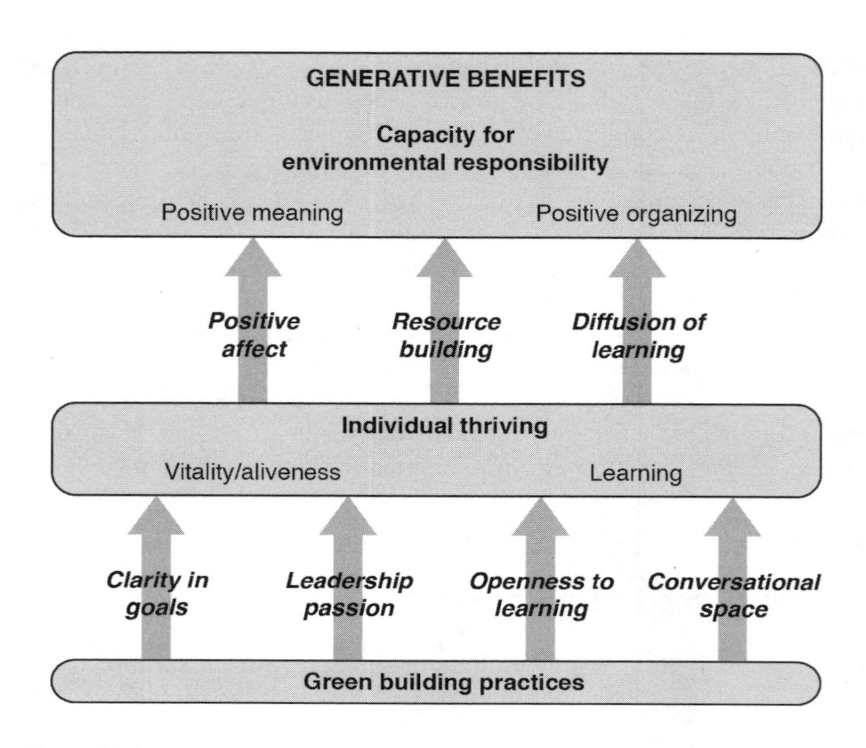

Figure 11.1

Generativity schematic as a transformative consequence of green buildings

Aliveness and Vitality

Green building practices are associated with a sense of aliveness and vitality achieved through clarity in goals and leadership passion behind those goals. Bringing together a diverse team and asking members to reconfigure their routines to meet the demands of green construction can be challenging. If not addressed, this diversity in composition and the magnitude of change required can create negative dynamics serving as barriers to successful execution. Clarity in goals can alleviate some of this ambiguity. Clarity can indicate that the goal is attainable and allows members to sense progress. This sense of progress creates positive emotions (Bagozzi, Baumgartner, and Pieters 1998). Clarity can also provide the opportunity to contribute meaningfully, generating energy within those associated with the project (Cross, Baker, and Parker 2003). A project manager for construction and remodeling explained the importance of goal clarity:

> We have educated practitioners over the years. It's hard to keep a carpenter from running over to Lowes and picking up unsustainable material—education at ground level is incredibly important. We have put lot of LEED intent [objective(s) of LEED guideline(s)] right on drawings, not buried in specification but in the drawings for people who are doing it. We spend lot of time with the site supervisor. This is important because you get the cost back by being clear about what you expect. Being ambiguous equals spending dollars, intents in LEED [specific guidelines] help in being clear, they spark conversations.

In addition, an organization's passion for creating a building that decreases its impact on the environment can generate positive emotions within the various constituents of the team. Passion is defined as "strong inclination or desire toward a self-defining activity that one likes (or even loves), that one finds important (high valuation), and in which one invests time and energy" (Philippe et al. 2010, 917). Such passion in organizational representatives or leaders of the team is associated with positive emotions (Mageau et al. 2005). When expressed, this energy and strong positive emotions can be contagious. The transfer of positive emotions from those leading to others comes from observing the leader's passion as well as feeling like "we are in this together" (Cardon 2008). Passion can attract and inspire others (Philippe et al. 2010), in turn creating a sense of aliveness. This is illustrated in the following comment by a sustainability manager:

> This was a sales office. In 2003 LEED was still new. There were lots of unknowns, very challenging. We were aiming for platinum [certification] so had no wiggle room. We put our commitment and passion upfront—architects, contractors knew that was our goal and we believed in it. People, internal and external, saw our commitment and knew we were one of the firsts so they were willing to invest in the learning curve. Lot of people donated their time. Contractors, architects, and other third part[ies] worked a lot more hours than they were paid for.

Thus, passionate leadership generates interactions through which people can understand the vision of the building, and can add value because of clarity in goals or purpose. Diffusion of this passion to others and presenting the opportunity to contribute meaningfully will generate aliveness and vitality (Cross, Baker, and Parker 2003) within those associated with the green construction.

Learning

Learning, as the second component of thriving, is given by a sense of movement and progress, whereby the individual experiences acquisition and application of knowledge and skills (Spreitzer et al. 2005). It can be described as not only seeing differently (Bushe 2005a) but also seeing with new possibilities that expand opportunities of knowing and doing (Dutton and Carlsen 2011, 217). Green building practices are associated with learning (see figure 11.1) through openness to ideas and creating a conversational space.

Green building practices, such as choosing materials that adhere to specifications while remaining within the budget, call for exploration and trial and error (Rerup and Feldman 2011). Exploration can be understood as the search for novel ideas or new ways of doing things, or reaching out to distant others in one's network of contacts (Mom, van den Bosch, and Volberda 2009). Organizational contexts that encourage experimentation influence individual actions toward self-generated initiatives (Ghoshal and Bartlett 1994). A context that encourages openness to learning provides trust and support to individuals to "do whatever it takes to deliver results" (Gibson and Birkinshaw 2004, 213). Given that most green building practices may be new for the focal organization and possibly other associated stakeholders, an organizational context that encourages exploration can provide a sense of movement by increasing the knowledge and skills acquired and applied (Spreitzer et al. 2005). This is illustrated in this quote from a facilities manager:

> Our CEO gave the go ahead and [the] only direction was to get the [LEED] certification. I found many great people within the organization who helped me. With those involved we brought other smaller players. We had no precedents within the organization but we figured a way to do it. For example, we started looking at used materials. We went to a demolition site where we marked what we needed, we were able to reduce cost by reusing things. It took a lot of effort but everyone was willing to do it.

The context of the team of diverse stakeholders executing the task is equally important to the organizational context for learning. The quality of the conversational space created within the team has an immediate influence on the energy of individuals and eventual influence on the arrangement and execution of organizational activities (Quinn and Dutton 2005). Such conversational spaces can be learning environments when shared beliefs and a sense of psychological safety (Edmondson 1999) enable members to collectively reflect on and integrate each other's experience for the team's way forward (Kayes, Kayes, and Kolb 2005).

Conversations in the spaces marked by ongoing reflection, asking questions, seeking feedback, discussing errors (Edmondson 1999) can bring to the surface opportunities for perspectives not seen before. Green building practices such as checklists for certification, if used for conversational learning where all members construct new meaning and knowledge together (Baker, Jensen, and Kolb 2005), can fuel the individual's sense of movement and learning. This quote by a project manager for building and remodeling exemplifies how discussion of the LEED certification checklist opened the space for conversation: "There are [LEED] credits and there are intents behind actions to achieve those credits. We focus on intents, not on the 'yes' in the checklist. We may not get credits for some of the things but we focus on intents because they are written in a broad-brushed way. This way ideas are generated, it sparks innovations, conversations."

In summary, a context for exploration and conversational spaces allows green building practices to stimulate an individual's sense of learning. In addition, the clarity of goals and passion behind those goals enable sense of aliveness and vitality. This combined sense of aliveness, vitality, and learning depict individual thriving as a generative outcome of green building practices.

Thriving has been related to individual-level outcomes such as positive health, wellness, and positive functioning (Spreitzer et al. 2005). As described previously, in the state of thriving individuals experience positive affect and engage in skill acquisition and application. Green building practices, through thriving, can generate more expansive thought-action repertoires (Fredrickson 2004) related to benefits such as high engagement, creativity, and building of resources. Next we describe how such individual-level outcomes relate to organizational-level benefits.

From Individual Thriving to Organizational Capacity Building

Green building practices, through organizational and team contexts, are associated with vitality and learning in individuals. However, work on organizational context and managerial action (e.g., Ghoshal and Bartlett 1994; Gibson and Birkinshaw 2004; Salvato and Rerup 2011) suggests that it is not only the work context that influences individual action, but an equally strong reverse pathway also exists. That is, individual action can lead to changes at an organizational level. We turn now to this dynamic of positive organizing and collective meaning.

Positive Organizing

Positive organizing is described as processes or acts that create not only upward spirals of learning and action for the organization, but also outward spirals infusing whole networks and communities with a sense of vitality and learning (Fredrickson and Dutton 2008). Learning, as a component of thriving, indicates building of individual resources such as knowledge (Spreitzer et al. 2005), positive relationships (Lilius et al. 2011), and energy (Quinn and Dutton 2005). These individual resources are subsequently converted to organizational resources. For example, the challenge of containing the cost premium encourages acts such as one described in an earlier quotation: the team's visit to a demolition site to procure materials that could earn LEED certification points. This builds knowledge resources for the individual by creating understanding of practices to follow in the "doing" of green building. The knowledge resource is generated through teamwork or interpersonal interactions. Such interactions also build relational resources for the individual (Spreitzer et al. 2005) and act as a conduit to diffuse the knowledge at an organizational level (Spender 1996; Lei, Hitt, and Bettis 1996). In addition, top managers can actively encode the tacit knowledge generated from experimentation (Lei, Hitt, and Bettis 1996) through successive meaningful dialogue that "enables team members to articulate their own perspectives, and thereby reveal hidden tacit knowledge that is otherwise hard to communicate" (Nonaka 1994, 20). These informal and formal interactions expand the base of organizational resources and allow the firm to not only refine its existing processes but also gather further resources (Cohen and Levinthal 1990; Spreitzer et al. 2005) in a continually expanding cycle of improved performance (Ray, Barney, and Muhanna 2004).

Moreover, successful execution can generate positive affect at the individual level. Internal and external recognition of this success through awards, certificates, and positive stories in the media can convert the individual's positive affect into a collective sense of efficacy or shared belief (Lindsley, Brass, and Thomas 1995; Bandura 2000) in the effectiveness of coworkers and existing processes. Thus, in the example quoted earlier, the act of facing a challenge and figuring out an innovative way to resolve it will build knowledge resources for individual team members that creates shared knowledge and belief that this can be done again. Such was expressed in different stories by different phrases such as: "Take the principles we learned in constructing a LEED-certified building and apply them at other places"; "Getting our feet wet showed us it can be done"; and "If we do it all over again we can do it for less." In this way, the knowledge and positive affect of individuals spiral upward to generate organizational know-how that can be leveraged for replicating, as well as creating, new outcomes.

But positive organizing is not only about these upward spirals for individual organizations. It is also about outward spirals that connect communities and network of organizations (Fredrickson and Dutton 2008). Such spirals can be created by actively diffusing the collective learning to other stakeholders. Green buildings are living symbols of an organization's values, and can create positive affect such as pride for the organization's members. The sense of achievement from accomplishing a challenging task, and the associated external recognition will encourage members to not only experience a heightened sense of collective efficacy as described, but also actively share their "narratives and war stories" (Nonaka 1994, 24) and mentor others going through the same process. This is exemplified in practices such as open houses and facility tours for the community, commonly conducted by many organizations occupying a green building. This active diffusion of learning is described in the following quote from the sustainability head at one of the organizations: "People feel proud of what they build. They want to share their stories with internal and external stakeholders. We have had several people come to us to inquire about LEED. They ask us to come and talk about it. We do open houses for people to come and ask questions, we mentor others in their LEED building constructions."

Positive Meaning

The generative benefits of green building are not restricted to the processes of organizing, but are also observed through changes in collective meaning and beliefs that shift the state of the system toward higher commitment for environmental responsibility (see chapter 9, this volume). In Gestalt theory's terms of "figure" and "ground," instead of changing the figure, individual-level positive affect can change the ground by providing a new lens to perceive old issues (Bushe 2005a). Finding positive meaning in work (Pratt and Ashforth 2003) is related to positive outcomes such as individual-level job and life satisfaction (Wrzeniewski 2003), engagement (Glavas and Piderit 2009), effective change implementation (Bushe 2005a) and sustained productive action (Dutton and Glynn 2008). Through the experience of constructing and/or occupying a green building, individuals can find such positive meaning and take actions that focus on the benefits of others, including the environment. The individual action for going the extra mile is elucidated in this quote by a plant engineer: "The lights in this new building automatically shut off. It made me more cognizant, now when I go to the older buildings and see lights left on in offices I go around and switch them off." This individual sense of responsibility is aggregated to the collective through conversations (Quinn and Dutton 2005) and vicarious learning by observing the behaviors of others. Such collective capacity is illustrated in this quote from a sustainability director: "We see a difference between the employees working in this (LEED) building and others in terms of volunteerism for other initiatives, keeping the plant clean and being more involved."

In addition to this expanded sense of individual and collective responsibility, vitality and aliveness at an individual level can influence people to infuse entire systems with significance and implications that green buildings are good, desirable, and beneficial for all stakeholders (Dutton and Glynn 2008). Positive emotions can foster positive interrelating (ibid.) with stakeholders; and more positive attitudes toward even those with whom the organization has had little contact (Brickson 2008), such as the generic entity called "community." Thus, aliveness and vitality at an individual level can broaden the collective meaning to the extent that the boundaries between the organization and stakeholders are blurred and "understanding the other's needs and expectations becomes essential to one's own sense of self." (ibid., 42). In other words, the understanding broadens from focusing on just the organization's own welfare, by drawing boundaries between the organization and other entities, to focusing on particular others or larger collectives by perceiving the organization—and its artifacts such as buildings—as part of a larger whole (Brickson 2005). This was described by a sustainability manager: "When people from the community drive around our facility they know it is managed differently. Our buildings don't belong to us, they belong to community since the community have to drive around and interact with them; we are part of the community."

Thus, by expanding the organizing processes and collective meaning of the organizational self in relation to others, the organization can build capacity for greater environmental responsibility.

Discussion and Implications

In this chapter we provide a new lens to understand the frequently mentioned benefits of green buildings such as health, productivity, and image, by suggesting generativity as a potential outcome. Generativity is reflected in organizational capacity for greater environmental responsibility given by enhanced processes of positive organizing and shifts in collective meaning generated from moments of individual thriving. We borrow from the positive organizational scholarship and appreciative inquiry perspectives to suggest that green building practices, associated with both construction and operations, can create positive outcomes at various levels—for individuals, organization, environment, and wider networks of stakeholders. We explain this by connecting green building practices to individual thriving, facilitated by organizational and team contexts. This thriving at the individual level spirals upward by building collective positive affect and organizational knowledge, and spirals outward by diffusion of learning to other stakeholders as well as expanding the definition of the organization's self in relation to others.

To delineate the implications of the proposed path we build on appreciative inquiry's fundamental principles (Bushe 2005b) and suggest the possibilities of change within the organization and its network. Appreciative inquiry is based on a principle that language and words form the building blocks of social reality (Cooperrider and Srivastava 1987); that human systems move in the direction of what they most frequently talk—and particularly ask questions—about. The narratives emerging within a green building team have the generative potential to change the organization's inner dialogue, not only leading to high-performance on strategic decisions (Bushe 2005b) but also accruing far-reaching consequences such as transforming the culture and vision of the organization. The implication for organizations is to bring to light such stories through the voice of the people associated with them. Intentional elaborations of the success stories about green construction create relational spaces that encourage learning by experimentation and further exploration of possibility. Simultaneous experience of positive affect generates real interest in shared possibilities (Barrett and Fry 2005, 98). Organizations can create spaces for this discovery of possibility such as project celebrations, open houses, factory tours, new employee orientations, mentorship programs, interactive websites, and social media spaces to share these examples and change the organization's narratives.

The outward spirals—shifts in collective meaning of the organization's self in relation to others—can be facilitated and amplified by what in appreciative inquiry terms is called as the "generative metaphor." Such metaphors "provoke new thought, excite us with novel perspectives, vibrate with multivocal meanings, and enable people to see the world with fresh perceptions not possible in any other way" (Barrett and Cooperrider 1990, 222). The emerging lens of the organization's relationship with others through its built environment can be underscored by images of what construction means to the organization in relation to the environment and its stakeholders. For example, describing the awards for green construction, a facilities manager hoped that "instead of an award for a green building there should be an award for avoiding construction; that we need to begin there in order to have the least impact on the environment and the communities we live in." Such images and metaphors can generate new thought repertoires for innovative actions. Execution teams can create the metaphors by taking a pause and generating an image of what the process means to them. It can also be generated in organization-wide settings. Working in a green building is a constant reminder to the employees of the values of stewardship and responsibility toward the others. A generative image for this experience can be co-created with stakeholders in settings such as open houses.

Conclusion

Generativity as a benefit from green construction enables us to explain consequences that go beyond cost saving, energy efficiency, and productivity gains. Research and theory development from the domains of positive organizational scholarship and the practice of appreciative inquiry, along with actual voices of actors in green construction projects, suggest that moments of individual thriving experienced during these projects can generate positive organizing and positive meaning, in turn creating greater capacity for environmental responsibility. Goal clarity, leadership passion, experimental disposition, and conversational spaces in these projects foster the individual's sense of aliveness, vitality, and learning. This state of thriving can generate new meaning, and positive organizing through the contagion of positive affect, resource building, and diffusion of learning. All this results in increased capacity for environmental responsibility. These linkages constitute generativity as a beneficial outcome of green construction. Most simply put, when individuals connect in a green building context, they come to identify more with the whole system(s) within which they reside. This connection with the whole is inherently generative at the individual, group, organization, and larger system levels because it fosters connections, learning, exploration, passion, clarity of purpose, positive affect, and increased responsibility for the whole.

Empirical evidence indicates that the mere perception that one's organization is serious about a sustainability agenda causes one to engage in more creative acts, to be more highly engaged as an employee, and to have more high-quality (deep) relations at work (Glavas and Piderit 2009). We see green construction as an obvious subset of sustainability and make a similar conclusion. Involvement in green building can generate a reconnection with the whole and thereby bring out the best in us—an inherent striving to do good for the environment around us.

Note

1. A similar path from practices to capability creation has been described by Lilius et al. (2011) as the relationship of everyday organizational practices to compassion capability.

References

Bagozzi, Richard P., Hans Baumgartner, and Rik Pieters. 1998. Goal-directed emotions. *Cognition and Emotion* 12 (1): 1–26.

Baker, Ann C., Patricia J. Jensen, and David A. Kolb. 2005. Conversation as experiential learning. *Management Learning* 36 (4): 411–427.

Bandura, Albert. 2000. Exercise of human agency through collective efficacy. *Current Directions in Psychological Science* 9 (3): 75–78.

Barrett, Frank J., and David Cooperrider. 1990. Generative metaphor intervention: A new approach for working with systems divided by conflict and caught in defensive perception. *Journal of Applied Behavioral Science* 26 (2): 219–239.

Barrett, Frank J., and Ronald E. Fry. 2002. Appreciative inquiry in action: The unfolding of a provocative invitation. In *Appreciative inquiry and organizational transformation: Reports from the field*, ed. Ronald E. Fry, Frank J. Barrett, Jane Selling, and Diana Whitney, 1–23. Westport, CT: Quorum Books.

Barrett, Frank J., and Ronald E. Fry. 2005. *Appreciative inquiry: A positive approach to building cooperative capacity*. Chagrin Falls, OH: Taos Publications.

Brickson, Shelly. 2005. Organizational identity orientation: Forging a link between organizational identity and organizations' relations with stakeholders. *Administrative Science Quarterly* 50 (4): 576–609.

Brickson, Shelly. 2008. Re-assessing the standard: The expansive potential of a relational identity in diverse organizations. *Journal of Positive Psychology* 3 (1): 40–54.

Bushe, Gervase R. 2005a. When is appreciative inquiry transformational?: A meta-case analysis. *Journal of Applied Behavioral Science* 41 (2): 161–181.

Bushe, Gervase R. 2005b. Five theories of change embedded in appreciative inquiry. In *Appreciative inquiry: Foundations in positive organization development*, ed. David L. Cooperrider, Peter F. Sorenson Jr., Therese F. Yeager, and Diana Whitney, 121–132. Champaign, IL: Stipes.

Cardon, Melissa S. 2008. Is passion contagious? The transference of entrepreneurial passion to employees. *Human Resource Management Review* 18:77–86.

Carlsen, Arne, and Jane E. Dutton. 2011. Research alive: The call for generativity. In *Research alive: Exploring generative moments in doing qualitative research*, ed. Arne Carlsen and Jane E. Dutton, 12–24. Copenhagen: Copenhagen Business School Press.

Cohen, Wesley M., and Daniel A. Levinthal. 1990. Absorptive capacity: A new perspective on learning and innovation. *Administrative Science Quarterly* 35 (1): 128–152.

Cooperrider, David L., and Suresh Srivastava. 1987. Appreciative inquiry in organizational life. *Research in Organizational Change and Development* 1:129–169.

Cooperrider, David L., and Diana Whitney. 2005. An invitation to the positive revolution in change. In *Appreciative inquiry: A positive revolution in change*, 1–3. San Francisco: Berrett-Koehler.

Cross, Rob, Wayne Baker, and Andrew Parker. 2003. What creates energy in organizations? *MIT Sloan Management Review* 44 (4): 51–56.

Dutton, Jane E., and Arne Carlsen. 2011. Seeing, feeling, daring, interrelating and playing: Exploring themes in generative moments. In *Research alive: Exploring generative moments in doing qualitative research*, ed. Arne Carlsen and Jane E. Dutton, 214–235. Copenhagen: Copenhagen Business School Press.

Dutton, Jane E., and Mary Ann Glynn. 2008. Positive organizational scholarship. In *The Sage handbook of organizational behavior*. vol. 1, ed. Julian Barling and Cary L. Cooper, 693–712. Thousand Oaks, CA: Sage.

Dutton, Jane E., Mary Ann Glynn, and Gretchen M. Spreitzer. 2006. Positive organizational scholarship. In *Encyclopedia of career development*, ed. Jeffrey H. Greenhaus and Gerard A. Callanan, 641–644. Thousand Oaks, CA: Sage.

Edmondson, Amy. 1999. Psychological safety and learning behavior in teams. *Administrative Science Quarterly* 44 (2): 350–383.

Erikson, E. H. 1950. *Childhood and society*. New York: Norton.

Frederickson, Barbara L. 2004. The broaden and build theory of positive emotions. *Philosophical Transactions of the Royal Society of London. Series B, Biological Sciences* 359:1367–1377.

Fredrickson, Barbara L., and Jane E. Dutton. 2008. Unpacking positive organizing: Organizations as sites of individual and group flourishing. *Journal of Positive Psychology* 3 (1): 1–3.

Fredrickson, Barbara L., and Marcial F. Losada. 2005. Positive affect and the complex dynamics of human flourishing. *American Psychologist* 60 (7): 678–686.

Fry, Ronald E. 2008. Business as an agent of world benefit: Transformative innovations for mutual benefit. *Develop* 3:8–18.

Ghoshal, Sumantra, and Christopher A. Bartlett. 1994. Linking organizational context and managerial action: The dimensions of quality of management. *Strategic Management Journal* 15:91–112.

Gibson, Cristina B., and Julian Birkinshaw. 2004. The antecedents, consequences, and mediating role of organizational ambidexterity. *Academy of Management Journal* 47 (2): 209–226.

Glavas, Ante, and Sandy K. Piderit. 2009. How does doing good matter?: Effects of corporate citizenship on employees. *Journal of Corporate Citizenship* 36:51–70.

Heerwagen, Judith H. 2002. Sustainable design can be an asset to the bottom line. *Environmental Design & Construction* 5 (4): 35–39.

Hoffman, Andrew J., and Rebecca Henn. 2008. Overcoming the social and psychological barriers to green building. *Organization & Environment* 21 (4): 390–419.

Kats, Greg E. 2003. *The costs and financial benefits of green buildings*. Sacramento: California Sustainable Building Task Force.

Kayes, Anna B., Christopher D. Kayes, and David A. Kolb. 2005. Experiential learning in teams. *Simulation & Gaming* 36 (3): 330–354.

Kibert, Charles J. 2003. Green buildings: An overview of progress. *Journal of Land Use* 19 (2): 491–502.

Kibert, Charles J. 2005. *Sustainable construction: Green building design and delivery*. Hoboken, NJ: Wiley.

Lei, David, Michael A. Hitt, and Richard Bettis. 1996. Dynamic core competences through meta-learning and strategic context. *Journal of Management* 22 (4): 549–569.

Lilius, Jacoba M., Monica C. Worline, Jane E. Dutton, Jason M. Kanov, and Sally Maitlis. 2011. Understanding compassion capability. *Human Relations* 64 (7): 873–899.

Lindsley, Dana H., Daniel J. Brass, and James B. Thomas. 1995. Efficacy-performance spirals: A multilevel perspective. *Academy of Management Review* 20 (3): 645–678.

Losada, Marcial, and Emily Heaphy. 2004. The role of positivity and connectivity in the performance of business teams. *American Behavioral Scientist* 47 (6): 740–765.

Mageau, Geneviève A., Robert J. Vallerand, François L. Rousseau, Catherine F. Ratelle, and Pierre J. Provencher. 2005. Passion and gambling: Investigating the divergent affective and cognitive consequences of gambling. *Journal of Applied Social Psychology* 35:100–118.

McAdams, Dan P., and Ed de St. Aubin. 1992. A theory of generativity and its assessment through self-report, behavioral acts, and narrative themes in autobiography. *Journal of Personality and Social Psychology* 62 (6): 1003–1015.

McNamara, David W., Brock Birkenfeld, Peter Brown, Nicole Kresse, Justin Sullivan, and Philippe Thiam. 2011. Quantifying the hidden benefits of high-performance building. *ISSP Insights* (December): 2–19. http://www.sustainabilityprofessionals.org/system/files/Valuing%20Green%20Building.pdf (accessed December 30, 2011).

Mom, Tom J. M., Frans A. J. van den Bosch, and Henk W. Volberda. 2009. Understanding variation in managers' ambidexterity: Investigating direct and interaction effects of formal structural and personal coordination mechanisms. *Organization Science* 20 (4): 812–828.

Nonaka, Ikujiro. 1994. A dynamic theory of organizational knowledge creation. *Organization Science* 5 (1): 14–37.

Paumgartten, Paul V. 2003. The business case for high-performance green buildings: Sustainability and its financial impact. *Journal of Facilities Management* 2 (1): 26–34.

Philippe, Frederick L., Robert J. Vallerand, Nathalie Houlfort, Genevieve L. Lavigne, and Eric G. Donahue. 2010. Passion for an activity and quality of interpersonal relationships: The mediating role of emotions. *Journal of Personality and Social Psychology* 98 (6): 917–932.

Pratt, Michael, and Blake Ashforth. 2003. Fostering meaningfulness in working and at work. In *Positive organizational leadership: Foundations of a new discipline*, ed. Kim Cameron, Jane E. Dutton, and Robert Quinn, 309–327. San Francisco: Berrett-Kohler.

Quinn, Robert W., and Jane E. Dutton. 2005. Coordination as energy-in-conversation. *Academy of Management Review* 30 (1): 36–57.

Ray, Gautam, Jay B. Barney, and Waleed A. Muhanna. 2004. Capabilities, business processes and competitive advantage: Choosing the dependent variable in empirical tests of the resource-based view. *Strategic Management Journal* 25:23–37.

Rerup, Claus, and Martha S. Feldman. 2011. Routines as a source of change in organizational schemata: The role of trial-and-error learning. *Academy of Management Journal* 54 (3): 577–610.

Salvato, Carlo, and Claus Rerup. 2011. Beyond collective entities: Multilevel research on organizational routines and capabilities. *Journal of Management* 37 (2): 468–490.

Schnackenberg, Andrew, Jagdip Singh, and James Hill. 2011. Theorizing capabilities of organizational agility: A paradox framework. Presented at the Academy of Management (AOM) Annual Meeting, San Antonio, TX.

Singer, Jefferson A., Laura A. King, Melanie C. Green, and Sarah C. Barr. 2002. Personal identity and civic responsibility: "Rising to the occasion" narratives and generativity in community action student interns. *Journal of Social Issues* 58 (3): 535–556.

Spender, J.-C. 1996. Organizational knowledge, learning and memory: Three concepts in search of theory. *Journal of Organizational Change* 9 (1): 63–78.

Spreitzer, Gretchen M., and Scott Sonenshein. 2003. Positive deviance and extraordinary organizing. In *Positive organizational scholarship: Foundations of a new discipline*, ed. Jane E. Dutton, Robert E. Quinn, and Kim Cameron, 207–224. San Francisco: Berrett-Koehler.

Spreitzer, Gretchen M., Katherine Sutcliffe, Jane E. Dutton, Scott Sonenshein, and Adam M. Grant. 2005. A socially embedded model of thriving at work. *Organization Science* 16 (5): 537–549.

Wilson, Alex. 2005. Making the case for green building. *Environmental Building News*. http://www.buildinggreen.com/auth/article.cfm/2005/4/1/Making-the-Case-for-Green-Building (accessed November 19, 2011).

World Inquiry's Innovation Bank. 2010. http://worldinquiry.case.edu (accessed November 19, 2011).

Wrzeniewski, Amy. 2003. Finding positive meaning in work. In *Positive organizational scholarship: Foundations of a new discipline*, ed. Jane E. Dutton, Robert E. Quinn, and Kim Cameron, 296–308. San Francisco: Berrett-Koehler.

IV

Perceptions, Frames, and Narratives

12

Conveying Greenness: Sustainable Ideals and Organizational Narratives about LEED-Certified Buildings

Beth M. Duckles

Introduction

Toward the end of a tour of a LEED- (Leadership in Energy and Environmental Design) certified corporate office building, my guide ran through a list of the building's sustainable features. He had already discussed the environmentally friendly carpeting, low VOC paints and adhesives, LED lighting, and furniture made from reused materials. When we passed a men's bathroom, he cracked a joke about waterless urinals. But as we finished, the guide kept saying his organization needed to "tell the building's story" better. While there was a plaque from the U.S. Green Building Council (USGBC) proclaiming the building's LEED status in the front lobby, no additional signs described green features. We walked to a pleasant outdoor courtyard and the guide explained that the pavers had been reclaimed from building waste, but you couldn't tell they were recycled. As we walked back inside, he mentioned again that the organization really needed to do better at telling the building's story.

By contrast, a tour at another LEED-certified building owned by a nonprofit organization started differently. Upon entry, I immediately noticed brightly colored signs, both large and small, describing the sustainable features of the building with graphics and statistics. The signs called attention to building features and how they reduced the building's environmental impact. Small plaques in the bathroom explained the operations of the dual-flush toilets and low-flow faucets. The tour guide was much more versed in the "story" of the building and took me through it as a part of a composed narrative. She rattled off the water and energy savings per year for this building compared to other buildings of its size. The tour ended on the roof with a discussion of solar panels; the guide explained that the technology was advancing rapidly. Telling the building's story, both visually and verbally, was a key part of this tour.

Both of these organizations built their sustainable buildings using the USGBC's LEED building certification system during the same time frame and in the same state. Each organization invested considerable effort and money to construct its building. As early adopters of the LEED certification system, project teams faced challenges in convincing stakeholders that green construction was the right choice. Yet the buildings have different public narratives. These organizations are conveying greenness in different ways.

Organizations use narratives to "position" themselves, to describe themselves in relationship to others. Narratives in organizations can be seen as a negotiation over meaning or the place where contestation and conversation occurs (Czarniawska 1997). Yet there is a limited understanding of the relationship between physical space and organizational narrative. While the field of architecture has a rich literature on the use of space as a narrative device, the field of organization theory and its narratives have not yet integrated the use of space as a narrative device. The stories that organizations tell around sustainable construction can help bridge this divide by giving a snapshot in time of how organizations frame their commitment to sustainability for stakeholders. In this chapter, I will look at the intersection of research on space, narrative, and sustainable construction. Then I will discuss four narratives and one counter-narrative that organizations use to legitimate green construction, exploring how five sectors—corporations, nonprofits, and education, government, and religious organizations—use those narratives.

Background

Scholars have long understood that how we construct buildings is a socially meaningful enterprise. The relationship between identity and the built environment has been explored in terms of political meaning (Rakatansky 1995), culture (Rapoport 1987), consumerism (Chase 1991), and cultural and ethnic identity (Atkin and Krinsky 1996). Gieryn (2000) discusses three attributes of space: geographic location, material form, and investment with meaning and value. Kamin (2001) comments, "The choices we make in creating our external world speak volumes about our inner values" (2001, xvi). Aldo Van Eyck advises architects, "What you should try to accomplish is built meaning" (Brown 1981, 44).

How this meaning is enacted in spaces is less clear. Though buildings are seen as cultural artifacts, the "reading" of the contested space and changing meanings of these artifacts is often difficult to ascertain (Goss

1988; Kraftl 2006). The influence of the political and regulatory environment on the construction of space can further complicate their reading (Faulconbridge 2009; Jones 2009). Gieryn notes that the construction of a science laboratory on a university campus is the site at which complex discussions among different interest groups and designers are finally solidified and literally structured. He writes, "The play of agency and structure happens as we build: we mold buildings, they mold us, we mold them anew. . . . In buildings, and through them, sociologists can find social structures in the process of becoming" (Gieryn 2002, 65). We know buildings are connected to meaning, but we do not clearly connect these meanings to organizations.

Organizational Narratives

One place where the organizational "process of becoming" can be examined is through the use of organizational narratives. Narratives are useful to understand the development of meaning around a collective phenomenon (Franzosi 1994, 1998; Reed 1989; Maines 1993). In particular, research on social movement narratives discusses how narratives help groups shift ideals and meanings (Polletta 1998, 2006; Polletta and Jasper 2001; Benford and Snow 2000). Organizational narratives within environmental movements have been used to explore the underlying structures within environmental negotiations (Troast et al. 2002), the framing of environmental debates (Hoffman and Ventresca 1999), and the development of recycling logics (Lounsbury, Geraci, and Waismel-Manor 2002). Analysis of organizational narratives can offer a linkage between meaning and organizational action (Stevenson and Greenberg 1998; Feldman et al. 2004; Boje 2008).

Stakeholder theory indicates that the organizational sector has an effect on the type of organizational narrative used. Because nonprofits, educational institutions, and public organizations work within a "nondistribution constraint," they are likely to be less affected by motives of seeking profit for their firm (DiMaggio and Anheier 1990; Frumkin and Galaskiewicz 2004; Weisbrod 1998; Hansmann 1980. However, nonprofits are still highly influenced by financial challenges and the need to balance budgets as their stakeholders expect that money be spent prudently. Since the organization is not legally allowed to accumulate capital for stakeholders, these organizations would be expected to appeal to a broader ideological frame that resonates with a specific set of stakeholders and uses a logic that goes beyond strict cost–benefit analysis in justifying the rationale for their choices (Rothschild and Milofsky 2006). By contrast, for-profit corporations would be more focused on appealing to their stakeholders in terms of market share, creating value, and attention to building coalitions and gaining influence (Scott 2007). While for-profit organizations could still be influenced by corporate social responsibility, this motivation would be tempered by a perspective that socially responsible actions must benefit the bottom line. The differences among sectors such as corporations, nonprofits, and education, government, and religious organizations become more pronounced as each organizational type mimics and creates normative homogeneity within its sector to institutionalize practices (DiMaggio and Powell 1983). Thus we could expect to see differences in how each sector conveys meaning to stakeholders for similar choices. We can see these dynamics in the narratives surrounding green buildings.

Organizational Narratives and Green Buildings

The green building movement could be characterized as a market-based movement. In other words, the movement aims to shift the market toward more sustainable construction. Unlike social movements that aim to change the state, research on market-based movements suggests that they rely both on "hot causes," or motivating reasons for action to encourage participation, and "cool mobilization," where actors form a relationship with the movement through collective experience (Rao 2009). In this sense, cool mobilization is a collective process that has less urgency than the "hot cause" but which creates communities with common values that shift the market. As of yet, there is little understanding of the mechanisms that create or sustain cool mobilization.

One way to examine organizational coalescence is by looking at narratives used to describe the creation and the space of the buildings as coolly mobilizing market narratives. Because the sustainable design and construction movement is focused on market change, the movement has an interest in developing new meanings that support the creation of more green buildings (see chapter 9, this volume). Recent research considers how to encourage more green building (Kingsley 2008), remove obstacles for greener buildings (Hoffman and Henn 2008), frame organizational change toward sustainable construction (see chapter 10, this volume) and consider the integration of sustainable ideals into professional labor (Duckles 2009).

The green building movement and its client organizations have created these new meanings through the organizational process of choosing and then following through on constructing and occupying green buildings. The choice of a green building is certainly not a singular one, involving many organizational stakeholders to design, finance, and construct the building. Yet a building has the unique quality of being a spatial outcome of an organization's decision-making process. Gieryn (2002) describes a constructed building as a way of concealing organizational conflict through the creation of an end product. Thinking of buildings as the final product that comes from a set of stakeholder discussions makes the building both a site of narrative and a place to create new narratives. Buildings become literal places where an organization can shape collective meaning. Looking at narratives of early green building adopters who had to legitimate their buildings to multiple stakeholders allows a better understanding of the relationship between the organizational sector and the complexity of sustainability as a concept. Narratives coolly mobilize the organizational market radicals or change agents by articulating the connection among building features, sustainable ideals, and organizational motives (Rao 2009).

As previously discussed, narratives will vary by organizational sector. In particular this chapter will focus on nonprofits, corporations, educational organizations, and government and religious organizations. Narratives on organizational greening among governmental buildings might focus on public stakeholders that are concerned with the use or misuse of taxpayer dollars, while environmentally focused nonprofits might focus on values-driven stakeholders who are concerned with educating the public over cost and for-profit organizations might focus on investment-oriented stakeholders who are concerned with net-present value and payback periods for green building features.

Building attributes could also affect these narratives. Some building attributes that are defined as sustainable through the LEED certification program are intentional architectural and space use choices, and have clear connections to the narratives about sustainable practices (e.g., solar panels or wind turbines). However, there are other attributes which may not be immediately visible (e.g., HVAC systems or insulation), or which may not offer a visible organizational narrative (e.g., recycled furniture or reclaimed materials). Building attributes can be seen as spatial and sometimes visual attributes of the organizational narrative and are a part of the way in which the organization tells its story.

Data and Analysis

To explore the processes by which organizational narratives are developed and employed, I conducted twenty-seven interviews with representatives of organizations that built a LEED-certified building. Interviews were conducted between April 2007 and November 2009 with the majority conducted between June and August 2007, discussing a total of twenty-four buildings.[1] This is a convenience sample, the choice of buildings is not representative of the approximately fifty buildings that were LEED for New Construction (LEED-NC) certified at the time and the sample is skewed toward Southern California.[2] I toured twenty-four LEED-certified buildings, all but one of which was discussed in interviews. I toured six corporate buildings, four education facilities, eight government buildings, three nonprofit buildings, and two religious buildings. In these tours, I learned how each building was conceived, how challenges were overcome, and choices were made in the process.[3] I augment these interviews and field observations with data from participant observation at green conferences as well as interviews with green building professionals and LEED Accredited Professionals (APs) working in a variety of building and design professions.

Early adopters can give the most robust, complete, and well-told explanations about why their organization built a green building. I offered respondents anonymity to encourage candor.[4] As a convenience sample, this data is not presented as a comprehensive account of all available individual organizational narratives. Instead, these are snapshots of narratives from a variety of building types and organizational forms. This data allows a discussion of a sample of narrative types that are available to early adopters seeking to legitimate building green. To analyze the data, I coded the transcribed interviews and fieldnotes using the software package Atlas.ti and used open codes to find thematic narrative elements.

Five Organizational Narratives for Green Building

The data revealed five prominent narratives that organizations employed to explain their green building initiatives: profit-driven green, practical green, deep green, green innovators, and hidden green.

Profit-Driven Green

By far, the most prevalent narrative constructed by those interested in green and sustainable building, terms that are typically seen as interchangeable, is that a green building is profitable; that the organization will receive a higher return on investment by constructing a sustainable building rather than a nonsustainable building. Respondents invoked this theme in two ways. First, respondents talked about the cost of the building itself and the tangible savings to the project over time due to green attributes. Second, respondents discussed possible future financial benefits that they had no sense of how to achieve by discussing sustainable successes in other sectors (see chapter 11, this volume). In building tours, organizations had creative ways to showcase the cost savings of their buildings such as using a window to display unique insulation.

Using the profit narrative is encouraged by the green building industry. Marketing books discuss ways to make it clear that the building will be cost effective and emphasize the need to make the green elements a central and integrated part of the project rather than an "add-on" (Yudelson 2007, 74). This is vital for green building advocates because research indicates that critics regard add-ons as making the process longer and more costly (Lutzenhiser et al 2001).

Respondents from all sectors discussed the economic savings they achieved due to sustainable choices. Some respondents expressed surprise at the economic viability of green buildings. One former building owner explained the building had exceeded expectations: "The fact of the matter is, I know I came in on that project $2 million under what it should have cost. And the water savings are 39 percent and the gas and electric savings are 38 percent. So it's going to pay for itself in two to three years. It's so exceeding its financial projections, it's not even funny."

The surprise at the fiscal success and the recognition that long-term savings would continue creating value for the project made respondents certain that the decision to build green had been the right one. Couching cost savings in terms, not only of the upfront cost of the project but also in terms of the ongoing reduction in water and utility costs was a common tactic. A lowered utility bill was a badge of pride, and owners joked about framing the bills and hanging them on the wall to prove cost savings. These claims are supported by a growing body of research on the fiscal viability of sustainable construction (Eichholtz, Kok, and Quigley 2009a, 2009b). It is notable that these studies were not available to respondents when they designed their buildings.

For this reason, respondents articulated a second, more challenging explanation of the financial impact of building green. Without hard data, respondents had difficulty making a case for the benefit of a building that appeared to be more costly in upfront costs. Instead, they drew from ideas of green thinkers such as Paul Hawken, Amory Lovins, or William McDonough (Hawken, Lovins, and Lovins 2000; McDonough and Braungart 2002; Hawken 1993). One of the often-told tales legitimating green buildings as financially viable was the story of Ray Anderson who founded Interface Carpeting (Anderson 1998). Anderson changed his manufacturing process to consider all aspects of how the company's behavior might affect the environment. The company accepted used carpeting and developed methods for recycling it. Interface now takes used carpeting as a raw material for carpet padding and creates products from the byproducts of the carpet recycling process. This creates additional revenue streams for the company and engages the company in "cradle-to-cradle" manufacturing (ibid.). A respondent stated: "And of course Interface, who you've probably heard of Ray Anderson, and his carpet, everybody thought he was going to lose his shirt and the company would go bankrupt. Literally, he's turned his company completely around to show that you can make money, a substantial amount of money, and not only that, but secure a very good place in the market for himself."

The function of this narrative is to allow individuals to believe they will make money if they move toward sustainable practices, but without having hard data to prove it a priori. In sustainable construction, there was no clear sense of how a profit would be made beyond energy savings, but respondents retold Anderson's story to have faith that thinking sustainably would yield a financial benefit. The story functions as an enabling myth to encourage a combining of innovation, sustainability, and capitalism to find new profit streams.

Such myths and stories can be particularly important as many components that contribute to cost savings for green buildings are hard to physically show because they are not easy to see. Efficient insulation or innovative heating, ventilation, and cooling (HVAC) systems are literally inside the walls. Tour guides were faced with telling the profit story while the most energy efficient components were hidden. To compensate, they discussed energy savings in comparison to other buildings of the same size. Some would show the energy efficiency of specialized systems with interactive touchscreens where visitors could see savings in real time. Buildings with unique insulating wall materials such as insulation made from recycled blue jeans or buildings constructed with straw bales might construct a window into the wall or have a display showing the material. While the renewable or recycled insulating material might not be cost effective in upfront costs, the window helps the tour guide discuss the energy efficiency of the building and the benefit of slightly higher upfront costs toward long-term energy savings. The challenge of telling this story was evident when individuals on the tour asked how much something like a photovoltaic array actually cost. Tour guides would give the appropriate figures while warning the questioner to look at the whole building cost rather than one component. However, some building attributes, particularly the low-cost, high-return attributes are more useful in telling the profit narrative than others.

Telling the profit story allows companies to legitimate their case to stakeholders who care about the bottom line and to make a case for the long-term fiscal stability of the building and by inference, the organization. The profit narrative counters the idea that green is more expensive and it helps to situate the organization when there is ambiguity. All organizations studied in all organizational sectors use this narrative in some manner, particularly in tours where questions of cost came up. These approaches are most employed by organizations whose stakeholders are most concerned with the bottom line and generating revenues for shareholders.

Practical Green

The practical green rationale indicates that building green just "makes sense"; a sustainable building is less about being eco-friendly and more about the practicalities of making a good building. There are two components to this rationale: first, the well-being of occupants derived from studies that highlight the health benefits of green construction. Second, commissioning, a relatively new step in the construction process, adds accountability and quality assurance.

Figuring prominently in the practicality narratives are studies on sustainable buildings creating a more effective workforce with more productive employees (Heerwagen 2000; Miller et al. 2009; Eichholtz, Kok, and Quigley 2009b) and with better indoor air quality (Paul and Taylor 2008; Abbaszadeh et al. 2006). Daylighting, one component of green building design, is seen as beneficial for occupants (Heschong 2003a, 2003c) as well as increasing retail sales (Heschong, Wright, and Okura 2002). Green buildings have been linked to increased student test scores in K–12 schools (National Research Council 2006; Kats 2006; Heschong 2003b). Respondents cite these sources as a single body of literature at conferences, in literature, and in conversations among green and sustainability-minded professionals stating that green buildings are better for occupants.

This narrative uses productivity studies to prove that a green building will make occupants happier and more productive. Personnel costs are substantial for most organizations and respondents argue that the well-being of employees and increased productivity will benefit the organization. One facilities manager pointed to improvements in his own employees: "The past three years, including the year that we were in the midst of construction and moving into this building, we delivered nearly 100 percent of all of our projects that made us . . . number one for project delivery [in our organization] . . . Health complaints, I don't have the stats on it, all I know is that when I hear from labor unions and employee groups of any type, and individuals, they are happier. They feel safer. They feel healthier."

These studies have had a tremendous impact on the green building industry. They indicate that green is not a high-minded ideal, but instead makes a practical difference for whoever uses the building. As productivity of workers increases, the company stands to benefit. This was referred to in tours when guides pointed out features such as operable windows, views out of workspaces, self-regulated temperature controls, and beautiful courtyard spaces.

A second way that the practicality narrative was invoked was through building commissioning. Basic commissioning is a LEED prerequisite and thus a requirement of LEED-NC certification (EA [Environment & Atmosphere] Prerequisite 1), and enhanced commissioning gives the building an additional LEED point (Energy and Atmosphere Credit 3). In commissioning, the design team hires an independent third party to examine the building after construction to ensure that building systems work properly. This is framed as the way to make certain the company gets the building it paid for by ensuring the systems work properly and that facilities management professionals are trained on the proper usage of the building systems. In the words of one respondent discussing commissioning: "It makes perfect sense. You don't buy a car without driving it. And really, that's all we're doing is making sure the mechanical systems and all of the different building components are working the way they were designed to." The comment "You don't buy a car without driving it" was a common theme. Respondents argued that the construction industry is one of the few industries where you can spend a lot of money and never check to see that you get what you paid for. Doing commissioning was tied with the client organization getting value for its money.

For the building owner, the practicality narrative frames the sustainable building as a commodity that should perform well and puts the building design team in the position of creating a product that is functional and high performing. This allows the organization's leaders to show that they value the well-being of their workers while also "getting their money's worth," thus arguing to stakeholders that they are using their money wisely. For some nonprofit organizations and government agencies, the practicality of having a good working environment for employees may be an argument that decreases employee turnover and increases retention as well as ensuring that the new building's value lasts longer.

Deep Green

Apart from the profit and the practicality narratives, some organizations frame the choice to be green from a personal, spiritual narrative. A green building is a sign of better stewardship of resources and a way to be a part of nature rather than harming it. Respondents had different ways to express these beliefs, from articulating the oneness of all life to invoking a specific spiritual tradition or a deep relationship to the earth. One respondent discussed her desire to create a building that was "feminine" and connected to the ecology of the site.

Organization owners and users articulated these decisions in light of their spiritual commitments. Six interviewees including Catholic Sisters and an employee at the Presentation Center in Los Gatos, California, discussed their reasons for building a welcome center made of straw bale and with a vegetative roof at their retreat facility.[5] One sister remarked:

> So what we're trying to do . . . as I say it's a retreat and conference center, and a variety of people come here for their own purposes. Spiritual purposes, community building purposes, social purposes. All of these different reasons. . . . We want them to leave here thinking, "Say, maybe I [should] start to rethink some of my values." To understand the connectiveness for example. . . . We're all one, we're all connected. . . . This isn't just words on a piece of paper. It's like trying to go being to live simply. To work with less, to have gentle footprints on Planet Earth.

Deciding to build green for these sisters was not a choice. They saw the decision to construct a green building as another way to enact and express their spiritual beliefs. The decisions they made in constructing the building were done in part to inspire others to think differently about their roles in stewarding the natural world.

Architects and engineers who worked with deep green organizations felt it was a unique process because clients placed a premium on green attributes if those attributes were a core part of reducing impact on the environment. In one building, the organization decided to cut the size of the building, reducing the number of bathrooms and even the size of the rooms from the original plans in order to pay for some of the more costly green attributes that the organization members believed were the right choice for their building. The architects expressed surprise at their client's willingness to forgo comfort in favor of a reduced environmental impact. During an interview with a nonprofit representative, I asked how the organization had decided which components to integrate into the LEED building. The respondent looked at me blankly, then responded that the organization had integrated every possible green component it could into the building; for this interviewee, the answer to my question was self-evident.

The story of such buildings was told as a lesson or as a way to encourage visitors to change their behavior. While signage existed in many LEED buildings, those with a deeper ecological focus had more prominent signage that extended past the front lobby and inside. These buildings had signage explaining low-flush toilets or waterless urinals in the bathrooms, but one organization with a deep ecology focus also had a sign explaining that the normal-looking tiles in the bathroom were recycled. For deep green buildings, it was important to show off even the less visible features designed to encourage sustainable behavior as way to both inform and educate visitors.

The deep green narratives in this study echo theoretical work on deep ecology with its focus on a spiritual and personal connection to the earth as a motivator for human behavior that takes the earth's well-being as central (Naess and Sessions 1995). Yet for deep ecologists who evoke these principles, constructing any building is problematic because they would advocate for not constructing one at all—green or not—to reduce the impact on the environment. This is challenging for organizations that need to build a new building, but wish to do so in a manner that fits with spiritual beliefs about the earth. While these organizations may not be true deep ecologists in the theoretical sense, the rationale that they use evokes stewardship of natural resources and the deep ecological and spiritual beliefs. This rationale is most clearly seen in religious and nonprofit organizations, but smaller for-profits and educational organizations also evoked this rationale.

Green Innovators

When an organization wanted to showcase its uniqueness, members talked about themselves as green innovators. Building owners who used this rationale often had buildings with newer or experimental systems such as reclaimed water systems or geothermal heating/cooling systems. Innovators would explain that they were setting an example for others, breaking new ground and demonstrating what is possible in building technology. In this narrative, respondents conveyed their hope that the building would be seen as a laboratory where other organizations could learn. The building was a "teaching tool" for new and, as yet, unproven techniques and tools.

While organizations wanted to show what was possible, there were often significant difficulties as early adopters of new products. Respondents mentioned a myriad of challenges with products and certification. From not enough fly ash in the concrete to a lack of knowledge about the chosen wood's chain of custody, almost all respondents had tales of difficulties with materials. While these problems are common in any building project, adding innovative materials is more likely to cause problems and delays. One respondent commented that challenges made them more aware of how innovative they were being. During a tour of a "demonstration" building, a guide showed several innovative building systems. He explained one feature that modeled thermal comfort in the building. When I expressed interest, the respondent said, "it's great stuff but we haven't done anything with it yet." Later in the tour, as we looked at another building system that was malfunctioning, the guide indicated that they had managed to get the system to work a few months earlier when a guest was visiting. There were challenges to getting most innovative systems to work consistently and most organizations seemed reflective about the trade-offs and their role as early adopters.

Respondents also proudly showed press coverage of their building, which often portrayed the organization as a leader. Most organizations welcomed this attention and incorporated the press about their building into their marketing strategy. Yet, there was a downside. Respondents commented that they had a higher workload due to newfound interest. A facilities manager for a government building said: "On the administrative side, as far as operating a building, it's created in many ways more work . . . now we're interacting with the public at the front door. We are giving tours, which is something in our previous location and the other district office, no one's coming to their door, knocking and asking for a tour of their facility . . . how do we prepare ourselves to actually have someone that can conduct tours?" Without a department designated for the new task of developing and leading tours, public relations offices struggled with the attention. Tour guides in these places were only partly familiar with the building systems and some remarked informally that they felt unprepared for the role. These firms struggled to respond to unintended consequences of public attention.

There are trade-offs with the use of the innovator narrative. On the one hand, having an innovative building makes the organization appear to be ahead of the curve and more visible through press and demand for tours. On the other hand, newer building systems and increased interest could create challenges. An organization's willingness to take on the cost of early adoption meant that they understood that they were taking on the "liability of newness" (Stinchcombe 1959). This narrative can be beneficial for governmental agencies and for-profit firms that are trying to show themselves as leaders in the field and as trying something different.

Hidden Green

Thus far, this chapter has detailed four narratives used by early adopters of green buildings to explain to stakeholders why they have built green: profit, practicality, deep ecology, and innovation. Yet some organizations carefully articulated what they were *not* doing with their buildings even as they showed off the green attributes. To understand this better, I looked at the counter-narrative or the way in which these organizations reacted to existing narratives through silence (Nairn, Munro, and Smith 2005) or the creation of a new narrative that conflicts with the dominant narrative (Lipsitz 1988). During a tour with a for-profit organization, the tour guide seemed proud to say that while some buildings "look green" their building did not. The respondent showed off furniture that was made from recycled material and told us "you couldn't tell" it was recycled. In contrast to the lengths other organizations took to ensure visitors knew the material or furniture was recycled, this organization specifically did not make this claim.

This counter-narrative produced a sense that the "greenness" of a building could be hidden and while the greenness was there, it was not put front and center. This was particularly true for corporate respondents who found green aesthetics unprofessional. In contrast to those who used the innovation rationale, one respondent said: "We don't want something that is experimental. We don't want to be guinea pigs for this. We want a building that's going to work. And the perception early on was that in order to get these higher levels of certification, you have to do crazy things. . . . So they didn't do anything crazy and yet they found, okay we were able to make a Platinum building out of something that is just a normal building. It's not a Birkenstock, you know. It's a normal developer type building."

Explaining that the building is not a "Birkenstock" building but rather a "normal" building, the speaker distinguishes their project from a building that is not appealing as a business setting. Green building advocates clearly understood this concern and attempted to address the negative image of sustainable buildings. A presentation from a U.S. Green Building Council in the early years showed a picture of a mud hut with the caption "green buildings do not have to look like this."

The counter-narrative of green as a hidden attribute combats the stereotypical drawback of a green building as something that is weird or experimental. In this way, the building itself becomes a part of the counter-narrative simply by not looking green and perhaps even looking mainstream or standard. The building addresses sustainability by LEED's definition but it looks appropriately professional. This is particularly important for corporations that may be ambivalent about being seen as adding on features (Lutzenhiser et al 2001). By making greenness less apparent, but still present if one looks closely, organizations can both articulate a need for green and not make that greenness central to the building's narrative.

Another reason for the counter-narrative is a direct reaction to the narrative of describing green buildings as healthier and more productive environments. In informal discussions, respondents reported it was difficult to describe one building as healthier because non-green buildings owned by that organization are by extension less healthy. Respondents were reluctant to address this overtly, however one consultant said:

> I think some clients are a little leery. I've heard before where publicly traded companies might be—there's a perception of—I think it's just changing somewhat—but if there's a perception in the market that LEED costs an inordinate amount [of] extra money, maybe your shareholders would be [saying] "Wait a minute. Your business is to be making whatever . . . not to be spending money on your head office". . . . If you own an existing portfolio of buildings, and you make a big deal about "Gee whiz, this brand new one is healthy and we have non-toxic finishes and daylight and fresh air," what are you implying about the other buildings? Are they unhealthy? Are they bad spaces to breathe? . . . But I could see why people may want to still do the right thing but be careful about how they describe that.

For large organizations with multiple buildings, sustainable construction had the capacity to create legal liability with repercussions (Galbraith 2009). Theoretically, workers compensation lawsuits could arise requiring justification for not updating older buildings to the same standard. In particular, health care organizations were fearful of such lawsuits as their business includes creating a healthy environment for patients.

The hidden green counter-narrative gives a context for the tour at the start of this chapter with the tour guide who wanted to "tell the story better." In a sense, the building itself tells the story that the building looks normal and "not green." The counter-narrative may silently portray sustainability as feasible and normal while responding to critiques that being too green gives the wrong image.

Conclusion

These five narratives are not mutually exclusive and organizations combine these to form their own composite narrative. However, the rationales are an attempt to respond to stakeholder concerns about building green and to explain the value of sustainability. For each rationale, the organization relates building attributes to the ideologies of the stakeholders by way of a sustainability narrative that most suits the task. Table 12.1 contains a summary of the relationships between narratives and organizational sector. While all rationales are available to all types of organizations, some narratives are more likely to be seen in certain sectors and are indicated as primary versus secondary rationales. Corporations or firms are more likely to employ the profit motive and to seek out ways in which their bottom line could be positively affected by the choice to build green. But in addition, for-profit firms would also engage in the legitimating rationales of green as a credential and arguments for practicality. Looking at a green building as a credential appeals to public relations needs and helps to make a "soft" case for green within a profitable enterprise. Particularly as an early adopter, these organizations will achieve positive attention in the press and secondarily use the green innovator narrative.

Table 12.1

Proposed relationship between green building narratives and sector

	Profit-driven green	Practical green	Green innovator	Deep green	Hidden green
Corporations	Primary	Primary	Secondary		Primary
Nonprofits	Secondary	Secondary	Primary	Primary	
Education		Primary	Secondary	Secondary	
Government		Primary	Primary		Secondary
Religious				Primary	

Nonprofits are more likely to have stakeholders who ask for sustainable ideals to be implemented in the organization's choice of building styles. For some nonprofit stakeholders, there may be a spiritual or deep ecological component. Other nonprofits are among the innovators, particularly nonprofits for whom the idea of innovating and being on the cutting edge of sustainability is related to their mission. Secondarily nonprofits may use the profit motive and the practical narratives to make a case for sustainability.

Educational organizations such as college campuses and K–12 schools use the practicality argument primarily responding to the research on better student learning in green schools. They may also employ deep green or innovator narratives to make their claims, depending on the background of the educational mission in relationship to sustainable practices.

Governmental organizations focus first on practicality, making the claim that they are using taxpayer dollars in a responsible and practical manner. But these organizations also present their buildings as being the early innovators where they are willing to try out new building systems. Secondarily, they may advocate for their buildings to be hidden green as they try not to call attention to spending what may be perceived of as excessive spending on the building. Religious organizations will use deep green rationales encourage visitors to think about their relationship with the earth.

There are important implications for future research from this project. Organizational narratives can and should be seen as not only a part of the organization, but also as a part of the built environment. By understanding the narratives told about sustainability and seeing their coding in buildings, we can identify different paths by which organizations choose to enact greenness. This allows institutional actors to better situate and frame arguments to support sustainable growth. Further, exploring how organizations already work with different narratives can show places where the dominant "green" narrative about saving the planet will be convincing for some organizations and not others. When not, the practicality and cost savings are vital narratives to consider.

This project opens a path toward future research that looks more closely at the way in which building attributes are able to articulate these narrative tropes. Future research could build on this work to explore how organizations that are not early adopters articulate their responses. More work can be done to strengthen the connection between organizational narratives and the building-as-narrative through the coding of building attributes and further understanding of the match between sustainable building projects and organizational missions.

Notes

This research has been supported by the National Science Foundation (#SBR-0727273) and the Marshall Foundation. I would like to thank Joseph Galaskiewicz, Don Grant, Charles Ragin, and the editors of this volume who gave generous editorial feedback on earlier drafts. In addition, I am grateful to all respondents who gave their time for this research.

1. Some individuals discussed multiple buildings and some buildings had multiple interview subjects.

2. The number of buildings increased as I collected data. I focused on early buildings but some later buildings were discussed in interviews.

3. The majority of buildings were certified under LEED-NC, though two were certified under LEED for Commercial Interiors (CI).

4. Several respondents felt strongly that others should know how their buildings were constructed and refused anonymity. I will note those cases unless there is a conflict that compromises the confidentiality of another source. I generalize both individuals' roles as well as the organizational details even when details might further develop the argument. For those who might engage in research with this population in the future, anonymity does not seem to be a necessity.

5. I spoke with Sister Patricia Mulpeters, Sister Rosemary McKean, Sister Joan Riordan, Sister Rosemary Campi, Sister Jacqueline Graham, and Cynthia Blain.

References

Abbaszadeh, S., L. Zagreus, D. Lehrer, and C. Huizenga. 2006. Occupant satisfaction with indoor environmental quality in green buildings. *Proceedings of Healthy Buildings* 3:365–370.

Anderson, Ray. 1998. *Mid-course correction: Toward a sustainable enterprise: The Interface model*. White River Junction, VT: Chelsea Green.

Atkin, Tony, and Carol Herselle Krinsky. 1996. Cultural identity in modern Native American architecture: A case study. *Journal of Architectural Education* 49 (4): 237–245.

Benford, Robert D., and David A. Snow. 2000. Framing processes and social movements: An overview and assessment. *Annual Review of Sociology* 26:611–639.

Boje, David. 2008. *Storytelling organizations*. Thousand Oaks, CA: Sage.

Brown, Denise Scott. 1981. With people in mind. *Journal of Architectural Education* 35 (1): 43–45.

Chase, John. 1991. The role of consumerism in American architecture. *Journal of Architectural Education* 44 (4): 211–224.

Czarniawska, Barbara. 1997. *Narrating the organization: Dramas of institutional identity*. Chicago: University Of Chicago Press.

DiMaggio, Paul J., and H. K. Anheier. 1990. The sociology of nonprofit organizations and sectors. *Annual Review of Sociology* 16 (1): 137–159.

DiMaggio, Paul J., and Walter W. Powell. 1983. The iron cage revisited: Institutional isomorphism and collective rationality in organizational fields. *American Sociological Review* 48 (2): 147–160.

Duckles, Beth M. 2009. The green building industry in California: From ideals to buildings. PhD diss., University of Arizona.

Eichholtz, Piet, Nils Kok, and John M. Quigley. 2009a. Why do companies rent green? Real property and corporate social responsibility. Energy Policy and Economics Working Paper EPE-024, Energy Institute, University of California, Berkeley. www.ucei.berkeley.edu/PDF/EPE_024.pdf (accessed January 29, 2011).

Eichholtz, Piet, Nils Kok, and John M. Quigley. 2009b. Doing well by doing good? Green office buildings. Center for the Study of Energy Markets Working Paper CSEM WP-192, Energy Institute, University of California, Berkeley. http://www.escholarship.org/uc/item/4bf4j0gw (accessed April 12, 2010). Also available at *American Economic Review* 100 (2010): 2494–2511.

Faulconbridge, James R. 2009. The regulation of design in global architecture firms: Embedding and emplacing buildings. *Urban Studies* 46 (12): 2537–2554.

Feldman, Martha S., Kaj Skoldberg, Ruth N. Brown, and Debra Horner. 2004. Making sense of stories: A rhetorical approach to narrative analysis. *Journal of Public Administration: Research and Theory* 14 (2): 147–170.

Franzosi, Roberto. 1994. From words to numbers: A set theory framework for the collection, organization, and analysis of narrative data. *Sociological Methodology* 24:105–136.

Franzosi, Roberto. 1998. Narrative analysis—or why (and how) sociologists should be interested in narrative. *Annual Review of Sociology* 24:517–554.

Frumkin, Peter, and Joseph Galaskiewicz. 2004. Institutional isomorphism and public sector organizations. *Journal of Public Administration: Research and Theory* 14 (3): 283.

Galbraith, Kate. 2009. The legal risks of building green. *New York Times Green Inc. Blog*, May 29. http://green.blogs.nytimes.com/2009/05/29/the-legal-risks-of-building-green/ (accessed January 2, 2013).

Gieryn, Thomas F. 2000. A space for place in sociology. *Annual Review of Sociology* 26:463–496.

Gieryn, Thomas F. 2002. What buildings do. *Theory and Society* 31 (1): 35–74.

Goss, Jon. 1988. The built environment and social theory: Towards an architectural geography. *Professional Geographer* 40 (4): 392–403.

Hansmann, Henry B. 1980. The role of nonprofit enterprise. *Yale Law Journal* 89 (5): 835–902.

Hawken, Paul. 1993. *The ecology of commerce: A declaration of sustainability*. New York: Harper Collins.

Hawken, Paul, Amory Lovins, and L. Hunter Lovins. 2000. *Natural capitalism: Creating the next industrial revolution*. Boston: Little, Brown and Co.

Heerwagen, Judith H. 2000. Green buildings, organizational success and occupant productivity. *Building Research and Information* 28 (5): 353–367.

Heschong, Lisa. 2003a. Daylight and retail sales. California Energy Commission Report P500-03-082-A-5. http://www.energy.ca.gov/2003publications/CEC-500-2003-082/CEC-500-2003-082-A-05.PDF (accessed October 10, 2008).

Heschong, Lisa. 2003b. Windows and classrooms: A study of student performance and the indoor environment. California Energy Commission Report P500-03-082-A-7. http://www.energy.ca.gov/2003publications/CEC-500-2003-082/CEC-500-2003-082-A-07.PDF (accessed October 10, 2008).

Heschong, Lisa. 2003c. Windows and offices: A study of office worker performance and the indoor environment. California Energy Commission Report P500-03-082-A-10. http://www.energy.ca.gov/2003publications/CEC-500-2003-082/CEC-500-2003-082-A-10.PDF (accessed October 10, 2008).

Heschong, Lisa, Roger L. Wright, and Stacia Okura. 2002. Daylighting impacts on retail sales performance. *Journal of the Illuminating Engineering Society* 31 (2): 21–25.

Hoffman, Andrew J., and Rebecca Henn. 2008. Overcoming the social and psychological barriers to green building. *Organization & Environment* 21 (4): 390–419.

Hoffman, Andrew J., and Marc Ventresca. 1999. The institutional framing of policy debates: Economics versus the environment. *American Behavioral Scientist* 42 (8): 1368–1392.

Jones, Paul. 2009. Putting architecture in its social place: A cultural political economy of architecture. *Urban Studies* 46 (12): 2519–2536.

Kamin, Blair. 2001. *Why architecture matters: Lessons from Chicago*. Chicago: University of Chicago Press.

Kats, Gregory. 2006. Greening America's schools: Costs and benefits. Capital E. http://www.leed.us/ShowFile.aspx?DocumentID=2908 (accessed October 10, 2008).

Kingsley, Benjamin S. 2008. Making it easy to be green: Using impact fees to encourage green building. *NYU Law Review* 83:532–567.

Kraftl, Peter. 2006. Building an idea: The material construction of an ideal childhood. *Transactions of the Institute of British Geographers* 31 (4): 488–504.

Lipsitz, George. 1988. Mardi Gras Indians: Carnival and counter-narrative in black New Orleans. *Cultural Critique*, no. 10: 99–121.

Lounsbury, Michael, Heather Geraci, and Ronit Waismel-Manor. 2002. Policy discourse, logics and practice standards: Centralizing the solid-waste management field. In *Organizations, policy, and the natural environment: Institutional and strategic perspectives*, ed. Andrew J. Hoffman and Marc Ventresca, 327–342. Stanford, CA: Stanford Business Books.

Lutzenhiser, Loren, Nicole W. Biggart, Richard Kunkle, Thomas Beamish, and Thomas Burr. 2001. Market structure and energy efficiency: The case of new commercial buildings. California Institute for Energy Efficiency Report. http://uc-ciee.org/additional-documents/8/423/121/nested (accessed January 2, 2013).

Maines, David R. 1993. Narrative's moment and sociology's phenomena: Toward a narrative sociology. *Sociological Quarterly* 34 (1): 17–38.

McDonough, William, and Michael Braungart. 2002. *Cradle to cradle: Remaking the way we make things*. New York: North Point Press.

Miller, Norm G., Dave Pogue, Quiana D. Gough, and Susan M. Davis. 2009. Green buildings and productivity. *Journal of Sustainable Real Estate* 1 (1): 65–89.

Naess, Arne, and George Sessions. 1995. Platform principles of the deep ecology movement. In *The deep ecology movement: An introductory anthology*, ed. Alan Drengson and Yuichi Inoue, 49–63. Berkeley, CA: North Atlantic Books.

Nairn, Karen, Jenny Munro, and Anne B. Smith. 2005. A counter-narrative of a "failed" interview. *Qualitative Research* 5 (2): 221–244.

National Research Council. 2006. Green schools: Attributes for health and learning. Washington DC: National Academies Press. http://www.nap.edu/openbook.php?record_id=11756 (accessed January 2, 2013).

Paul, Warren L., and Peter A. Taylor. 2008. A comparison of occupant comfort and satisfaction between a green building and a conventional building. *Building and Environment* 43 (11): 1858–1870.

Polletta, Francesca. 1998. Contending stories: Narrative in social movements. *Qualitative Sociology* 21 (4): 419–446.

Polletta, Francesca. 2006. *It was like a fever: Storytelling in protest and politics*. Chicago: University of Chicago Press.

Polletta, Francesca, and J. M. Jasper. 2001. Collective identity and social movements. *Annual Review of Sociology* 27 (1): 283–305.

Rakatansky, Mark. 1995. Identity and the discourse of politics in contemporary architecture. *Assemblage* 27:9–18. doi:10.2307/3171424.

Rao, Hayagreeva. 2009. *Market rebels: How activists make or break radical innovations*. Princeton, NJ: Princeton University Press.

Rapoport, Amos. 1987. On the cultural responsiveness of architecture. *Journal of Architectural Education* 41 (1): 10–15. doi:10.2307/1424903.

Reed, John Shelton. 1989. On narrative and sociology. *Social Forces* 68 (1): 1–14.

Rothschild, Joyce, and Carl Milofsky. 2006. The centrality of values, passions, and ethics in the nonprofit sector. *Nonprofit Management & Leadership* 17 (2): 137–143.

Scott, W. Richard. 2007. *Institutions and organizations: Ideas and interests*. Thousand Oaks, CA: Sage.

Stevenson, William B., and Danna N. Greenberg. 1998. The formal analysis of narratives of organizational change. *Journal of Management* 24 (6): 741–762.

Stinchcombe, Arthur L. 1959. Bureaucratic and craft administration of production: A comparative study. *Administrative Science Quarterly* 4 (2): 168–187.

Troast, John G., Andrew J. Hoffman, Hannah C. Riley, and Max H. Bazerman. 2002. Institutions as barriers and enablers to negotiated agreements: Institutional entrepreneurship and the Plum Creek habitat conservation plan. In *Organizations, policy, and the natural environment: Institutional and strategic perspectives*, ed. Marc Ventresca and Andrew J. Hoffman, 235–261. Stanford, CA: Stanford Business Books.

Weisbrod, Burton A. 1998. Institutional form and organizational behavior. In *Private action and the public good*, ed. Walter W. Powell and Elisabeth S. Clemens, 69–84. New Haven: Yale University Press.

Yudelson, Jerry. 2007. *The green building revolution*. Washington, DC: Island Press.

13

Challenging the Imperative to Build: The Case of a Controversial Bridge at a World Heritage Site

Olivier Berthod

Introduction

Certification schemes—most notably the LEED (Leadership in Energy and Environmental Design) certification—dominate the debate on environmental sustainability in construction projects. This debate, however, concerns itself with construction projects whose necessity has already been demonstrated. In other words, LEED asks the question "How do we build green?" and ignores the question "Do we *really* have to build in the first place?" Donovan Rypkema, a dedicated advocate of heritage preservation and sustainable restoration, goes so far as to translate LEED into "Lunatic Environmentalists Enthusiastically Demolishing." Joking aside, this provocative statement alludes to a crucial issue in the nascent debate on construction and environmental sustainability: how do we determine whether building is necessary at all?

The problem behind this question is that every construction project provides a different answer at different points in time. This question is a process and it keeps on evolving, for it tackles a paradox that may concern any type of construction project. When we decide to build, this is based on the identification of specific needs (for housing, for new infrastructures, etc.). Based on this information, we decide to sacrifice portions of our natural environment, or not. From then on, greening the construction is nothing but a way of apologizing to the nature by minimizing our impact. But how are we sure that the need we identified in the first place will remain as solid as the concrete solutions that we proposed to build?

The rapidity with which our social, economic, institutional, and natural environments change constantly challenges the imperative to build. The decisions behind a construction project engages various actors and organizations in a project that can extend over a very long period of time until the artifact (a building, a road) reaches completion. And yet the problem that had motivated the decision to build can take a different turn than initially expected, and render irrelevant the rationales behind the decision to build in the first place. The imperative to build is thus never a certainty. Instead, it is typically an assumed solution to a spatially related problem. The decision to build often relies on the capacity of some "issue sellers" to gather a critical mass of public support and to sustain it. Interestingly, this form of coordination is more often driven by effective communication and persuasive storytelling than by technical necessity (Throgmorton 1992; Suchman 2000; Sandercock 2003). Therefore—and quite unsurprisingly—some have gone so far as to argue that concerns for sustainability in urban projects, however prominent they may be in official discussion, are often supplanted by matters of prestige, politics, special interests, and power behind the curtain (Flyvbjerg 1998). This perspective proposes a more fragmented and conflicting view of the imperative to build, and suggests a much messier understanding of sustainable urban development. Sustainability, in this interpretation, takes on the elusiveness of an ever-evolving issue and is subject to conflicts over priorities, definitions of collective interest, and visions about what an appropriate future is supposed to look like (in this vein, see Hoffman 1999 on corporate environmentalism; Buttel 2003 on environmental reform; or Lounsbury, Ventresca, and Hirsch 2003 on waste management). Hence we face the following two dimensions. On the one hand, we decide to build with little certainty about the very relevance of the project in the long run. We bet about the future—nothing more certain than that that. And on the other hand, we engage in processes of communication; we sell projects like issues and try to define collectively a certain idea of sustainability for our cities and our relation with the surrounding natural environment. I suggest that these two dimensions may collide. It is precisely this tension that I propose to illustrate in this chapter.

I present a case study to examine the ways in which actors *determined*, *stabilized*, and *defended* the imperative to build a four-lane bridge across a wide meadow within a (now former) UNESCO World Heritage Site. To do so, I will analyze the public debates that took place surrounding the activities of planning and construction. The case depicts the collision of competing views of collective interest and sustainable urban evolution. In a first phase, the city administration identified a traffic problem. It then discussed different conceptions of a desirable future in terms of traffic and urbanism. This resulted in the domination of one technical solution (the so-called Waldschlößchenbridge—henceforth: WSB) and a division between two movements: the *supporters* and the *opponents*. Over the years, however, the initial traffic problem took a surprising turn, challenging the very necessity to build. Eventually, in 2003, the city successfully applied to the UNESCO-World Heritage Site accolade. These elements challenged the very necessity to build that informed the technical solution proposed. Consequently, the supporters entered a conflict with various constituencies and sources of institutional influence, not least with the bodies of the UNESCO Convention for the Protection of the World Heritage. Nonetheless, they managed to defend their solution all the same, thus pushing the city into building a solution that is in line with its original motivation but decoupled from current needs. For the purpose of this case study, I make use of a narrative-based analysis and shed light on how supporters and opponents of the project spoke about the project and related planning activities. This chapter first offers a brief discussion on technical rationality and the imperative to build, and then presents the case study. In the conclusion, I relate the case study to the questions I just raised.

Shedding Ambiguity on the Imperative to Build

A construction project is as much a matter of bricks and concrete as it is of human and organizational coordination (Weick 2004) (see chapter 3, this volume). While technical rationality alone can help define what dimensions a wood truss must take if it is to carry the weight of a roof, the very decision to build depends more often on what decision makers are convinced about, or on what vision they want to convey for their city, than on technical facts and data (Flyvbjerg 2009).

In fact, sociological approaches have long advocated a departure from theories that promote technical rationality and efficiency as drivers of decision making. In this perspective, members of a group (be it an organization, a project, a team, an industry, or a community) tend to *define* collectively what they consider to be a rational action in the face of a problem, thus creating socially what we commonly call "reality" (Berger and Luckmann 1966; Meyer and Rowan 1977). Instead of postulating people's "naïve" adherence to "hard" data and information, this line of argument puts dynamics of collective consensus making to the fore, and focuses on what we actually do with what we know and do not know (Hacking 2000). In this respect, DiMaggio and Powell (1983) write about the emergence of *collective* rationalities that may differ from one social arena (i.e., a community, an industry, etc.) to the other. For example, environmentalists are likely to defend priorities in urban development (and consider them as absolutely rational) that are much different from the priorities defended by urban planners (who also consider their priorities as rational), or by citizens who are stuck in traffic jams every morning. Decision makers usually capture the essence of these societal expectations and may decide to adopt them (or not) in order to increase their apparent legitimacy and thus secure their access to support or resources or both (Meyer and Rowan 1977; Oliver 1991).

Typically, the abstract notion of "sustainability" faces such a dilemma. Deeply imprinted by the Brundtland report on sustainable development (World Commission on Environment and Development 1987), concerns for sustainability in research and practice focus on the influence of today's decisions on future generations. This abstract definition further suggests the defense of a "common future" (ibid., 3) and the willingness "to meet the needs of the present without compromising the ability of future generations to meet their own needs" (ibid., 24). On examining construction projects, however, we find little consensus about what "our" common future should look like. In fact, every project represents one more arena for stakeholders from different constituencies to debate the vision of this common future. In this respect, sustainability is nothing but a vague idea that changes from case to case, depending on how people mobilize and how they discuss potential futures. Hence the imperative to build will depend largely on the way actors define a desirable urban evolution for their habitat, and on their capacity to do so convincingly. Such debates often resemble democratic *tours de force* within interdependent webs of resources (e.g., local development funds, public opinion, electorates).

Different approaches and analytical devices have been put forth to unpack this pluralism. In this chapter, I observe actors' communication efforts as they perform actions and make decisions (Morrill and Owen-Smith 2002; Zilber 2009). In this view, planning is seen as a storytelling activity by which actors more or less persuasively talk about past problems, present solutions, and potential futures (Throgmorton 1992; Sandercock 2003). For matters of cohesion, I use the word "narrative" to refer to the dominant themes underlying communication over time (see chapters 10 and 12, this volume). The evolution of the overall narrative contributes to the dissemination of a shared mental model within the community. In other words, talking about the project and its planning is seen as the carrier of a cultural influence that is used to justify action and further impart a sense of rationality to what is being done (Phillips, Lawrence, and Hardy 2004).

Narratives provide both the collective *and* the rational dimensions that are central to our discussion. The collective aspect arises from the fact that the dominant narrative is made of multi-authored stories and of elements that are understandable to most of the audience concerned. Rationality is provided by the very nature of narratives: the seemingly logical flow of events, starting with a beginning and ending with certain results, and making sense of events while constructing logical links between them. Narratives convey continuity and traditions and may thus become a source of inertia and resistance to change in groups and organizations (Geiger and Antonacopoulou 2009), for example by coordinating actors' opinions on specific issues (Boyce 1995). The way in which stories are understood shapes mental models in actors' minds (Bower and Morrow 1990), allowing dominant narratives to rise above subversive stories (Ewick and Silbey 1995), and can thus contribute to making one definition of the imperative to build drown out other critical voices.

The Dresden Case Study

This chapter reports on a qualitative case study of the controversial construction of a four-lane bridge in the German city of Dresden. Dresden is the capital city of Saxony, a sovereign state of the Federal Republic of Germany, at the borders with Poland and the Czech Republic. The city is located on both sides of the Elbe River and stretches in a narrow strip along the water. WSB is the name of a new bridge over the Elbe, 2.5 km east of the city center, which was still under construction in 2012. The project started in the early 1990s, as planners in the city administration delivered a new traffic plan to the city council.[1] The city was in the process of renewing its infrastructures as a result of the German reunification. The WSB figured in the plan as one project among many others. Different actors within the city administration and in the city council took positions for the WSB or for other options. This fostered the constitution of the two camps: the *supporters* and the *opponents* (two categories that eventually gained a very strong political dimension in the city). The opponents especially criticized the location picked for the WSB (the largest meadow in town, down a hill, Waldschlösschen was a vast and popular recreational area). They advocated building somewhere else (i.e., *not to build* at Waldschlösschen) and sparing the area. Nonetheless, the city council, led by a majority of councilors in favor of the WSB (center-right wing), voted in favor of the project in 1996. The city administration planned the bridge from 1994 to 2004 (construction to finish in late 2012). As hinted upon in the introduction to this chapter, the facts that motivated the 1996 decision changed over time. First, around 1999 traffic started decreasing—and has been doing so, smoothly but steadily, since then. Second, in 2003 the city applied successfully for the UNESCO World Heritage Site accolade, and therewith committed itself to a new set of official priorities in terms of urbanism and construction projects, as defined by the operational guidelines of the World Heritage Center.[2] Little by little, the opposition grew even larger. Numerous opponents—including former employees of the city administration, members of the city council, local environmentalists, all later joined by the UNESCO World Heritage Committee and national politicians and famous architects and intellectuals—started challenging the very necessity of the bridge and suggested abandoning the project (i.e., *not to build at all*) or building a tunnel instead.[3] Nonetheless, the supporters of the WSB prevailed, and the project went forward. The case burst into the German press as the World Heritage Committee threatened the withdrawal of Dresden from its seminal list. The committee did so in 2009.

For the purpose of my investigation, I collected official documents, studies, and other textual material (press releases, content from the websites of related administrations and social movements, feasibility studies and reports on various aspects of the project, legal documents, official decisions, internal documents and memos), covering the years 1994 to 2009. This body of data encompasses more than six thousand pages of documents. I also collected visual data such as maps and 3D visualizations and pictures, and I conducted seventeen interviews with program specialists from the World Heritage Center, local politicians and environmentalists, and professionals from diverse fields who were involved to some degree in the construction project. This collection of data further included informative material on World Heritage policies and site management as well as local political issues and the traffic situation in Dresden, and information and observations gleaned from repeated visits to the construction site. The aim of this chapter is to show how actors determined, stabilized, and defended the imperative to build. For this purpose, I retrieved from this database all documents that were issued for public use by the local administrations and political actors involved in the construction project, in which events and decisions were told and retold and interpreted.

I analyzed this material to detect the ways in which actors involved in the project "produce[d] structures, constraints, and opportunities that were not there before they took action" (Weick 1988, 306). Defending a certain conviction implies postulating the superiority of the chosen course of action over other alternatives. Looking for implicit assertions allows for more consideration of the meaning between the lines and for nuances in arguments. This type of analysis seeks "to surface and analyze dualities embedded in the text" (Brown 2004, 98). I followed this reasoning in order to systematically examine how major narrative themes emerged from the argumentation and to observe how one main understanding of the imperative to build arose and took precedence over time.

The Identification of a Problem and the Emergence of Two Competing Views

The project idea emerged in the early 1990s as part of a broader traffic study, as urban planners in the city administration prepared a new traffic plan for the city. Shortly after German reunification, Dresden faced a growing volume of traffic and needed to cope with different traffic flows. The plan proposed renovating several older bridges, diverting some of the main thoroughfares to improve the traffic situation in the city center, and constructing a new bridge in the eastern part of the city in order to lead traffic toward the northern industry area. At this point in time, the WSB was one option among many. It would soon become one of the most polarizing issues in Dresden.

The city first identified a specific problem that made necessary the decision to build—whether green or not was then a secondary question. As suggested in the introduction to this chapter, in the early days of the project, no one would have thought to challenge the very imperative to build a new bridge. In point of fact, the traffic situation was clearly problematic. Instead, city administration and city council members were debating the questions of *where* and *what* was necessary to build. Debates on locations and technical solutions revealed two major solutions, each based on a different philosophy and supported by two different sets of actors:

1. The mayor, together with planners and other members of the city administration—mostly longtime employees—and a political majority center-right in the city council, saw in the WSB the best solution. Even though this group evolved over the years in its membership basis, I refer to this set of actors as the "supporters." A basic postulate behind their work was the promotion of Waldschlößchen as the optimal area for construction.

2. The members of the team for strategic urban development (also part of the city administration) defended the construction of three smaller bridges (with two lanes instead of four), scattered from east to west, with a focus on sparing the Waldschlößchen area from any construction incursion. As we will see, this constituency grew much bigger eventually, joined by a majority center-left, various environmental associations, German

intellectuals, and, later, the World Heritage Committee. I refer to this set of actors as the "opponents." A basic postulate behind their work was the refusal to build at Waldschlößchen.

Supporters and opponents developed two competing narratives about the necessity of the project, the relevance of their solutions, and a specific opinion about sacrificing the Waldschlösschen area. These narratives evolved over the years along a total of three phases, as summarized in table 13.1. In phase 1, we will see how supporters and opponents debated the options with respect to where to build and what to build until the city council voted the official start of the project. In phase 2, we will see how the supporters stabilized the decision taken and made it legitimate in the eyes of the community. In phase 3, we will see how supporters and opponents confronted each other when the very rationale behind the project (i.e., the traffic argument) started to gain in prominence and when the project endangered Dresden's World Heritage status.

Table 13.1

Main arguments of competing green building narratives

	Phase 1 Determining (1990–1996)	Phase 2 Stabilizing (1996–2004)	Phase 3 Defending (2004–2009)
Supporters' narrative	Traffic forecasts, traffic bundling, environmental sacrifice, tradition	Traffic forecasts, tradition, conformity with environmental regulations, subventions	Tradition, direct democracy, localness
Opponents' narrative	Short-term flexibility, traffic distribution, environmentally friendly design, technical progress as a form of ideology	Weak traffic impact, environmental impact, environmentally unfriendly design, financial priorities	Compromise, human rights (UNESCO), global

Phase 1: The Competition of Potential Futures

This chapter examines the ways in which actors *determined*, *stabilized*, and *defended* the imperative to build a four-lane bridge across a wide meadow in Dresden. In the first phase (starting in the early 1990s until 1996), diverse actors worked to define the original imperative to build. In this respect, the major camps pleaded for two different futures for the city and its relation with its natural and urban environment.

The arguments about "not to build" in phase 1 were about not building at Waldschlößchen. Again, at this point in time, most stakeholders were still convinced by the project rationale (i.e., the traffic issue). The question of whether to build or not to build was a *geographical* one. The supporters bypassed this opposition. They claimed that the merging of vehicles over one large road would represent a more efficient way to process the increasing traffic volumes. Waldschlösschen was indeed the best option to do so. Furthermore, the supporters saw in the WSB a way to link residential areas in the south with industrial areas in the north. Needless to say, they recognized that such a bridge would have to cross through large recreational areas (which had been under protection through zoning since 1908). In their view, however, it was preferable to build big at one critical location and hence to spare other areas.

The opponents, on the contrary, basically refused building at this location. To promote their refusal, they defended a decentralized distribution of traffic instead. In their opinion, this strategy would contribute to support more lines of public transportation (and thus diminish the prominence of single-occupant auto traffic in the city) and spare the Waldschlößchen area. Also, opponents claimed, constructing more bridges would entail diachronic construction efforts (i.e., building one bridge after the other), thus rendering the city administration more flexible in its decision making with respect to potential changes in traffic volumes. Finally, opponents felt that their alternative locations and smaller-scale designs (i.e., two travel lanes instead of four) would offer a friendlier integration of the projects into the natural and urban landscapes.

While the opponents often presented the multi-bridge solution as synonymous with environmentally friendly design and sustainable public transportation (they would often refer to Scandinavian car-free cities), the supporters of the WSB answered more pragmatically, insisting on the dire traffic situation in projected in the official forecasts and highlighting the urgency to act. Supporters did not eliminate the issue of preservation of the natural environment. Instead, they presented their project as a sacrifice of nature deemed necessary for the good of the citizens. The supporters of the WSB further mobilized narrative elements of local history; bridges had been proposed on numerous occasions in the past (from the early 1900s to the 1980s), in more or less the same location as the WSB, and had been abandoned for various reasons such as costs or war efforts. The supporters made large use of this narrative element, rooted in the notion of tradition, to buttress the suggested relevance of their favored solution. A typical claim was: "In studies in the 20s, the 30s, the 60s and the 80s, a traverse at WS was always considered as the best option" (a city councilor and supporter of the WSB, in a declaration to the local press, 1996).

In response, the WSB opponents who urged the multi-bridge concept pursued a narrative rooted in critical thinking and progressivism. They pleaded for the need to rethink these old concepts, to develop a long-term vision for traffic issues, and to save on construction costs. This argument, however, proved rather ineffective. The stories told by the WSB supporters (and especially the narrative of tradition) helped the citizens make sense of the WSB as the dominant solution to traffic problems. This can be observed in citizens' statements to the local press:

> I am very pleased with the solution with the two bridges. But I also know that the planners have been favoring the WSB since the 30s (A resident, 1996)

> We taxi drivers don't understand this discussion. The WSB was projected and considered necessary as long ago as the 30s. When will they finally build? (Spokesman, local union of taxi drivers, 1996)

Although the supporters based their justifications for the WSB option on the familiarity of the solution, studies later showed (in 2006) that previous planning in the city's history had no imprinting effect on the street network and that the WSB could not be considered as a missing link. Nonetheless, this narrative, rooted in the history of the bridge, took on a decisive role in May 1996, as the city administration organized a "bridge-workshop" to debate the options and elect an official course of action for Dresden. During this workshop, it became clear to all participants that the state would help finance the solution with the largest potential for societal acceptance and traffic resolution, independent of any strategic vision. Hence the supporters began putting other aspects to the fore, namely: the potential for a streetcar line over the WSB (despite the absence of cost/use data), the familiarity of the WSB compared to other new ideas, and the perceived (and widely communicated) increase in traffic. Based on these aspects, the city administration reached a majority in favor of the narrative told by the WSB supporters. As a result, the city council officially voted for the WSB project.

Phase 2: Stabilizing the Imperative to Build

A second phase (starting in 1996 and lasting until 2004) took over. During this period the supporters stabilized the imperative to build in the form of a collective consensus, rendering the solution legitimate to most stakeholders. From 1996 to 2004, the city council voted on subsequent phases of technical implementation. The city administration, (as mandated by the council) began developing a concept that would address all of the requirements necessary to apply for plan appraisal and comply with construction laws.

It became necessary for the supporters of the WSB to neutralize opposition to the project and stabilize the citizens' expectations. At first, they did so by continually evoking the aforementioned arguments about traffic and the bridge tradition at Waldschlösschen and undermining the importance of nature. Project supporters maintained the notion of the traffic increase as their main narrative argument. They stressed the project's necessity with the help of traffic forecasts, and communicated widely the financial subventions that state authorities had promised as indications of WSB's technical relevance. These two elements served as explanations in their pressing for a quick start of the project:

> We must decide at last and stop the endless discussions, or else we'll have only beautiful plans and no bridge. . . . If I must find a balance between the impact on residential districts and on nature, I will always be on the people's side. (Herbert Wagner, mayor of Dresden, in a statement to the local press, 1996)

> In this time [the nineteenth century], the idea of an external ring was born . . . along the path of the current Fetscher Street—Stauffenberg Avenue. . . . As yet, this bridge has never been realized. . . . The bridge construction shall become the task of the century for Dresden. (Planners from the city administration, in a document directed at potential architects, 1997)

As the project went forward, the city administration (mandated by the vote of the city council) and supporters began making use of more technical and informed arguments in telling the story of the WSB, stressing its compliance with the local environmental regulations. For example, on February 10, 2000, the city communicated about the construction plans via a press release and provided the community with details on forthcoming renovations and compensatory measures, depicting (among other things) the future construction of new spaces for public gardens, rehabilitation of the land surrounding the infrastructure, and the creation of a protected island for local flora. However, on July 14, 2000, the regional directorate (the local agency in charge of examining plans and granting permission to begin construction) rejected the plans due to shortcomings with respect to noise protection and the construction of soundproofing walls. On February 18, 2003, the city submitted new plans (with new connections to the street network), in which it retold the story of the increasing traffic trend and set forth environmental compensatory measures. This time it used new forecasts that simulated data for up to the year 2015. On February 26, 2004, the regional directorate granted permission to have the plan appraised.

Needless to say, during this same phase, the WSB opponents (back then: members of the city administration mostly involved in the multi-bridge solution, but also local activists and center-left wing politicians) continued to discredit the WSB. In their eyes, the location of the bridge remained critical. The opponents relied heavily on the problems concerning the bridge's potential negative impact on its environment. While they still recognized the imperative to build (as related to traffic congestion), they kept asking questions about where it should be built (proposing other locations closer to Dresden center) and what form it should take (e.g., a smaller bridge or a tunnel). Over these phase 2 years, project opponents began a critical comparison of changing traffic perspectives with the technical dimensions of the WSB, initially planned as a regional road. The questions of "what to build" and "why build" started colliding in the supporters' discussions, thus raising questions about the very necessity to build and sacrifice the Waldschlösschen area. Furthermore, in order to include the bridge in the landscape, urban planners needed to make it particularly low. This resulted in costly solutions to integrate the bridge into the road network along the shore. Hence the opponents attacked the costs of the project (today estimated at over 180 millions euros, most likely more) as well as its factual necessity by challenging the traffic volumes expected in the forecasts compared with the dimensions of the solution proposed at Waldschlösschen, and by pointing out the potential damage to a particularly large recreational area. Politicians opposed to the WSB used these elements to challenge spending public money on the WSB project and started allocating it differently.

In direct relation to this latter argument from opponents (and particularly interesting for our discussion), the public debate over the WSB plan escalated in the years 2004 and 2005. In 2004, Dresden citizens elected a new city council. Up until this point, the political parties from the center-right wing had a majority that actively supported the WSB. After the election, the council switched in favor of a center-left coalition, which had made a tradition of opposing the project. Quickly, the debates in the council became a source of frustration for all parties. The new coalition tried to unbind the financial resources dedicated to the bridge in order to allocate them for other municipal problems (e.g., rehabilitation of kindergartens, street renovations). In response, the supporters initiated a referendum, asking the citizens whether or not they would like to see the WSB built.

Supporters and opponents (both represented by political parties and various social movements) began a harsh communication campaign, both in print and online. The political party at the far-left even issued a leaflet resembling a newspaper dedicated to the bridge, in which it reiterated its classical critiques, namely that the WSB would not solve traffic issues, that the WSB was too expensive, and that it would destroy protected meadows. Eventually the city administration distributed an official brochure to residents' households in February 2005. In its more than twenty pages, supporters and opponents compared their concept with the other camp's vision for Dresden's future, thus revealing the system of meaning that had emerged over the years. In this brochure, the supporters put forth a final plea for the WSB project. Their text addressed the negative expectations about the project and aimed at reassuring the citizens. Back then, the traffic had already started to decrease. Concealing this, the supporters made clear that most individual interests were being addressed in the bridge project, reminded them that environmental compensation measures would balance the negative impact of the construction on the meadow, and reassured them that the project would have little impact on the municipal budget (thanks to the subventions). In the brochure, the supporters further linked the necessity for a bridge with what we could interpret as liberal ideas by stressing the importance of an urban policy that would grant equal rights to car drivers and users of public transportation without restrictions: "Mobility is decisive for the freedom to choose one's social life and an important contribution to social integration. . . . The importance of all means of transportation, particularly of the car, for the daily organization and planning of life of numerous people must be accepted."

The opponents took the exact counterpoint. They argued more pragmatically and with less symbolism. They deplored the irreversible impact of the project on the meadow, invoked the weak effect of the bridge on traffic issues, criticized the confidence in forecasts, and especially stressed the financial implications of such a substantial construction project for a city with an unstable budget and weak capacities for financing other issues of social necessity. In the brochure, they stated: "A lot of citizens ask us: 'And what would be the alternative to the WSB?' We then ask: 'Why such a gigantic effort to build a bridge that does not solve our problems and brings with it a lot of new ones?' The WSB is here and now completely out of place. It is hardly useful. It damages a lot. And it is only costly."

In the end, the opponents met with little success. Almost fifteen years had passed since the discussion about this bridge had begun. Most citizens were puzzled by the never-ending debate and demonstrated their desire to see something happen: on February 27, 2005, in the referendum on the bridge project, 67. 9 percent of the voters chose the WSB over not building at all. This result bound the city to the project until February 27, 2008. From a legal point of view, with this victory in the polls, the supporters secured a total commitment of both the city administration and the city council to the project.

Phase 3: Defending the Imperative to Build

During the third and last phase (starting in 2004 and lasting until 2009), the opponents took a more aggressive turn in their challenging of the very motivation behind the project, attacking the traffic issue and the environmental impact even more directly.

As mentioned in the introduction, large infrastructure projects suffer one classical flaw: the particularly long time between the decision to build and the actual completion of the project. During this time span, many things can change, and situations that had proven problematic can evolve positively and thus cast doubt on former solutions. In Dresden (in spite of the referendum) two major changes contributed to the opponents again questioning the imperative to build at all at Waldschlößchen, thus forcing the supporters to actively defend the imperative.

The first change was a measured decline in traffic volume since 2000, as was common in many other cities in Europe. This new trend cast doubt on the forecasts on which the decision to build was based. For example, in 1999, the three bridges closest to the future WSB were transporting 58,500, 47,000 and 34,500 cars daily, respectively. Forecasts for the years 2000 to 2015 postulated an increase in traffic. Calculations for 2015 included the simulated impact of the WSB on traffic and predicted 51,500, 37,000, and 33,500 cars daily over the same bridges, respectively. In sharp contrast, real traffic calculations in 2009 (still decreasing since 2000) yielded figures that were below the ones from 1999 (in fact even closer to the figures from the early 1990s), with 45,500, 38,000, and 32,500 cars daily, respectively.

A second change came from abroad. On December 9, 2002, the city decided to apply for UNESCO World Heritage status and submitted its application in January 2003. The International Council on Monuments and Sites (the ICOMOS, commissioned by the World Heritage Center, is the NGO in charge of inspecting the cultural sites) inspected both the application and the site in 2003. The ICOMOS delegation saw the future location of the construction (still untouched in 2003) and the plans for the bridge. Eventually Dresden and the Elbe Valley entered this famous list of heritage sites on July 2, 2004. However, applying for the World Heritage label entailed numerous changes for the management of the city. Most particularly, in the case of the construction project, the city administration was required to keep the World Heritage Center informed of any developments and to seek approval before proceeding (World Heritage Center 2005, paragraph 172). In other words: the listing as a World Heritage Site did not automatically mean the acceptance of the WSB project.[4]

In the summer of 2005, representatives of social movements against the WSB sent officials to the World Heritage Center in Paris, where they expressed their concerns about the bridge, stressing both the traffic situation and the fact that the city had been planning the construction in the middle of the protected area. On November 4, 2005, the World Heritage Center sent two letters to Dresden through diplomatic channels, asking the city to take a position on the matter.

The city administration (now legally obliged to proceed with the WSB project in virtue of the referendum) increased its communication activities. First, the then mayor defended the necessity to find a solution in the form of a compromise and to avoid a conflict with the World Heritage Center. Nonetheless, this new situation influenced the two competing narratives about the bridge. The city administration issued several leaflets and brochures on the matter and distributed them widely in the community. In these documents, the administration left out the argument of forecasts about traffic increase. Instead it evoked heavily the weight of tradition and explained the imperative to build as the result of the repeated postponements of the project over the last decades. It further linked the imperative to build with the direct will of the citizens, as echoed by the referendum. Based on this, the administration suggested temporal projections, sketching potential futures, and rooting the bridge in past experiences:

> City bridges often provide a panoramic view. The space which a bridge spans in the city generally allows viewers to see and take in a larger part of the city. These panoramic views are an experience rather like a bird's eye view. For this reason, bridges are inviting places. . . .
>
> At the end of the 1880s, after visiting the Gotthard Tunnel construction site, Friedrich Nietzsche described it in almost euphoric terms as an engineering mas-

terpiece. The train trip through the completed tunnel, however, made him feel anxious. . . . A tunnel is, and will always be, a claustrophobic experience. (City administration, in a leaflet distributed to the citizens, 2006)

Finally, the administration did not propose to rethink the project. It presented it as a local affair, best debated by the citizens and their administration. Instead it pleaded for the environmental friendliness of the WSB and its role for the enhancement of the World Heritage Site:

On their own, neither the landscape nor the man-made structures are of such high value that they would qualify for the World Heritage seal of approval. Only their co-relationship qualifies them for World Heritage status. . . .

If it was not accepted that a bridge could be integrated into the cultural landscape, then its World Heritage status would not have been justified. (City administration, in a leaflet distributed to the citizens, 2006)

The World Heritage Committee, based on expert studies ordered by the World Heritage Center, issued numerous criticisms regarding the inclusion of the bridge into the site. Furthermore, on July 11, 2006, during its yearly session, the World Heritage Committee put Dresden and the Elbe Valley on the list of endangered World Heritage sites. This measure of list displacement is supposed to shed a blaming light on those sites in need of specific attention and assistance due to critical evolution in their conservation.

Meanwhile, numerous external actors (famous German architects, politicians, artists, Nobel Prize laureates, etc.) began joining the movement of the opponents, extending over the local level to include actors with national and international stature. The opponents made use of the weight of the UNESCO and of the World Heritage convention in their efforts to gain more prominence. They made clear that fighting the World Heritage program was about refusing the basic idea that World Heritage sites are shared by humankind, and that all citizens, even outside of Dresden, were now concerned with the conflict. The opponents, attempting at slowing down the project, used this state of confusion to advocate other solutions (most notably the construction of a tunnel or, as for matter of compromise, a smaller bridge). For example, Wolfgang Tiefensee, the Federal Minister of Transport, Building and Urban Affairs, wrote that his ministry might consider unfreezing supplements in financial aid for a solution that would satisfy the World Heritage Center.

Addressing this challenging turn of events, the supporters elevated their storytelling and began to link the imperative to build with the very idea of direct democracy, thus decoupling the debate from any technical and natural-environmental arguments. For example, they stressed the fact that the bridge had initially been pushed by those same persons who fought the GDR dictatorial regime in 1989. Considering the history of the city, playing on the respect due to democratic rights had enormous potential. Dresden had been under various dictatorial regimes for most of the twentieth century: the National Socialists from 1933 to 1945, the communist regime imposed by the Russian occupation until 1949, and the German Democratic Republic from 1949 to 1990. The supporters echoed implicitly this experience in their storytelling. For example, a local politician (from the party supporting the bridge project) compared the local intellectuals and environmentalists that were against the bridge with a "totalitarian elite":

> When the citizens stop bending under the dictate of their [i.e., the opponents'] pathetically uplifted forefinger, then they [i.e., the opponents] look in panic for new partners anywhere in the world. The lower elite by definition ask for help from the upper elite, in this case UNESCO. . . .
>
> When Dresden was still part of the GDR and Ulbricht [Chairman of the Council of State, 1950–1971] had the Saint Sophia's Church demolished, UNESCO showed no interest in this city. It is not until one becomes a democracy that one gains the right to be patronized by them and kept on a leash. (Local politician, supporter of the WSB, in an open letter)

The only way to democratically unfreeze the WSB project would have been a new decision by the city council with a two-thirds majority. The political parties supporting the bridge, however, refused the arrangement. The city council (still with a majority against the bridge) and the regional directorate of Dresden (in charge of monitoring legal enforcement within the city) eventually initiated a legal battle. The city made extensive use of the aforementioned democratic angle in its communication, focusing on the illegitimate character of any decision that would have postponed immediate construction. On March 9, 2007, the Higher Administrative Court declared that the city council of Dresden should indeed follow the decision to which it was bound by way of referendum. In its 2008 session, the World Heritage Committee decided to retain Dresden on the list of World Heritage sites in danger. In the following months, the construction proceeded in a straightforward manner. On June 30, 2009, the World Heritage Committee voted to withdraw the natural landscape "Dresden and the Elbe Valley" from the list of World Heritage sites, thus reducing the chances for opponents to succeed. This was the first time in Europe and the second time worldwide that the committee had to go as far in sanctioning. The bridge should be operational by late 2013.

Conclusion

In the WSB case in Dresden, we saw the messiness inherent in how we commonly define "sustainability" and our relations to the natural and urban environments. Instead of a shared and well-defined goal, we observed a struggle to define urban policies and their concrete manifestations. In so doing, we have followed the recursive interplay between what has been done (decisions, actions) and what has been told (narrative themes surrounding the project). We were able to divide this process into three phases: the determination, the stabilization, and the defense of the imperative to build.

I started this chapter by asking: How do we determine whether building is necessary at all? This question challenges the imperative to build in the first place. I suggested that the imperative to build was never a certainty since it was always subject to changes in the economic, institutional, social, and natural environments. In sharp contrast, I noted that while the facts and motivations behind construction projects may evolve, the concrete buildings and roads remain.

Challenging the imperative to build has often been something of a cliché. Indeed, this concern has long been a source of inspiration for artists and poets. Hermann Hesse (1990), observing the growing urbanization around his home, deplored his growing inability to preserve the hilled landscape of Montagnola for future generations. Charles Baudelaire (1869/2006) saw in the "enormous cities" of his time a reason for artists to strive for even more poetic prose. As a result, challenging the imperative to build is often linked to romantic nostalgia and to people's natural aversion to change. As Friedmann puts it (1998), it is easier to dismiss someone's earnest efforts by labeling them utopian.

This is where the Dresden case becomes interesting. In Dresden, challenging the imperative to build took on various forms, which provided us with a finer appreciation of such debates. In fact, we can divide this issue into four intertwined questions: *What* to build, *Where* to build, *How* to build, and *Why* build. A certification program such as LEED can help answer the first three questions. In this respect, the "why" question remains more critical, as conventional cost/use calculations do not work. In Dresden, this was so for three reasons that are typical of large-scale projects (see Flyvbjerg, Bruzelius, and Rothengatter 2003). First, project leaders (in this case, the city) tend to extrapolate levels of demand that communities most likely never experience. Second, they often submit plans with cost forecasts that are lower than in reality. Third, such projects often span long periods of time from start to completion, thus fostering rigid webs of commitments, power relations, and interests that are highly resistant to change (Miller and Hobbs 2005). Hence, the planning work on what to build, where to build, and how to build (and however "green" the solutions may be) has little relevance for the environment or for any kind of sustainable development policy as soon as the very necessity to build becomes decoupled from the actual project. Or to put it another way: why do we need to discuss efficiency if we are not even effective with what we do? Hence, one critical lesson explores how we can improve the process of project realization so that one also keeps control of the objectives of the infrastructure as well as the ability to correct decisions and trajectories.

The most obvious resistance to such highly disruptive moves can easily be tracked to institutions, such as local agencies which award construction permits, or decisions voted on by municipal councils. If we believe Berchicci and King (2007), the good news is that this same institutional context can be mobilized to improve environmental performance. In the Dresden case, we saw how opponents invoked the very idea of World Heritage—to which the city itself had initially subscribed—in order to jeopardize the project. According to Berchicci and King, the nature of such institutions as LEED or the World Heritage program is self-regulating, since actors may decide to opt in or out, and because they exert little central enforcement. The influence of the World Heritage program has been strong enough to make actors comply in many other conflicting situations (von Schorlemer 2009). And yet in Dresden, this had little effect. After all, losing a label is not the end of the world. However strong their effect may be, self-regulating institutions have a limited range of influence that stops where local interests prove stronger (e.g., electorate, local subventions, etc.) and where resource dependencies are already set up. This leads us to the paradox I mentioned in the introduction: the very question of the necessity to build often emerges after several years (as it did in Dresden, when traffic started to decrease on its own), once other institutions have long validated the "what," "where," and "how" questions. Hence we should reflect on how to unfreeze such constellations and on how to allow for more critical evaluations of the motivations behind the projects *during* the implementation of construction projects instead of one-shot evaluations, as is usually the case.

To conclude, challenging the imperative to build is not good advice for making friends. And yet such studies, however utopian and idealistic they may sound, do us a favor by repeatedly pointing out potential dysfunctionalities so that we may progress toward improvements in practice and regulations.

Acknowledgments

I am grateful to the German Research Foundation (DFG) for providing me with financial support, and to the members of the doctoral program Research on Organizational Paths at the Freie Universität Berlin for their helpful discussions. I would also like to thank the editors of this volume as well as the MIT Press reviewers for decisive feedback on earlier versions of this manuscript.

Notes

1. All planners evoked in the case are employees of the city administration. They are, however, not autonomous. The city administration acts as the executive body of the city's management structures. The mayor manages the city administration and its divisions (finances, urbanization, security, culture, economic development). Decisions, however, are made by the city council, which entails missions to the mayor and her/his administration via voted motions.

2. An independent international secretariat, the World Heritage Center coordinates the day-to-day management of the World Heritage Convention of 1972, organizes the yearly sessions of the World Heritage Committee, supervises the implementation of the decisions made by the committee, and coordinates the reporting on the sites.

3. Composed of twenty-one state parties, the World Heritage Committee establishes the list of World Heritage sites and makes decisions with regard to the implementation of the Convention of 1972.

4. This obligation is directly related with the origins of the program. The concept of World Heritage first came into the spotlight in 1959 as the construction of the Aswan High Dam across the Nile River threatened Egypt's pharaonic temples by flooding from what is now known as Lake Nasser. UNESCO triggered an international campaign to gather financial support for dismantling and rebuilding the temples at the shores of Lake Nasser. Eventually UNESCO initiated or supported a multitude of similar projects, ending in the Convention Concerning the Protection of the World Cultural and Natural Heritage, ratified in 1972 (UNESCO 1972).

References

Baudelaire, Charles. 1869/2006. *Petits poèmes en prose*. Paris: Pocket.

Berchicci, Luca, and Andrew King. 2007. Postcards from the edge: A review of the business and environment literature. *Academy of Management Annals* 1:513–547.

Berger, Peter, and Thomas Luckmann. 1966. *The social construction of reality, a treatise in the sociology of dnowledge*. New York: Penguin Social Sciences.

Bower, Gordon, and Daniel Morrow. 1990. Mental models in narrative comprehension. *Science* 247:44–48.

Boyce, Mary. 1995. Collective centring and collective sense-making in the stories and storytelling of one organization. *Organization Studies* 16 (1): 107–137.

Brown, Andrew. 2004. Authoritative sensemaking in a public inquiry report. *Organization Studies* 25 (1): 95–112.

Buttel, Frederick. 2003. Environmental sociology and the explanation of environmental reform. *Organization & Environment* 24 (2): 306–344.

DiMaggio, Paul, and Walter Powell. 1983. The iron cage revisited: Institutional isomorphism and collective rationality in organizational fields. *American Sociological Review* 48:147–160.

Ewick, Patricia, and Susan Silbey. 1995. Subversive stories and hegemonic tales: Toward a sociology of narrative. *Law & Society Review* 29 (2): 197–226.

Flyvbjerg, Bent. 1998. *Rationality and power: Democracy in practice*. Chicago: University of Chicago Press.

Flyvbjerg, Bent. 2009. Survival of the unfittest: Why the worst infrastructure gets built—and what can we do about it. *Oxford Review of Economic Policy* 25 (3): 344–367.

Flybjerg, Bent, Nils Bruzelius, and Werner Rothengatter. 2003. *Megaprojects and risk—An anatomy of ambition*. Cambridge, UK: Cambridge University Press.

Friedmann, John. 1998. Planning theory revisited. *European Planning Studies* 6 (3): 245–253.

Geiger, Daniel, and Elena Antonacopoulou. 2009. Narratives and organizational dynamics: Exploring blind spots and organizational inertia. *Journal of Applied Behavioral Science* 45 (3): 411–436.

Hacking, Ian. 2000. *The social construction of what?* Cambridge, MA: Harvard University Press.

Hesse, Herman. 1990. *Mit der Reife Wird Man Immer Jünger—Betrachtung und Gedichte über das Alter*. Frankfurt: Insel Verlag.

Hoffman, Andrew. 1999. Institutional evolution and change: Environmentalism and the U.S. chemical industry. *Academy of Management Review* 42 (4): 351–371.

Lounsbury, Michael, Marc Ventresca, and Paul Hirsch. 2003. Social movements, field frames and industry emergence: A cultural-political perspective on US recycling. *Socio-economic Review* 1 (1): 71–104.

Meyer, John, and Brian Rowan. 1977. Institutionalized organizations: Formal structure as myth and ceremony. *American Journal of Sociology* 83:340–363.

Miller, Roger, and Brian Hobbs. 2005. Governance regimes for large complex projects. *Project Management Journal* 36 (3): 42–50.

Morrill, Calvin, and Jason Owen-Smith. 2002. The emergence of environmental conflict resolution: Subversive stories and the construction of collective action frames and organizational fields. In *Organizations, policy, and the natural environment. Institutional and strategic perspectives*, ed. Andrew Hoffman and Marc Ventresca, 90–118. Stanford, CA: Stanford University Press.

Oliver, Christine. 1991. Strategic responses to institutional processes. *Academy of Management Review* 16:145–179.

Phillips, Nelson, Tom Lawrence, and Cynthia Hardy. 2004. Discourse and institutions. *Academy of Management Review* 29 (4): 635–652.

Sandercock, Leonie. 2003. Out of the closet: The importance of stories and storytelling in planning practice. *Planning Theory & Practice* 4 (1): 11–28.

Suchman, Lucy. 2000. Organizing alignment: A case of bridge-building. *Organization* 7 (2): 311–327.

Throgmorton, James. 1992. Planning as persuasive storytelling about the future: Negotiating an electric power rate settlement in Illinois. *Journal of Planning Education and Research* 12:17–31.

UNESCO. 1972. Convention Concerning the Protection of the World Cultural and Natural Heritage, English text. 17th Session of the General Conference, November 16, Paris.

von Schorlemer, Sabine. 2009. Compliance with the UNESCO World Heritage Convention: Reflections on the Elbe valley and the Dresden Waldschlösschen bridge. *Jahrbuch fur Internationales Recht. German Yearbook of International Law* 51:321–390.

Weick, Karl. 1988. Enacted sensemaking in crisis situations. *Journal of Management Studies* 25 (4): 305–317.

Weick, Karl. 2004. Rethinking organizational design. In *Managing as designing*, ed. Richard Boland and Fred Collopy, 36–53. Stanford, CA: Stanford University Press.

World Commission on Environment and Development. 1987. Report of the WCDE: Our common future. United Nations, General Assembly, 14th Session, 8–19 June, Nairobi.

World Heritage Center. 2005. Operational guidelines for the implementation of the World Heritage Convention. Paris, FR: World Heritage Center. http://whc.unesco.org/en/guidelines/ (accessed January 7, 2013).

Zilber, Tamar. 2009. Institutional maintenance as narrative acts. In *Institutional work: Actors and agency in institutional studies of organizations*, ed. Tom Lawrence, Roy Suddaby, and Bernard Leca, 205–235. Cambridge, UK: Cambridge University Press.

14

Incorporating Biophilic Design through Living Walls: The Decision-Making Process

Clayton Bartczak, Brian Dunbar, and Lenora Bohren

Introduction

Over the past several thousand years people have constructed increasingly complex structures to shield themselves from the outside world. However, in recent times this shielding seems to have come at a price to their physical and psychological well-being. In many instances, buildings trap high concentrations of contaminants, imprisoning occupants in an atmosphere of toxins that often make them sick (Hansen and Burroughs 1999; Hines et al. 1993; Wolverton 1996). Some indoor environments also detach occupants from the world's natural elements by confining them to uninspiring indoor spaces, proven detrimental to psychological health (Heerwagen and Hase 2001; Joye 2007; Lohr, Pearson-Mims, and Goodwin 1996; Pearson 1989; Seignot 2000; Wener and Carmalt 2006; Wolverton 1996).

Incorporating natural elements in the built environment—a strategy referred to as biophilic design—promotes both physical and psychological health (Darlington et al. 2000; Heerwagen and Hase 2001; Joye 2007; Liu et al. 2007; Lohr, Pearson-Mims, and Goodwin 1996; Orwell et al. 2004; Pearson 1989; Seignot 2000; Wener and Carmalt 2006; Wolverton 1996; Wood et al. 2006; Yoo, Kwon, and Son 2006). Historians believe that the Chinese used plants as interior decoration as long as three thousand years ago, and archeologists confirm evidence of interior plant use from at least two thousand years ago in the perfectly preserved ruins of Pompeii, Italy (Manaker 1987).

Technology's sword cuts a wide beneficial swath with biophilic design (see chapter 16, this volume). Although the benefits of biophilic design are well documented, the absence of innovative green building technologies—such as living walls (see figure 14.1)—in most buildings lends itself to investigation. Why is there not widespread adoption of living walls and

Figure 14.1

A living wall. University of Guelph-Humber, Toronto, Canada.
Photo courtesy of Nedlaw Living Walls, Inc.

other biophilic technologies? What factors weight the decision to incorporate biophilic design elements—specifically living walls—into a building? Awareness of the decision-making process to incorporate biophilic design elements into buildings provides valuable insight for green technology advocates, property owners, building designers, and architects.

Psychological Benefits of Plants and Biophilic Design

The humanistic social-psychologist Eric Fromm coined the term "biophilia" and defined it simply as humans' "love of life" (Fromm 1964, 45). Edward O. Wilson later popularized the biophilia concept in 1984 in his innovative book by the same name. Wilson defines biophilia as "the innate [human] tendency to focus on life and lifelike processes" and describes it as an inherent love of nature and need to associate with other life forms and natural processes (1984, 1). Wilson suggests that a living organism is infinitely more interesting than any imaginable type of inanimate matter. As an example, Wilson states that "no one in his right mind looks at a pile of dead leaves in preference to the tree from which they fell" (84). Simply put, our connection with living things is part of what makes us human.

Wilson argues that because the natural world has had such a tremendous impact on human evolution, we have an ingrained need to associate with other living things and experience beautiful places. Furthermore, we are psychologically fulfilled and physically healthier when this happens. Biophilia suggests that the satisfaction we experience in beautiful places is embedded in the neural framework of the human brain. Moreover, our concept of beauty evolved from our ancestors' need to find places that provided food, shelter, and security, and consequently improved their overall well-being.

To some extent, contemporary building designers respond to biophilic tendencies. However, proponents of biophilic architecture argue that virtually all areas of our built environment should incorporate our inherent need to connect with nature and living things (Heerwagen and Hase 2001; Joye 2007; Wilson 1984). Results of a recent study conducted with occupants of a high-rise building lend credence to this assertion. The tenants' primary concern was the lack of living vegetation in the building and their sense of disconnection from the outside. Occupants explicitly requested that the designers integrate "outside" features by additional use of plants and natural lighting (Wener and Carmalt 2006).

However, even if we agree about human enjoyment of natural settings (or those that mimic nature), does this provide sufficient leverage to justify the added line item to the budget? Does popular opinion warrant the additional resources required to incorporate these elements into the built environment? Randolph Hester answered these questions best in his book *Design for Ecological Democracy* by saying "when a person exclaims that being in nature makes her feel healthy, it is typically a purely emotional response, but increasingly it is backed by scientific evidence" (2006, 303). Emotions may spark the desire but solid reasoning must justify the need. Indeed, numerous studies have demonstrated the various physical and psychological benefits of buildings that incorporate natural features and plant life (Darlington et al. 2000; Heerwagen and Hase 2001; Joye 2007; Liu et al. 2007; Lohr, Pearson-Mims, and Goodwin 1996; Orwell et al. 2004; Pearson 1989; Seignot 2000; Wener and Carmalt 2006; Wolverton 1996; Wood et al. 2006; Yoo, Kwon, and Son 2006).

Worker Productivity, Stress Reduction, and Student Performance

In homes, offices, schools, and hospitals, studies indicate the positive psychological benefits to occupants of buildings with biophilic-inspired elements. Workers in offices with views of nature tend to feel less frustrated and more patient. They report higher levels of overall satisfaction and well-being, and are—not surprisingly—more productive (Heerwagen and Hase 2001). Even plants in portable containers within the workspace decrease stress levels and increase productivity. Specifically, in 1996 Lohr, Pearson-Mims, and Goodwin conducted a study in which they asked participants to perform a computer task either in a room with several potted plants (incidentally, not in direct line of sight but within the participant's peripheral vision) or in a room without plants (Lohr, Pearson-Mims, and Goodwin 1996). While performing the task, participants' blood pressure rose, signaling increased stress levels, whereas, interestingly, participants in the plant-adorned room experienced a significantly smaller rise in blood pressure and a faster return to normal than those in the plant-less room.

In 1991, Ulrich and colleagues conducted a similar study which showed stress "recovery was faster and more complete when subjects were exposed to natural rather than urban environments" (Ulrich et al. 1991, 201). Considering that a large portion of the world's population live in urban environments and have relatively brief encounters with nature (e.g., observing trees from office windows and interior potted plants in the workplace; having lunch in the park; passing by a living wall; or viewing a green roof), this finding is significant since it implies that even these brief encounters can be beneficial to our psychological well-being.

In 2005 researchers conducted a study measuring the connection between exposure to photographs of natural settings and the ability to remain attentive to a task. In this study, after participants completed a task that caused mental fatigue, researchers exposed them to images of a natural setting or an urban environment. Participants shown the natural settings consistently regained a notable degree of attention capacity enabling them to perform well on successive tests (Berto 2005).

A more recent set of studies from Japan (Li 2010; Park et al. 2010) demonstrates significant positive health results from "forest bathing" and human exposure to nature. While the authors studied subjects in parks and forests, they suggest that even short exposures to places with the characteristics of nature (plants and trees with soil, humidity, scents, moving leaves, variety of colors and shapes) provide discernible physical and mental health benefits.

In addition to tangible biophilic design elements such as living walls and other plant installations, natural light (also known as daylighting) is another example of biophilia in the built environment. Kellert, Heerwagen, and Mador concisely explain the connection between biophilia and daylighting in their book *Biophilic Design*: "Another carryover of evolution is that modern humans are psychologically and biologically attuned to light and changing cycles of light and darkness. Daylight and sunshine enabled visual surveillance of surroundings, finding food and water, locating and pursuing game, and avoiding threats such as predators that would be concealed in darkness" (2008, 90–91).

Studies show that students in schools with enhanced levels of daylighting and views of nature are happier, able to concentrate better, retain information more readily, and score better on standardized tests (Lovins 2004). Similarly, in an impressive nine-year study, researchers demonstrated that hospital patients in rooms with a view of nature exhibit reduced stress levels, shorter hospital stays, require less pain medication, have fewer postsurgical problems, and generally recover better than patients without a view (Joye 2007; Wener and Carmalt 2006; Ulrich 1984). Another study demonstrated a similar association among the presence of plants and flowers and the length of surgical patients' hospital stays, decreased need for analgesics (pain medication), more positive feelings, and higher overall satisfaction with their hospital rooms (Park and Mattson 2009).

Human Behavior

Exposure to natural elements even has a beneficial effect on aggressive behavior. Increasing the amount of and access to vegetative areas in public housing developments decreases violent behavior and crime rates (Wener and Carmalt 2006). Similarly, in a study of public housing, Kuo and Sullivan (1996) found overall rates of violence and of women hitting children were lower among those women living in buildings that had trees included in the landscape design than those without.

Overall Environmental Impact

The potential to indirectly decrease the negative impact people have on the natural environment is yet another psychological benefit of including natural elements in built environments. Biophilia proponents argue that when people are more connected to living things and natural processes, they recognize their interdependence with nature and will correspondingly respect it by trying to limit their environmental impact. Pyle states that "if people were closely acquainted with the plants and animals they live among and depend upon, the eco-crimes they permit and abet and conduct and sponsor daily would not be so easy to bear" (2003, 211; Joye 2007). In fact, a study conducted on outdoor enthusiasts implies a strong association between a person's level of participation in outdoor activities and a heightened environmental concern and corresponding "pro-environmental" behavior (Teisl and O'Brian 2003). Likewise, other studies demonstrate that experiences of nature are powerful predictors of behaviors associated with our desire to protect nature (Finger 1994; Kals, Schumacher, and Montada 1999; Langeheine and Lehmann 1986). Whether we learn to love nature through outdoor adventures or conversely, actively engage in outdoor activities due to our love of nature is not immediately clear, but an undeniable association exists between the two.

Physical Effects of Poor Indoor Air Quality

Detrimental as the psychological impact of poorly designed indoor environments can be, the physical effects can pose even greater hazards to the health of those living and working within their walls. Dr. B. C. Wolverton reports "indoor air pollution is now considered by many experts to be one of the major threats to human health." In support of Wolverton's assertion, indoor air is often ten times more contaminated than outdoor air (1996, 7; Hines et al. 1993). Furthermore, the United States Environmental Protection Agency (EPA) has ranked indoor air pollution as one of the top five threats to public health (Wolverton 1996).

The energy crisis of the 1970s spurred construction of airtight buildings in an effort to limit energy loss. But unbeknownst to building designers at that time, airtight buildings often create more problems than they solve. Subsequent research demonstrated that ventilated buildings that allow some air-conditioned or heated air to escape—and fresh air to enter—support a healthier indoor environment. Otherwise, dangerous pollutants trapped inside can spawn health problems such as Sick Building Syndrome, a term used to describe a condition that occupants of poorly ventilated spaces may develop and can include symptoms such as dizziness, nausea, headaches, eye irritation, and fatigue, among others (Hansen and Burroughs 1999; Hines et al. 1993; Wolverton 1996).

The wide variety of indoor air pollutants include (1) volatile organic compounds (VOCs), often found in aerosols, paints, furnishings, finishes, clothing, paper products, cleansers, and electronic appliances; (2) combustion gasses that result from incomplete combustion of indoor fuel equipment; (3) bioeffluents, the product of normal biological processes in which people and animals produce odors, carbon dioxide, moisture, and microbes; (4) cigarette smoke; and (5) soil gases such as radon that come from the ground underneath buildings (Hansen and Burroughs 1999; Hines et al. 1993; Seignot 2000; Wolverton 1996).

Thankfully, designers' informed choices can reduce the use of materials that "off-gas" carcinogenic chemical pollutants. However, pollutants such as VOCs (volatile organic compounds) are hard to avoid because of their prevalence in so many products used by consumers (Darlington et al. 2000; Seignot 2001; Wolverton 1996). Formaldehyde, one of the most common VOCs, is a conventional component of paper products, pressed fiber board, adhesives, sealants, paints and other finishes, fabrics, insulation, and clothing (Seignot 2001).

Another VOC, benzene, a by-product of cigarette smoke, glue, paint, furniture wax, and detergent, often contaminates the workplace. Fortunately, numerous studies demonstrate plants' ability to remove benzene from indoor air. One study found that plants were effective in benzene mitigation both in the presence and absence of light. Even unused potting soil has some benzene removal capabilities, though not as much as soil with plants (Orwell et al. 2004). For those seeking a plant with superb benzene mitigation properties, a 2006 study of seventy-three plant species, found *Dracaena deremensis* cv. Variegata was the most effective (Liu et al. 2007). Yet another study demonstrated that "potted-plants can provide an efficient, self-regulating, low-cost, sustainable, bioremediation system for indoor air pollution, which can effectively complement engineering measures to reduce indoor air pollution, and hence improve human well-being and productivity" (Wood et al. 2006, 163).

Proven, documented, and now widely known, various indoor air contaminants can have a profound negative impact on occupant health. William Fisk, in charge of the Indoor Environment Department at Lawrence Berkeley National Laboratory, suggested that each year in the United States alone, improved air quality in the workplace could lead to as many as sixteen to thirty-seven million fewer cases of the common cold and the flu, creating a savings of $6 to $14 billion from lost time and productivity (Wener and Carmalt 2006).

Furthermore, there is a direct correlation between indoor environmental quality and worker absenteeism, illness, and productivity. Because worker salaries account for 35–40 percent of many companies' budgets, even small investments to improve employee health, productivity, or retention have an enormous impact on overall profitability (Hansen and Burroughs 1999). Could something as simple as incorporating plants into the workplace environment improve indoor air quality enough to increase worker productivity and decrease absenteeism due to sickness? Studies indicate a resounding yes. In fact, a 2007 study proved that indoor plant use was a significant factor related to increasing worker productivity and decreasing sick leave (Bringslimark, Hartig, and Patil 2007).

Physical Benefits of Plants

Without a doubt, the use of plants in the built environment not only can provide a much needed connection to natural elements in the built environment, but also can help mitigate some of the problems associated with poor-quality indoor air. Blending function and aesthetics, plants in the built environment can positively impact the physical health of a building's occupants. Although complex, the photosynthetic process wherein plants absorb carbon dioxide and release oxygen furnishes sufficient reason for their installation in living and working areas (Wolverton 1996). In addition, plants help regulate humidity and temperature via transpiration, a process by which plants draw moisture through the soil and expel it as water vapor (Pearson 1989; Seignot 2000; Wolverton 1996). Lohr and Pearson-Mims demonstrated in a 1996 study that plants can effectively reduce the accumulation of particulates such as dust, dirt, soot, or smoke on horizontal surfaces. Since particulate matter can become airborne and often contains contaminants that cause health and comfort issues, plants essentially clean the indoor air.

NASA (National Aeronautics and Space Administration) investigated how plants improve indoor air quality in an attempt to find solutions to life support systems in space. NASA developed the BioHome, a small living space with high concentrations of various indoor pollutants. When researchers introduced plants into the BioHome, tests showed a significant reduction in chemical concentrations. The study "concluded it was possible to remove as much as 87 percent of indoor air pollutants within a matter of hours" (Seignot 2001, 108). Of further interest, the BioHome study revealed that many pollutant mitigation processes take place below the soil level, thereby demonstrating that a number of the benefits from plants are due to microbes living in and around their root systems (ibid.).

There are, however, conflicting opinions on the efficacy of plants to filter and purify indoor air. In a study published in 1989, Godish and Guindon determined it is not necessarily the plants themselves that are able to remove harmful contaminants from indoor air as had been previously concluded (Godish and Guindon 1989). While their study also showed a reduction in airborne contaminants (specifically formaldehyde) by the introduction of potted plants, they were also able to show that a greater contaminant reduction occurred when the plants were defoliated. Their studies helped advance the understanding that plants alone do not filter and purify indoor air, but rather that their entire mini-ecosystem of plant, soil, and soil microorganisms combine to create a functional air purification process.

Current Installations of Living Walls

To create the benefits of biophilia just described, a number of companies have recently developed systems to more fully integrate plants into the built environment. Whether through interior plant installations, outdoor plant installations, or interior living walls, the focus remains constant: to improve air quality and to bring attractive natural elements into a built environment. Living walls are typically large-scale installations consisting of a variety of vegetation and are permanent features of buildings' interior spaces. In contrast, although potted plants can provide ample vegetation, they are more easily removed or relocated than a building's living-wall installation.

One of the most successful and well-known living-wall installations is located at the University of Guelph-Humber in Toronto, Canada. Designed by Alan Darlington, this four-story bio-wall is visible from nearly every occupied space in the building. Darlington's company, Nedlaw Living Walls (previously known as Air Quality Solutions) focuses on designing and installing large-scale "bio-filter walls" that not only contribute to interior aesthetics, but directly integrate into the building's mechanical system as a means to help filter and purify air. The quantity of plant life in the bio-wall is sufficient to supply all the building's fresh air intake needs with the added benefit of cooling the building during the summer and providing humidity in the winter (Pliska 2005). Since maintenance of indoor air quality typically introduces fresh outside air, this system reduces energy consumption and associated costs by decreasing the amount of outdoor air that subsequently needs to be heated or cooled.

Nedlaw Living Walls' unique product is a hybrid of two well-known remediation technologies: biofiltration (the natural decontamination processes of beneficial microbes) and phyto-remediation (the use of plants to assist in the remediation of contaminated air and soils). Spanning North America from Nova Scotia in the east to California in the west, there are currently approximately one hundred of these bio-filter-type walls in use today.

Toronto-based Genetron Systems, founded by Dr. Wolfgang Amelung, is another industry leader in living-wall installation. Genetron's mission statement exemplifies the company's concern for improving human connections with nature: "To be the possibility for human transformation of humanity through harmony with nature" (Genetron Systems 2006). Genetron created the unique "Breathing Wall Ecosystem" which not only filters and purifies indoor air, but is as close to a natural self-sufficient ecosystem as technology currently allows. These breathing wall ecosystems, designed with specific plants, animals, and water, create what Genetron terms a "nutrient desert." A system with very little extra nutrients, a nutrient desert is unlike most human-made aquatic environments where algal growth is often a problem. One of Genetron's most impressive installations, the breathing wall ecosystem installed at the University of Toronto, includes a large variety of tropical plants, several species of fish, snails, and electric grow lights (ibid). Genetron has designed and installed approximately fifteen to twenty breathing wall ecosystems, most if not all in Canada.

Unfortunately, these two companies still occupy a niche in the green building market. The questions remain: Why is there not more widespread adoption of a technology with such varied benefits? What factors are involved in the decisions to incorporate biophilic elements—specifically, living walls—in buildings? To answer these questions, we conducted a study to examine the factors that influenced the decisions to install living walls in several buildings.

Studying the Living-Wall Decision Process

We selected a group of fifteen participants of varying demographics, prequalified for their exposure to work environments enhanced by living walls. Our goal was to determine what factors, if any, have influenced the decision to include living walls in a building's design and how those factors rank in terms of importance. We requested feedback from the study participants on several subjective factors such as initial cost and maintenance, as well as factors related to the physical and psychological benefits of exposure to living walls, and we collected objective information such as the living-wall locations and the building types where the living walls were installed.

Factors that Influence the Living-Wall Decision Process

We presented thirteen factors that may have influenced the decision to invest in a living-wall installation to each participant and asked them to indicate whether each factor was *Not important, Somewhat important,* or *Very important*. These factors were developed through discussions with living-wall professionals to ensure all relevant factors were included. A chart detailing the respondents' answers is seen in figure 14.2.

In most cases, we conducted interviews with people who had been involved in past living wall projects. We explicitly told the participants that we sought to understand the various factors that influenced their decisions to install a living wall as opposed to learning about factors or opinions they may have developed after their living-wall installation.

Economic Factors

One of the categories of questions we asked interviewees relates to the various economic factors that may have influenced their decision to install a living wall. Although not linked to the aforementioned psychological or physical benefits of living walls, we determined that understanding the economic aspects of the living-wall decision was critical since building industry decisions are often heavily influenced by economics. What

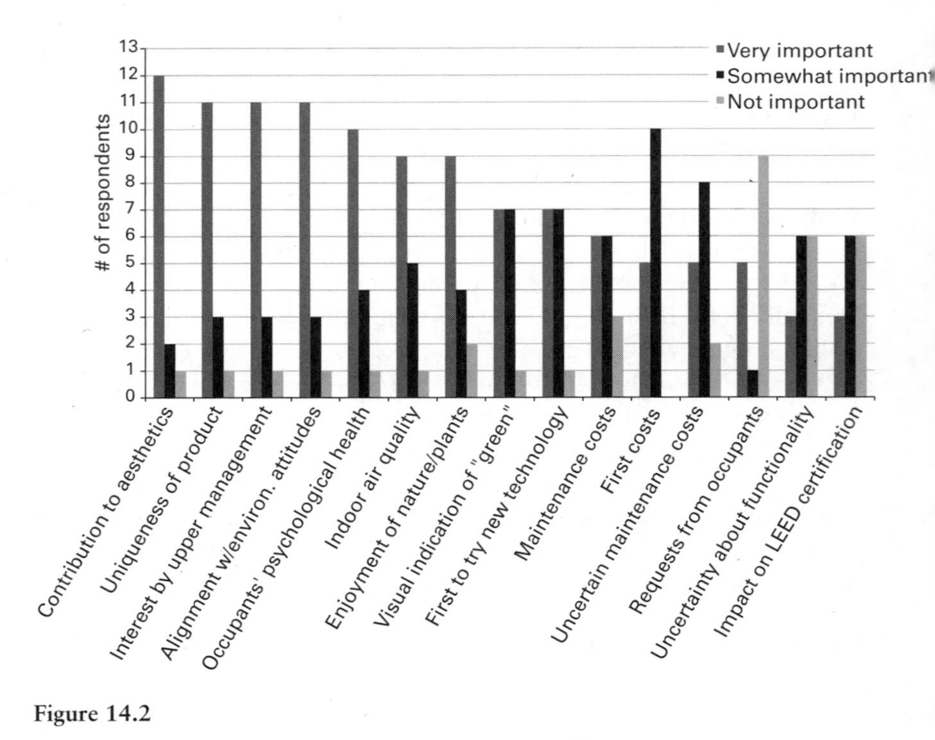

Figure 14.2

Factors influencing the decision to install a living wall

factors influence the decision when considering a living-wall installation? Is initial cost most important? What about maintenance costs? Does being viewed as a green building or green company warrant added expense?

We asked our study participants to rank the following economic factors in terms of importance: (1) first costs of a living wall, (2) maintenance costs, (3) uncertainty about maintenance costs, and (4) a visual way to recognize the company/building as green.

Although not easily quantifiable, recognizing a building as green plays a vital role in marketing—therefore, a contributing economic factor. As evidence, one respondent stated her organization's use of its living wall as a green marketing tool was of key importance to their decision to invest resources in its installation (for more on narrative frames, see chapter 12, this volume). Another respondent stated their facility's living wall "provides a good marketing opportunity."

Three of the four economic factors listed above had considerable importance to the respondents with the exception that "uncertainty about maintenance costs" was not considered as important. In fact, two interviewees responded with *Not important*. While all respondents ranked "First costs" as *important* in varying degrees, only one thought that the visual impact of recognizing a company or building as green was *Not important*. To be fair, although visual cues that a building is green are not imperative to marketing, they can provide a clear indicator to occupants and visitors who may be otherwise ignorant of the building's green features. That said, the green marketing factor was of secondary importance to initial cost.

Unsolicited comments about economic factors were all positive toward living walls. For example, one person observed that the capital cost of a living wall was small when compared to total construction cost. Furthermore, this respondent pointed out that the capital cost of their living wall was equal to less than one tenth of one percent of the project's total construction cost.

One participant offered that maintenance costs are not much more than the recurring cost to paint the building. Another interviewee stated, "the living wall is definitely worth the maintenance cost." Although this participant was not directly responsible for the building's maintenance budget, as vice provost academic of the university where the living wall was installed, he was privy to information regarding the building's operational budget and could therefore make an informed comment.

Another economic issue that surfaced relates to the impact of a living wall on a building's operational costs. Four of the sixteen study participants stated that because a living wall can filter air, therefore decreasing ventilation requirements, operation costs could be viewed as an energy-cost saver. Since we did not specifically request this information from the interview participants, we did not include it in the figures of the study's data.

Psychological and Physical Benefits of Living Walls

Indicative of the value placed on improved air quality, aesthetics, and psychological health, the benefits of living walls (as perceived by interview participants) ranked higher in overall importance than economic factors (see figure 14.2). We asked participants to ascribe value to four benefits: (1) improvement to indoor air quality, (2) enjoyment of nature/plants, (3) contribution to building aesthetics, and (4) impact on occupants' psychological health.

Figure 14.2 illustrates that the benefits of living walls were of considerable importance to all respondents. Since the interview participants all made decisions to build living walls it was not surprising that when compared to the four economic factors, the benefits were more important than the economic factors. In fact, for the questions about the benefits of living walls, eight of the respondents described all four factors as *Very important* and only one respondent described the factor "Enjoyment of nature/plants" as *Not important*—this being the only factor about the benefits that was described in this way.

Novelty of Living Walls

The uniqueness or novelty of living walls was presented as a plausible factor in decision making. We asked participants to rank the importance of two factors: (1) uniqueness of the product, and (2) desire to be one of the first companies to try this new technology.

Results, as shown in figure 14.2, indicate the "Uniqueness of the product" factor was most influential for respondents. Only "Contribution to building aesthetics" exceeded the number of respondents who said it was *Very important*.

The second question relating to the novelty of living walls addressed how important it was to be "One of the first companies to try this new technology." One respondent described this factor as *Not important* and all others described it as either *Somewhat important* or *Very important*. This suggests that although being one of the first companies to try this new technology was not nearly as important as the overall uniqueness of the product, it was still at least somewhat important for fourteen of the interviewees.

Several interview participants provided unsolicited comments regarding the novelty of living walls that highlighted the ability of living walls to affect corporate image and visitor attraction. For example, one respondent stated that he wanted a living wall so people would recognize the organization and its building as unique. Another said her business wanted a living wall partially because it would attest to how progressive and different her company was compared to its competitors. Another person observed that since his organization facilitates entrepreneurial activities, the living-wall installation was seen as a way to identify the organization as progressive and offering fresh, new ideas for its client businesses. Other respondents stated that because of the living wall, "visitors immediately see the building as different" and that because it is distinctive, "people come to visit our living wall all the time."

An intriguing and unexpected factor that may have influenced decision makers also relates to the novelty of living walls. This factor is what we are calling the "Event center factor." Five of the interviewees made unsolicited comments about planning the location of the living wall (in many cases a large communal space such as a lobby, foyer, or atrium) for use as a community event space. We did not anticipate this factor. Thus, since five interviewees volunteered this information as an influential factor in the decision to install a living wall, it was at least somewhat important. These comments suggest the spaces where living walls are installed have a remarkable capability to encourage social interaction and attract visitors.

Link to Environmental Attitudes

We asked interview respondents to describe the importance of the living walls' "Alignment with the organization's environmental attitudes." This suggests respondents saw an association between installing living walls in their buildings and reducing the building's environmental impact. As we mentioned in the analysis of the economic factors, four of the interviewees stated that because a living wall can filter air, the building's ventilation can be reduced and consequently the energy needed to condition outdoor air is decreased (though this may be subject to interpretation of local building codes). If true, this can be interpreted as an environmental factor as well because a reduction in a building's energy use will most likely lead to a reduction in fossil fuel consumption and associated greenhouse gas emissions.

Interest Shown by Upper Management

A similar factor that affected almost all respondents' decision to install a living wall was the interest level that upper management showed in the proposal to include a living wall in the construction project. This relates to the "Alignment with organization's environmental attitudes" factor because in many cases, if an organization's upper management has environmental inclinations, the organization as a whole may as well. Eleven respondents described this factor as *Very important*, three as *Somewhat important*, and only one as *Not important*. One respondent stated that during the cost analysis of the construction project the project team did not fully understand the value of the living wall, and had it not been for upper management's strong interest, the living wall would have been removed from the building plans.

Influence on Green Building Certification

Of the fifteen respondents, six described "Influence on LEED certification" or another green building certification program as *Not important*, six described it as *Somewhat important*, and three described it as *Very important*. Although the interview participants were not specifically asked, four people also stated that they earned credits under the LEED rating system for their living walls, all under the Innovation in Design (ID) category.

When we conducted this study, green building and the LEED rating system had only recently gained widespread recognition in the commercial construction industry. Therefore, it is possible that many respondents either did not have the option to pursue LEED certification or did not know it was available when their living walls were designed and built. Furthermore, all the respondents who described this factor as *Not important* also stated that their project did not pursue LEED certification or any other green building certification.

The Most Important Factors

After asking respondents to describe each potential factor in terms of importance, we then asked them to list the three most important factors that affected their decision to install a living wall (not all respondents listed three factors as the most important—some only listed two).

Because project teams often make design decisions based on a cost–benefit analysis, we thought it was important to directly compare the economic factors with those factors related to the benefits of living walls. Figure 14.3 shows how the three economic factors compare to the top three identified benefits of living walls. Based on participants' responses, these top three benefits were considerably more important than the three economic factors.

The "Contribution to building aesthetics" factor was described as the most important factor by the highest number of people. The next two highest scoring factors are also benefits of living walls—namely, "Improvements to indoor air quality" and "Effects on psychological health of occupants." The fourth highest scoring factor was "A visual way to recognize building or company as green." No other factor was described as one of the three most important factors by more than two people, thereby leading to the conclusion that the more easily quantifiable economic factors were of lesser importance than the benefits of living walls or the green marketing factor.

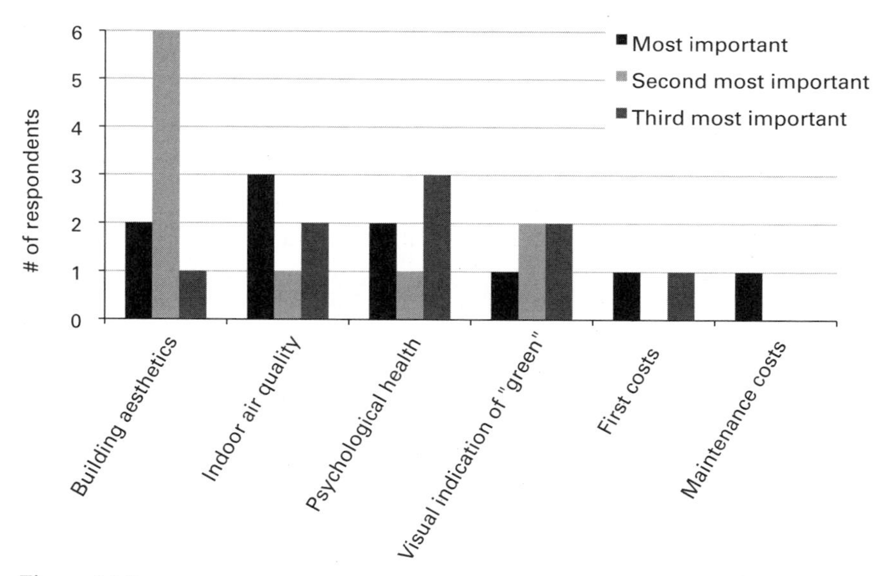

Figure 14.3

Comparison of three economically related factors to three benefits of living walls

Objective Attributes

All of the factors discussed to this point can be considered subjective in nature since respondents were asked to express their opinion on the importance of each factor in the decision-making process. We also asked questions which respondents did not have to evaluate in terms of importance, and are therefore considered objective attributes. The three most interesting objective attributes are (1) location of the living wall, (2) building tenant type, and (3) construction type. It should be noted that although we interviewed a total of fifteen people, in some case we interviewed more than one person about the same living wall and so there were only twelve different living walls discussed.

Ten of the twelve living walls discussed are in Canada and eight of those are located in the Toronto metropolitan area. Of the two living walls in the United States, one is in New Jersey and one is in New Mexico. It is important to note that many living walls we studied were designed and built in Canada; there were more established living-wall companies and noted living walls in Canada at the time of this study.

We asked respondents to classify the tenant of the building in which the living wall is located and gave them numerous options. Participants identified five of the twelve different living walls as having a higher education institution as the building tenant while no other tenant type was selected by more than two respondents (other options included education K–12, commercial leased space, commercial owner-occupied, nonprofit, residential, and health care). Although it is possible that at the time of the study there were more living walls located in higher education buildings, it is also possible that we were simply able to identify more people associated with higher education institutions to participate in this study. Thus, the fact that we discussed more living walls located in buildings owned by higher education institutions does not necessarily signify that more living walls overall are located in higher education buildings.

One study respondent stated that the living wall they installed in their university facility was primarily a research initiative to study its effects on indoor air quality. Furthermore, she stated that whether the wall was successful or not, the research project would be a success and thus, it would be worth the risks and costs the university would incur.

A final objective attribute was whether the project was new construction or a building renovation. In ten of the twelve living walls examined, the installation of a living wall was included in a new construction project while only two were part of a building renovation. This is likely because the size, complexity, and potential impact of a living wall on a building's systems makes it a less viable concept (and likely much more expensive) when attempting to retrofit an existing building.

Praise for Living Walls

Although we did not explicitly ask for participants' opinions on the living-wall systems, several made comments on the pleasure that building occupants derive from the walls. In spite of maintenance issues, one participant declared they "have no regrets in the installation of the living wall." Furthermore, she believes the living wall is advantageous to the building and its occupants. Another person related that when water leaks were discovered, the building tenants complained and said they wanted the wall to be repaired, not removed.

Concerns about Living Walls

Several respondents also made comments about concerns and issues they have encountered with living walls. One person explained that the city building officials almost refused to allow the living wall installation until the project team took the time and effort to educate them on how the wall would function, including precautions that would be taken to avoid water leaks and other potential problems.

Most of the other issues participants brought up were about the unique maintenance required for a living-wall system. Two respondents stated that the sheer size of their living walls make access difficult for maintenance personnel. Two other respondents said the time and cost for maintenance was more than they expected. Similarly, three people explained they had to train their maintenance staff extensively to care for plants on a vertical surface. Others mentioned the plants occasionally die and have to be replaced, and another described an initial problem with insects. Last, one person said she thought there would be fewer mechanical problems (pumps, lighting, leaking, etc.), another said they were still worried about mold growth (but had not discovered any yet), and another said the initial maintenance was high until the plants were established.

Several solutions were offered to address maintenance concerns. Several building managers hired the same company that designed and built the living walls to maintain them while one team went beyond that by contracting a monthly fixed cost for maintenance. Thus, monies are held in escrow, so to speak, for when maintenance is needed.

Practical Suggestions for Future Living-Wall Projects

In addition to the praise and concerns expressed about living walls, several people also offered suggestions for future project teams considering living walls for their buildings. Most important, two respondents said that ideally, living walls should be incorporated into the project at the concept design phase of the project to facilitate a holistic design plan. Both respondents believed this would limit installation defects and scheduling problems since the wall would be integrated with the building design from the start. Another person suggested that future projects should use the most experienced living-wall company available in their area to minimize potential for problems. A different respondent advocated for more natural lighting near the living wall as a way to decrease electric lighting requirements and increase the likelihood of plants thriving. A final participant recommended that someone design and install a living wall that is not only beautiful and improves indoor air quality, but is also useful for food production, thereby encouraging urban agricultural practices.

Conclusion

From the research data analysis, three definitive conclusions stand out. First, although economic factors had considerable impact on the decision-making process, the benefits were notably more important to respondents. This tells us that people who have installed living walls determined the benefits outweighed the costs to install and maintain living walls. Second, we found that "contribution to building aesthetics" was the most important factor. Mentioned as one of the three most important factors by more people than any other factor (a total of nine), it was described by more respondents as *Very important* than any other factor (a total of twelve). Therefore, the overall visual impact of a living wall must be of chief importance in influencing people's decisions to install them in their buildings. Last, the fact that living walls are a relatively new and uncommon technology in the sustainable building industry played a considerable role in almost every respondent's decision-making process. Only one person described either "Uniqueness of the product" or "One of the first companies to try this new technology" as *Not important*, while the remaining respondents described them as either *Somewhat important* or *Very important*. This implies that if living walls were a more common green technology in use today, they might not hold the same level of appeal for project teams. In addition, from the information gathered, we drew the following conclusions about living walls in general: first, the fact that five respondents mentioned (without being asked) that their living wall is located in a space regularly used as a community gathering place alludes to the ability of living walls to attract people through their physical beauty and hints at the relevance of the biophilia hypothesis. Since we did not specifically ask participants about this characteristic of living walls, we cannot generalize these opinions to all the participants but it is nonetheless an interesting finding. Second, ten of the twelve living walls discussed in our interviews were installed as part of a new construction project, indicating that they may be more feasible (or at least more likely) for new construction projects than for building renovations. This also suggests the possibility of complications when installing living walls as part of building renovation projects.

In conclusion, previous studies on plants and other natural elements suggest living walls and other biophilic design features can provide numerous benefits to the built environment, and our study showed that those benefits were of chief importance in the decision to install living walls in their buildings. Specifically, we found that living walls' contribution to overall building aesthetics stood out as important to respondents, as was the perception that living walls are a unique and relatively uncommon technology. In addition, respondents' statements that people are often drawn to their living walls and the building spaces they occupy suggests living walls can indirectly enhance social interactions, promote community engagement, and increase exposure for the buildings and organizations where they are located. Living walls have the potential to provide incredible benefits to built environments. Green building supporters should advocate for the expanded use of living walls and other biophilic features in sustainable construction.

References

Berto, Rita. 2005. Exposure to restorative environments helps restore attentional capacity. *Journal of Environmental Psychology* 25 (3): 249–259.

Bringslimark, Tina, Terry Hartig, and Grete Grindal Patil. 2007. Psychological benefits of indoor plants in workplaces: Putting experimental results into context. *HortScience* 42 (3): 581–587.

Darlington, Alan, M. Chan, D. Malloch, C. Pilger, and M. Dixon. 2000. The biofiltration of indoor air: Implications for air quality. *Indoor Air* 10:39–46.

Finger, Matthias. 1994. From knowledge to action? Exploring the relationships between environmental experiences, learning, and behavior. *Journal of Social Issues* 50 (3): 141–160.

Fromm, Erich. 1964. *The heart of man*. New York: Harper and Row.

Genetron Systems. 2006. Genetron Systems—The nature of the beast. http://www.breathingwall.com

Godish, Thad, and Carlos Guindon. 1989. An assessment of botanical air purification as a formaldehyde mitigation measure under dynamic laboratory chamber conditions. *Environmental Pollution* 62 (1): 13–20.

Hansen, Shirley, and H. E. Burroughs. 1999. *Managing indoor air quality*. Lilburn, GA: Fairmont Press.

Heerwagen, Judith and Betty Hase. 2001. Building biophilia: Connecting people to nature in building design. *Environmental Design & Construction*. March 8.

Hester, Randolph T. 2006. *Design for ecological democracy*. Cambridge, MA: MIT Press.

Hines, Anthony, Tushar Ghosh, Sudarshan Loyalka, and Richard Warder. 1993. *Indoor air: Quality and control*. Englewood Cliffs, NJ: Prentice Hall.

Joye, Yannick. 2007. Architectural lessons from environmental psychology: The case of biophilic architecture. *Review of General Psychology* 11 (4): 305–328.

Kals, Elisabeth, Daniel Schumacher, and Leo Montada. 1999. Emotional affinity towards nature as a motivational basis to protect nature. *Environment and Behavior* 31 (2): 178–202.

Kellert, Stephen R., Judith Heerwagen, and Martin Mador. 2008. *Biophilic design: The theory, science and practice of bringing buildings to life*. Hoboken, NJ: John Wiley & Sons Inc.

Kuo, Frances, and William Sullivan. 1996. Do trees strengthen urban communities, reduce domestic violence? Forest Report R8-FR 55, Technical Bullitan No. 4. USDA Forest Service Southern Region, Athens, Georgia.

Langeheine, Rolf, and Jurgen Lehmann. 1986. *Die bedeutung der erziehung für das umweltbewußtsein* [The importance of education for ecological awareness]. Kiel: Institut für die Pädagogik der Naturwissenschaften (IPN).

Li, Qing 2010. Effect of forest bathing trips on human immune function. *Environmental Health and Preventative Medicine*. http://www.ncbi.nlm.nih.gov/pubmed/19568839 (accessed January 15, 2012). doi: 10.1007/s12199-008-0068-3.

Liu, Y., Y. Mu, Y. Zhu, H. Ding, and N. Arens. 2007. Which ornamental plant species effectively remove benzene from indoor air? *Atmospheric Environment* 41: 650–654.

Lohr, Virginia, Caroline H. Pearson-Mims, and Georgia Goodwin. 1996. Interior plants may improve worker productivity and reduce stress in a windowless environment. *Journal of Environmental Horticulture* 14 (2): 97–100.

Lohr, Virginia, and Caroline Pearson-Mims. 1996. Particulate matter accumulation on horizontal surfaces in interiors: Influence of foliage plants. *Atmospheric Environment* 30 (14): 2565–2568.

Lovins, Amory. 2004. Lightfair: Economics of daylighting and occupant productivity. D04–04. RMI. http://www.rmi.org/Knowledge-Center/Library/D04-04_EconomicsOfDaylighting (accessed December 28, 2012).

Manaker, George. 1987. *Interior plantscapes: Installation, maintenance, and management*. Englewood Cliffs, NJ: Prentice-Hall Inc.

Orwell, Ralph, Ronald Wood, Jane Tarran, Fraser Torpy, and Margaret Burchett. 2004. Removal of benzene by the indoor plant/substrate microcosm and implications for air quality. *Water, Air, and Soil Pollution* 157:193–207.

Park, Bum Jin, Yuko Tsunefsugu, Tamami Kasetami, Takahide Kagawa, and Yoshifumi Miyazahi. 2010. The physiological effects of Shinrin-yoku (taking in the forest atmosphere or forest bathing): evidence from field experiments in 24 forests across Japan. *Environmental Health and Preventative Medicine*. http://www.ncbi.nlm.nih.gov/pubmed/19568835 (accessed December 28, 2012). doi: 10.1007/s12199-009-0086-9.

Park, Seong-Hyun, and Richard Mattson. 2009. Therapeutic influences in hospital rooms on surgical recovery. *HortScience* 44 (1): 102–105.

Pearson, David. 1989. *The new natural house book*. New York: Simon and Schuster.

Pliska, Shane. 2005. Biophilia, Selling the love of nature. *Interiorscape Magazine*. http://www.planterra.com/research/article_biophilia.html (accessed December 28, 2012).

Pyle, Michael. 2003. Nature matrix: Reconnecting people and nature. *Oryx* 37 (2): 206–214.

Seignot, Lynne. 2001. *Interior planting: A guide to plantscapes in work and leisure spaces*. Hampshire, UK: Gower Publishing.

Teisl, Mario, and Kelly O'Brian. 2003. Who cares and who acts? Outdoor recreationists exhibit different levels of environmental concern and behavior. *Environment and Behavior* 35 (4): 506–522.

Ulrich, Roger S. 1984. View from window may influence recovery from surgery. *Science* 224 (4647): 420–421.

Ulrich, Roger S., Robert F. Simons, Barbara D. Losito, Evelyn Fiorito, Mark A. Miles, and Michael Zelson. 1991. Stress recovery during exposure to natural and urban environments. *Journal of Environmental Psychology* 11 (3): 201–230.

Wener, Richard, and Hannah Carmalt. 2006. Environmental psychology and sustainability in high-rise structures. *Technology in Society* 28:157–167.

Wilson, Edward O. 1984. *Biophilia*. Cambridge, MA: Harvard University Press.

Wolverton, B. C. 1996. *How to grow fresh air: 50 Houseplants that purify your home or office*. New York: Penguin Books.

Wood, Ronald, Margaret Burchett, Ralph Alquezar, Ralph Orwell, Jane Tarran, and Fraser Torpy. 2006. The potted-plant microcosm substantially reduces indoor air VOC pollution: I. Office field study. *Water, Air, and Soil Pollution* 175:163–180.

Yoo, Mung Hwa, Youn Jung Kwon, and Ki-Cheol Son. 2006. Efficacy of indoor plants for the removal of single and mixed volatile organic pollutants and physiological effects of the volatiles on the plants. *Journal of the American Society for Horticultural Science* 131 (4): 452–458.

V

Perspectives on the Future

15

Constructing Green: Challenging Conventional Building Practices

Monica Ponce de Leon

From the onset of industrialization, material production in the various design fields—from industrial design to the design of environments and buildings—has had and continues to have a devastating effect on the planet. Given the magnitude and complexity of the problem, how can designers participate in the solution? The answer to this question is complex. Building is at the center of many disciplines, and therefore no one field can provide effective alternatives. Addressing the environmental impact of buildings will demand that educators and practitioners in many fields think differently about their missions, their histories, and their purposes. Voluntary programs are not enough; we need regulatory mechanisms that ensure change in the marketplace. Access to innovation and information is critical so that building designers, owners, and users can make informed choices. The technology and capacity exist, but we are not moving fast enough. A design and construction revolution is necessary to move our society quickly to the next level of sustainability.

Let's begin by reviewing some facts that have now become common knowledge: buildings are one of the heaviest consumers of natural resources and account for a significant portion of the greenhouse gas emissions that affect climate change. In the United States alone, buildings account for 38 percent of all CO_2 emissions (EIA 2008b). Buildings represent almost 40 percent of U.S. primary energy use and 75 percent of all U.S. electric consumption; they consume 14 percent of potable water (EIA 2008a).

Perhaps less discussed, but no less significant, is buildings' share of material consumption and waste output. The U.S. Environmental Protection Agency (2009) estimates that 170 million tons of building-related construction and demolition debris are generated annually, with 61 percent coming from nonresidential and 39 percent from residential sources. This represents 30 percent of all waste output in the United States (2003 numbers reported in 2009). Globally, buildings use 40 percent of raw materials (Lenssen and Roodman 1995).

Today, many designers see third-party certification systems as the only viable solution to reduce the negative environmental impact of buildings. Third-party certification systems and organizations are increasingly streamlined, recognized, and respected. For example, the LEED (Leadership in Energy and Environmental Design) rating, assigned by the USGBC (U.S. Green Building Council), has undergone a revolution in just over fifteen years. Since its inception in 1994, the program has grown from six volunteers on one committee to hundreds of volunteers on more than twenty committees (USGBC 2012). As of January 2012, there are now seventy-nine USGBC chapters and 130,049 registered and certified projects (ibid.).

Despite its success, LEED has fundamental flaws that expose the limit of third-party certification. For example, negative points are not part of the system. In theory, one could do something terrible for the environment, garner points through other means, and still have a LEED-certified building. In addition, LEED compounds the very complex issue of transportation of materials to the site: local production is not always best, despite its immediately apparent benefits. There are some strong arguments in favor of large-scale, centralized production in some industries. For instance, energy-intensive production required in such industries as steel and glass manufacturing are more efficient and less polluting in midsize-to-large plants than they tend to be in small, community-scale operations. Similarly, the manufacture of some energy-saving products, such as window glass with low-emittance (low-e) coatings, requires highly developed equipment that is not economically viable on a small scale.

Despite these limitations, LEED remains an effective tool for informed designers who want to make a difference. I have experienced this first-hand in my practice with the Macallen Building. Completed in 2007, this is the first LEED Gold-certified residential building in Boston. Its innovative green-building technologies will save over 600,000 gallons of water annually while consuming 30 percent less electricity than a conventional building. The project has won numerous awards, including an AIA/COTE Top Ten Green Project in 2008 and an AIA Institute Honor Award in Architecture in 2010. But in that project (http://www.themacallenbuilding.com/), the pursuit of LEED certification enabled us to incorporate environmentally sensitive strategies into the design, not only because our client wanted to build a better building, but also because LEED certification allowed him to send a clear message to critical stakeholders (see chapter

12, this volume). In this case, LEED certification served as a marketing tool. Without LEED, the incentive might not have been there.

The real problem with LEED lies elsewhere. As of January 4, 2012, the USGBC boasts that the count of LEED-certified residential and commercial projects is 27,529 (chart on http://www.usgbc.org/Default.aspx). While this number might seem impressive, it should be cause for dismay. LEED certification needs to be placed in a larger context, one that compares total buildings constructed with total LEED-certified buildings. One would find there is quite the disparity.

The challenge is clear. Despite the hype, the excitement, and the best intentions, a voluntary system will never be as effective or have as much impact in transforming the way we build and manage buildings as actual legislation. Despite its extraordinary rate of success, the environmental impact of LEED certification at this time is negligible. Only legally bound regulatory statutes have proven effective in producing tangible, widespread change.

The solutions to many of the environmental challenges that pertain to buildings are close within our reach, particularly on the energy front. Advances in the design of building envelopes place within our grasp a substantive reduction in the environmental impact of new and existing buildings. The digital revolution gives us the tools to address both energy and water consumption in buildings. We must design buildings that acknowledge the computer revolution and use software as a means to control environmental quality, energy, and water consumption within existing and new buildings. The North House, by Team North/RVTR, is a great model of what is currently possible (see chapter 8, this volume). The North House is a research prototype that competed in the U.S. Department of Energy's Solar Decathlon in 2009, where it placed fourth. It employs two primary technological systems using feedback and response mechanisms: a Responsive Envelope, which reconfigures in response to changing weather conditions, and a custom-developed Adaptive Living Interface System (ALIS), which provides detailed performance feedback and system control to the inhabitants, equipping them with informed control of their home. Over the course of one year, the home is able to produce up to 6,600 kWh of electricity beyond what it consumes, and it is able to feed this additional power back into the energy grid.

We must examine why the means to reduce energy and water consumption in buildings has advanced in research institutions but lags behind in the construction industry. While the digital revolution moves forward at previously unimaginable speed and digital technology is now deeply embedded in our daily lives, its application in buildings remains in the dark ages. Smart buildings have been the subject of sophisticated research, but few have actually been built. Traditional market mechanisms have failed. It is evident that only performance requirements reinforced by legislation will ensure a fundamental transformation of the industry.

The issues of material consumption in buildings and waste output by the construction industry represent complex problems with difficult histories affecting the environment. For example, consider the selection and use of particular species of wood in American furniture. The use of oak as a common material at the end of the nineteenth century is traditionally linked to the Arts and Crafts movement's call for honest indigenous materials. But a closer look at material history reveals a different picture. Until the 1890s, most designers disliked oak as a primary wood because of its open coarse grain. Instead, walnut was the material of choice. However, at the end of the century walnut was practically depleted from Midwest forests, and therefore it was scarce and expensive. The resurgence of oak was due to this scarcity. Oak was abundant, so American cabinet makers developed methods to treat oak that deemphasized the grain and darkened it, making it close to the look of walnut. By the 1910s, as the supply of large oaks declined in the Midwest, oak lost its prominence. The furniture industry began to harvest trees in the South making poplar, maple, and birch the woods of choice. This cycle of "popularity" and "depletion" is one that persists even today, but now on a global scale. This story illustrates how the various fields included under the rubric of design have had a tremendous impact on the degradation of the environment. Whether in the production of objects, buildings, or interiors, more often than not, design has been at the center of human interventions that have negatively affected particular ecosystems on a global scale.

How to address the problem of material consumption remains one of the most pressing problems of our time. Today, despite the proliferation of information available to us about materials, the right solution in material selection is hardly apparent. Let's take plywood, a very common material. If you look at its embodied energy, plywood remains one of the most sustainable materials for the production of buildings and furnishings; it is a dramatically more efficient use of material than wood. Its laminate structure is engineered to provide great strength with very little matter. Plywood is typically 1.8 percent binder by weight. Very few types of glue can be used to bind thin layers of veneer together, and only those with higher molecular weights can be used without penetrating the ply layers. Until the mid-twentieth century, vegetable and animal glues were used. With the advent of formaldehyde-based binders, plywood manufacturers almost exclusively switched to the new, cheap, and waterproof phenol formaldehyde (PF) and urea formaldehyde (UF) resins. PFs are dark in color and utilized primarily for exterior and sheathing applications. Off-gassing is minimal once the curation process is finished. UFs are light in color and used in furniture-grade plywood. UFs continue to off-gas throughout their life cycle, posing a health risk to both laborers on the production end and to end-users, but are far cheaper than PFs for the manufacturer. In short, like all formaldehyde-based products, PFs and UFs are known carcinogens and both types of plywood are potentially harmful to human health. Current alternatives to these glues include soy-based binders and polyvinyl acetate (PVAc) adhesives (like Elmer's glue). A number of manufacturers and research institutes remain in the research phase of product development, but at least one major plywood company, Columbia Forest Products, carries a soy-based adhesive, which it sells at a premium. This is just one example of the complexity of material selection in building.

Now let's consider the use of wood as a primary material in building. Wood products compose 47 percent of the manufactured materials in the United States and consume only 4 percent of the total energy used for manufacturing raw materials (APA—The Engineered Wood Association 2012). However, an analysis of the same material from another point of view gives us a potentially different perspective. Deforestation accounts for one-fifth of total global carbon emissions and is most threatening in the tropics. Furthermore, industrial logging accounts for almost one-third of deforestation worldwide. Since export and domestic timber account for only 16 percent of that one-third, on the surface it does not seem that the building and product manufacturing industries are contributing greatly to deforestation. However, the furniture, interiors, and building industries are the primary users of tropical hardwoods, and it is through tropical deforestation that damage of global significance threatens species diversity, environmental health, and cultural heritage.

To combat deforestation, third-party organizations have again developed standards for responsible practice. The most widely recognized of these organizations is the FSC (Forest Stewardship Council). FSC certification allows consumers to know that the wood products they are purchasing have been logged from a sustainably managed forest. As a designer, FSC wood is an easy choice, but one with limited impact. As of July 2009, FSC forests account for only 5 percent of the world's productive forests (FSC International Center GmbH 2010). As successful as the

FSC has been, voluntary, market-driven initiatives are slow to take hold and effect change. We need legislation that will ensure that compliance with standards for best practice becomes the norm, rather than remains the exception.

Parallel to the implementation of statutory laws, addressing the challenge of waste and consumption in buildings will require the fundamental transformation of construction practices. Today's building industry came to fruition during the industrial revolution and has remained virtually unchanged since. At the turn of the twentieth century, construction methods were conceived for ease of assembly at a time when materials seemed abundant. It was in the nineteenth century that standardization of materials across large geographic areas came into being, forever transforming the way that buildings are produced. The consistency of dimensional lumber or "modern" brick sizes and their implications for construction are very much part of the reality of building today. These techniques came into being without the critical input of those outside the building industry, propelled almost exclusively by short-term economic forces with unexpected societal and environmental consequences. For example, the efficiency of assembly of dimensional lumber, enabled by the widespread use of platform framing, resulted in the boom of the lumber industry, but by the end of the nineteenth century most forests in the American Northeast had been depleted.

Despite their enduring history, conventional building practices can no longer be sustained. Standardization and efficiency must be reconceptualized in the service of material economy. A new building industry must emerge with ease of disassembly as its primary goal. New material standards and construction techniques must be developed that will enable us to easily transform our buildings as our needs change and will allow materials to be reused readily when a building is taken down. Recent advances in digital technology have laid a fertile ground for this wholesale transformation. Parametric software allows for complex variations to emerge out of repetitive systems, and digitally guided fabrication enables material efficiency. We already have the capacity to design for disassembly—this should become the catalyst for new building practices that address contemporary needs while anticipating environmental consequences.

References

APA—The Engineered Wood Association. 2012. Panel design specification. Form no. D510C. Revised May 2012. American Plywood Association. http://www.apawood.org/level_c.cfm?content=pub_searchresults&pK=510c&pT=Yes&pD=Yes&pF=Yes&pubGroup= (accessed December 13, 2012).

EIA. 2008a. Annual energy outlook 2008 with projections to 2030. U.S. Department of Energy Report DOE/EIA-0383. Energy Information Administration. http://www.eia.gov/oiaf/archive/aeo08/ and http://www.eia.gov/oiaf/aeo/pdf/0383(2008).pdf (accessed January 19, 2012).

EIA. 2008b. Assumptions to the annual energy outlook 2008. U.S. Department of Energy Report DOE/EIA-0554. Energy Information Administration. http://www.eia.gov/oiaf/archive/aeo08/assumption/index.html and http://www.eia.gov/oiaf/archive/aeo08/assumption/pdf/0554(2008).pdf (accessed January 4, 2012).

FSC International Center GmbH. 2010. Forest Stewardship Council fact sheet. http://us.fsc.org/ (accessed February 3, 2013).

Lenssen, Nicholas, and David M. Roodman. 1995. Worldwatch paper #124: A building revolution: How ecology and health concerns are transforming construction. Washington, DC: Worldwatch Institute.

U.S. Environmental Protection Agency. 2009. Estimating 2003 building-related construction and demolition materials amounts. Washington, DC: U.S. EPA. http://www.epa.gov/wastes/conserve/imr/cdm/pubs/cd-meas.pdf (accessed January 4, 2012).

USGBC. 2012. U.S. Green Building Council. http://www.usgbc.org (accessed February 3, 2013).

16

Constructing the Biophilic Community

William Browning

Nature is part of our humanity, and without some awareness and experience of that divine mystery man ceases to be man. When the Pleiades and the wind in the grass are no longer part of the human spirit, a part of very flesh and bone, man becomes, as it were, a kind of cosmic outlaw, having neither the completeness and integrity of the animal nor the birthright of true humanity.

—Henry Beston, *The Outermost House*, xxxv

Introduction

Although the effects of our communities on ecosystems cannot be understated, we must recognize that the design of human communities also affects human health and productivity. An increasing body of research on *biophilia*, the concept first framed by biologist E. O. Wilson (1984) to describe humanity's innate need for and connection to nature and natural systems, has shown that the current design of many of our buildings and communities negatively affects human health. By separating humanity from natural systems, we have harmed both. In the continuing evolution of the sustainable design movement, we need to expand our scope beyond how buildings impact the environment to include how buildings and nature can interact to improve the human experience. Drawing on the experiences of William Browning and colleagues at the research and consulting firm Terrapin Bright Green, this chapter discusses how, with careful planning, we can find holistic solutions that can enhance ecological and human health.

Background

At Terrapin Bright Green, we came to the biophilia conversation somewhat indirectly. In 1994, the Rocky Mountain Institute (RMI) published *Greening the Building and the Bottom Line* (Romm and Browning 1994).

This report documents eight case studies in which businesses that pursued efficient lighting, heating, and cooling to reduce operating costs found one or more of these positive side effects: measurably increased worker productivity, decreased absenteeism, and improved work quality. Based on research done at Western Electric's Hawthorne plant in Chicago from 1929 to 1932, it had been general business consensus until that point that *any* change in a worker's environment would increase productivity. It was thought that changes in working conditions affect productivity only because they signal management's concern—the so-called "Hawthorne effect." Any gains were believed to be only temporary (for studies on the Hawthorne effect see Brill, Margulis, and Konar 1984; Dickson and Roethlisberger 1966; Parsons 1974, 1978).

Although the original experiment was flawed—the research pool included only five subjects, who could monitor their own production rate on an hourly basis and who were, along with their supervisors, being rewarded for gains in productivity—it still led researchers to ignore the effects of building design on worker productivity for sixty years. One such ignored study was from 1984, by the Buffalo Organization for Social and Technological Innovation (BOSTI), which found direct correlations between specific changes in the physical environment and worker productivity (Brill, Margulis, and Konar 1984). Energy-efficient design can produce productivity gains as high as 6–16 percent, providing savings far in excess of the energy savings. Daylighting and efficient lighting design, in particular, can measurably increase work quality by reducing errors and manufacturing defects.

For example, in 1986 the mail sorters at the Main Post Office in Reno, Nevada, became the most productive sorters in the western region of the United States—an expanse stretching from Colorado to Hawaii (Lee Windheim, Robert McLean, pers. comm.). What happened? In one portion of the facility, the Reno Post Office implemented a lighting retrofit that lowered the ceiling, and installed more efficient direct/indirect lighting systems. Over the following twenty weeks, productivity in the retrofitted portion increased by more than 8 percent, whereas workers in the unmodified area showed no change in productivity. A year later, after the retrofit had been extended to the whole building, productivity had stabilized at an increase of about 6 percent. The energy savings projected for the building came to about $22,400 a year, with an additional savings of $30,000 a year from reduced recurring maintenance costs associated with the new ceiling. Combined, the energy and maintenance savings came to about $50,000 a year: a six-year payback. The productivity gains, however, were worth $400,000 to $500,000 a year. In other words, the productivity gains alone would pay for the entire renovation in less than a year.

Similarly, a tilt-up concrete warehouse—as unexciting and unattractive a building type as one could imagine—in Costa Mesa, California, also shows how retrofits can lead to impressive productivity gains (Todd Jersey Architecture, pers. comm.). VeriFone, a company that makes credit card verification keypads, converted its warehouse into its worldwide distribution headquarters. The building had only three windows, yet because of seismic codes they could not add more. Consequently, the company decided to daylight the building via the roof. After this retrofit, the remanufacturing floor required no electric lights 80 percent of the time, with the employees working almost entirely in daylight. The building exceeds the state's strict energy code regulations, using half as much energy per square meter as the code allows. The natural lighting decision more than slashed energy costs; it also dropped absenteeism, which was 47 percent lower than among employees in the previous building.

In 1993, Walmart opened the "Eco-Mart" in Lawrence, Kansas. Designed in part by Rocky Mountain Institute's Green Development Service, the store used new experimental features to make the store more energy efficient and environmentally sustainable. One of the main energy-saving features was an array of light-monitoring skylights that optimized the amount of daylight admitted into the store. To cut costs, Walmart decided to install the skylights in only one half of the store. To their surprise, managers discovered that sales per square foot were significantly higher for those departments located in the daylit half. Furthermore, employees in the artificially lit half began requesting that their departments be moved to the daylit side (Tom Seay, pers. comm.).

Productivity gains are financially significant, but in many ways are just placeholders for a more important issue: people's well-being. These productivity gains were realized by improving the physical working environment, a side effect of attempting to increase energy efficiency. As we look toward creating a healthier, more sustainable society, the question of how to design buildings and places that maximize human well-being emerges. What can we do to make the built environment the best possible place for people to live and work?

The Basis of Biophilia and Design

Following the publication of *Greening the Building and the Bottom Line*, the U.S. Department of Energy initiated a study of the new Herman Miller

SQA factory in Zeeland, Michigan, to determine what conditions would lead to gains in productivity (Wise et al. 1996). This 300,000-square-foot daylit facility, designed by William McDonough & Partners, features extensive skylights and windows with a view to the surrounding restored prairie landscape. Studying the factory's three shifts—day, swing, and night—researchers found an overall increase in productivity from the previous building, but the only conclusive results occurred during the daytime shift. The night shift, although well lit, had access to neither natural daylight nor an outside view.[1]

The framework for this experiment was based on principles of biophilia or "the innately emotional affiliation of human beings to other living organisms" (Wilson and Kellert 1993, 31). This emerging field of research combines work from evolutionary psychology, anthropology, archeology, geography, neuroscience, and a number of other disciplines to assess how the built environment affects human well-being. One of the most intriguing early pieces of research was a 1984 study by Roger Ulrich on the recovery times of cholecystectomy surgery patients, in which some of the patients had a view of trees and shrubs and others had a view of a brick wall (Ulrich 1984). The patients with the view to nature had a shorter average recovery period, took fewer painkillers, and had fewer nursing calls. This research, and other subsequent studies on healing times and stress recovery, led to the use of "healing gardens" as elements in many new hospitals (Marcus and Barnes 1999).

There are now several hundred papers that document various aspects of this desire of humans to have contact with nature, and the psychological and physiological effects of this interaction. How this research is translated into the built environment is the core of biophilic design. For designers seeking a biophilic response to their work, researchers have developed three categories of design interventions: *Nature in the space*, *Natural analogs*, and *Nature of the space*.

Nature in the Space

Nature in the space, or bringing plants, water, and animals into the built environment, is something that humans have done since time immemorial. Flowers, potted plants, gardens, aquariums, and fountains are all examples of bringing nature into the space. Bartzcak and Dunbar (see chapter 14, this volume) offer a great in-depth look at how this can be achieved with living walls. But bringing nature into the space can also mean designing buildings so that indoor occupants have a view of outdoor landscapes. We see the intrinsic value of nature at play every day, for example, in how we price real estate: the Manhattan apartment with the view onto Central Park is more expensive than one with a view onto treeless Lexington Avenue; the hotel room with an ocean view is more expensive than the one looking onto the parking lot.

Part of the relaxation response linked to having a view onto natural spaces is due to the biomechanics of the human eye. When we focus on a desktop computer screen or read a paper, the muscles in the eye contract, curving the lenses into a convex form. Looking at a distant object, our muscles relax and the lenses flatten. In other words, occasionally looking at a distant view provides intermittent periods of relaxation. Recent research by Lisa Heschong of Heschong Mahone Group indicates that nonrepetitive movement in the far visual plane, such as the bird flying past the window or a single leaf trembling on a tree, has a unique capacity to garner our attention (Lisa Heschong, pers. comm.). We are also particularly attuned to movements within fractal patterns. Fractals are nested shapes that remain the same as they scale up or down; fern leaves are the classic example. Flames and waves also typically occur as fractal patterns, which may explain why the flames in a fireplace and waves in water can both transfix and relax us (Joye 2007; Hagerhall et al. 2008). Further, the neuroscientist Irving Biederman at the University of Southern California has found that settings rich in interpretable visual information (such as a complex natural landscape) activate dense areas of mu-(opioid) receptors in the brain, which indicate that the brain is having a pleasurable experience (Biederman and Vessel 2006).

Natural Analogs

Natural analogs are materials and patterns that evoke nature and can be characterized into three broad types: ornamentation, biomorphic forms, and the use of natural materials. Cultures around the world ornament their buildings with representations of nature such as carvings and paintings of fruit, flowers, leaves, acorns, seashells, birds, and other animals. Biomorphic forms mimic natural shapes. The columns and ribbed vaulting in a cathedral remind us of trees in a forest. The Finnish architect Eero Saarinen and the Spanish architect Santiago Calatrava are famous for creating explicitly biomorphic buildings that look like trees, shells, bones, and wings. Finally, natural materials that expose their inherent qualities are another form of a *Natural analog*. Wood where the grain can be seen, stone where the texture and veins are apparent, and highly textured fabrics are all examples. Less directly, details like the remnant patterns of wood grain in board-formed concrete offer traces of natural life in nonliving materials. Although *Natural analogs* elicit a biophilic response, the experiences of *Nature in the space* and *Nature of the space* tend to elicit stronger responses.

Nature of the Space

Nature of the space deals with how we respond psychologically and physiologically to different spatial experiences. In numerous experiments, people from around the world show a cross-cultural preference for certain types of landscapes—typically characterized by a ground plain of grasses, flowers, and low shrubs, punctuated by copses of spreading shade trees, rock outcroppings, and small rivers, with views to distant mountains and herds of grazing animals and birds. Judith Heerwagen and Gordon Orians (Heerwagen and Orians 1986, 1993) postulated the "Savannah Hypothesis": because evolutionary evidence indicates that humans arose on the savannahs of Africa it would make sense that we would prefer to find ourselves in these savannah-like conditions. It is a landscape that humans artificially create all over the world—parks, suburban lawns, and golf courses are all essentially recreated savannahs.

English geographer Jay Appleton was perhaps the first modern researcher to attempt to codify which landscape elements people found most appealing, but *Nature of the space* touches on a human legacy found in many traditional geomancy systems (Appleton 1975). Buried among the layers of mysticism in feng shui and Ayurvedic design, there are insights about local climatic responses and occupant psychology. Most of these patterns are based on long-term observation within a specific cultural context. The research into the spatial patterns in preferred landscapes attempts to reach a deeper, non-culturally specific understanding of spatial patterns. Today, researchers cite between seven and sixteen patterns, such as experiences of what Appleton referred to as "enticement" or "peril." For example, an elevated view across a distant landscape is a highly preferred spatial pattern that Appleton called "prospect." It's a view made still more appealing if it contains shade trees and water. In a contrasting but also preferred spatial pattern that Appleton called "refuge," one's back is protected with shelter overhead so the space embraces and nurtures the occupant, such as in an inglenook next to a fireplace. The biophilic experience is stronger when patterns occur together, such as a space that has both prospect and refuge, as occurs when sitting under the overhanging roof on the raised porch of a bungalow.

Biophilic Design in Practice

Combining elements of *Nature in the space*, *Natural analogs*, and *Nature of the space* can seem daunting, but with some forethought it is possible to mix these features and in the process create buildings and communities with a powerful positive impact. For example, The Johnson Wax Company (now SC Johnson) Administration Building, designed by Frank Lloyd Wright successfully incorporates all three biophilic design elements. A water feature at the building's entrance, in conjunction with planters and potted trees, provide *Nature in the space*, whereas the columns in the great work room are a *Natural analog* with slender tapering forms that are then capped with a massive spreading disk, creating a biomorphic grove of shade trees in a space that itself taps into elements of *Nature of the space* by evoking elements of savannah landscapes. *Nature of the space* is also represented with a low entry sequence that does not provide a direct view of the central space, evoking the pattern of refuge. As one comes to the great workroom in the center of the building, the space expands, allowing full view of the width of the building, evoking the patterns of enticement and prospect. The building's successful biophilic design is evident in the happiness of the SC Johnson staff who work in a building that is essentially unchanged from its original design in 1939.

Similarly, the new LEED Platinum Bank of America Tower at One Bryant Park in Manhattan was designed with biophilia in mind. A daylit building in which more than 90 percent of the occupants have a view to the outside, the building is shaped to focus views to Bryant Park, and longer vistas toward Central Park and the Hudson River, thus providing *Nature in the space*. From the shape of the building itself, inspired by a quartz crystal, to the noticeably grained door handle at the building's entrance, the Bank of America Tower features numerous *Natural analogs*. Even the elevator lobby, with its walls of dark reddish leather begging to be touched, features a strong *Natural analog*. *Nature of the space* in the forms of prospect and refuge are found in the balance between the glass-fronted offices and perimeter meeting spaces, each of which features floor-to-ceiling glass with views of the outdoors.

The previous two buildings had biophilia integrated into their initial design. Showing the potential of biophilic elements to transform an existing space, the offices of CookFox Architects and Terrapin Bright Green are a strong example of a biophilic retrofit to the top floor of an old department store on Manhattan's 6th Avenue. The designers combine *Nature in the space* with *Nature of the space* to create a dynamic workspace.

The space is set back from the parapet, exposing 3,600 square feet of roof surface, covered by a green roof with eight different species of succulents. There are a few enclosed spaces along an interior wall; the rest of the occupants sit in an open-plan space with forty-two-inch tall partition walls. This ensures that everyone has access to daylight and 95 percent of the occupants have a view out to the skyline or the green roof.

Bringing biophilic experiences into an existing urban setting is about more than just buildings, however. Urban pocket parks can create a biophilic experience in otherwise densely built environments. Paley Park in Manhattan is a tiny park that features sidewalls covered in ivy, a waterfall as a back wall, and a grove of shade trees. This incredible little park brilliantly weaves together prospect and refuge, and *Nature in the space*. Meanwhile, in Portland, Oregon, small sidewalk drainage wetlands have been created in the strip between the street and the sidewalk. These systems provide a function (cleaning storm water) while also helping to bring *Nature in the space*.

Hannoversch Munden, a medieval German town, not only illustrates how biophilic design can make cities into better human habitats, but also how it can serve as a mechanism to increase urban cohesiveness. Composed of a large paved plaza with a cathedral on one side, the town hall on the other, and the remaining space surrendered to bus parking, the center of Hannoversch Munden periodically flooded with storm water. To address the storm water issue and help revitalize the area, planners decided to repurpose the plaza. Artist/engineer Herbert Dreiseitl led a community design process that transformed the plaza into a large water feature. Storm water is collected, filtered, and cleaned. Some is then stored in cisterns and slowly introduced to the groundwater, while the rest flows across a series of shallow stepping planes. During the day, Turkish and German families play and splash together (helping to speed the process of evaporation)—a unique activity given the often strained relations between these cultural groups. Three adjoining light stanchions are mounted with a mirror that bounces light onto vibrating metal shapes in the water where both children and adults play. At night the vibrating metal shapes cause reflections of the light to dance across the facades of the surrounding buildings. Dreiseitl says that dealing with storm water is an opportunity to connect people with water as art and life; it is also the essence of designing a biophilic community.

Too often, designers focus their attention on the structure they are building, and neglect its effect on the people who use it. Researchers at the University of Michigan found that just *being* in an urban environment impairs our basic mental processes—research participants who took a walk through nature scored 12 percent better on tests of memory and cognition than their urban counterparts (Berman, Jonides, and Kaplan 2008). One outcome of this mental fatigue, posits professors Frances Kuo and William Sullivan of the University of Illinois, Urbana-Champaign, is shorter fuses, which lead to increased acts of violence (Kuo and Sullivan 2001).

Conclusion

In the future, we expect successful designers to increasingly rely on their ability to connect humans back to nature. Demographics are one reason for this conclusion; as of just a few years ago, more people now live in urban than in rural environments for the first time in history. This trend is expected to continue through the next century, which may be good news ecologically. Cities typically create a smaller environmental footprint per person and thus provide one avenue to conserve natural resources. However, this may be bad news for people's health. Besides established concerns about pollution, as more people move toward cities, we risk losing touch with the natural planet around us. As demographics push us toward increasingly denser habitats, designers will be required to perform more planning and deliberate effort to incorporate natural systems into urban environments. Although not fully appreciated, these systems are important for our health, happiness, productivity, and well-being. We would be well served to build biophilic communities where people can thrive as they live, work, and play.

Note

1. This study was initiated by Rocky Mountain Institute, the U.S. Green Building Council, the U.S. Department of Energy, and Herman Miller. The study involved site visits to the Herman Miller SQA building, and personal communications with the Battelle researchers, architect William McDonough, and Keith Winn and Joseph Azzerello of Herman Miller.

References

Appleton, Jay. 1975. *The experience of landscape*. New York: Wiley.

Berman, Marc, John Jonides, and Stephen Kaplan. 2008. The cognitive benefits of interacting with nature. *Psychological Science* 19 (12): 1207–1212.

Beston, Henry. 1949. *The outermost house*. New York: Henry Holt and Company.

Biederman, Irving, and Edward A. Vessel. 2006. Perceptual pleasure and the Brain. *American Scientist* 94 (3): 247–253.

Brill, Michael, Stephen T. Margulis, and Ellen Konar. 1984. *Using office design to increase productivity*, vol. 1. Buffalo, NY: Workplace Design and Productivity.

Dickson, William J., and F. J. Roethlisberger. 1966. *Counseling in an organization: A sequel to the Hawthorne researches*. Cambridge, MA: Harvard University Press.

Hagerhall, Caroline M., Thorbjörn Laike, Richard P. Taylor, Marianne Küller, Rikard Küller, and Theodore P. Martin. 2008. Investigations of human EEG response to viewing fractal patterns. *Perception* 37 (10): 1488–1494.

Heerwagen, Judith H., and Gordon H. Orians. 1986. Adaptations to windowlessness: A study of the use of visual decor in windowed and windowless offices. *Environment and Behavior* 18 (5): 623–639.

Heerwagen, Judith H., and Gordon H. Orians. 1993. Humans, habitats, and aesthetics. In *The biophilia hypothesis*, ed. Stephen R. Kellert and Edward O. Wilson, 138–172. Washington, DC: Island Press.

Joye, Yannick. 2007. Fractal architecture could be good for you. *Nexus Network Journal* 9 (2): 311–320.

Kuo, Frances E., and William C. Sullivan. 2001. Aggression and violence in the inner city: Effects of environment via mental fatigue. *Environment and Behavior* 33 (4): 543–571.

Marcus, Clare C., and Marni Barnes. 1999. *Healing gardens: Therapeutic benefits and design recommendations*. New York: Wiley.

Parsons, H. M. 1974. What happened at Hawthorne? *Science* 183 (4128): 922–932.

Parsons, H. M. 1978. What caused the Hawthorne effect? A scientific detective story. *Administration & Society* 10 (3): 259–283.

Romm, Joseph J., and William D. Browning. 1994. *Greening the building and the bottom line: Increasing productivity through energy efficient design*. Snowmass, CO: Rocky Mountain Institute.

Ulrich, Roger S. 1984. View through a window may influence recovery from surgery. *Science* 224 (4647): 420–421.

Wilson, Edward O. 1984. *Biophilia*. Cambridge, MA: Harvard University Press.

Wilson, Edward O., and Stephen R. Kellert, eds. 1993. *The biophilia hypothesis*. Washington, DC: Island Press.

Wise, James A., Judith Heerwagen, David B. Lantrip, and Michael Ivanovich. 1996. Protocol development for assessing the ancillary benefits of green building: A case study using the MSQA building. In *NIST Special Publication 908, Third International Green Building Conference and Exposition*, ed. A. H. Fanney and P. R. Svincek, 63–80. Gaithersburg, MD: National Institute of Standards and Technology.

Contributors

Editors

Rebecca L. Henn is an assistant professor at The Pennsylvania State University and a doctoral candidate at the University of Michigan, holding doctoral fellowships granted at both the Erb Institute for Global Sustainable Enterprise and the Graham Environmental Sustainability Institute. She earned her architectural degree from Carnegie Mellon University in 1994, master in design studies degree from Harvard University's Graduate School of Design in 2006, and practiced architecture at Bohlin Cywinski Jackson, Gluckman Mayner Architects and Celento Henn Architects + Designers for the intervening twelve years, learning first-hand the joys and challenges of creating material, social, and cultural artifacts.

Andrew J. Hoffman is the Holcim (US) Professor of Sustainable Enterprise at the University of Michigan, with joint appointments at the Stephen M. Ross School of Business and the School of Natural Resources & Environment. Hoffman also serves as director of the Frederick A. and Barbara M. Erb Institute for Global Sustainable Enterprise. Before entering academia, Hoffman was a construction supervisor for T&T Construction & Design in southwestern CT and Nantucket MA.

Contributors

Lauren Barhydt completed her master of architecture degree with commendation from the University of Waterloo in 2010. While working on her degree, she was project manager for the North House prototype. Barhydt now lives and works in Toronto, where she continues to investigate sustainable buildings and the role of the occupant.

Clayton Bartczak is a sustainability consultant with experience on new construction and existing building projects and is a LEED AP BD+C and O&M and an IFMA Sustainability Facility Professional (SFP). He holds a master of science degree in construction management from Colorado State University. Bartczak lives in Denver, Colorado, with his wife, Kim, and his son, Miles.

Lyn Bartram is an associate professor in the School of Interactive Arts & Technology at Simon Fraser University. Bartram's interests lie in the design and deployment of human-centered interactive systems, particularly in the realms of conservation and sustainability. She founded and leads the Human-Centered Systems for Sustainable Living research group. Bartram holds a PhD from Simon Fraser University.

Olivier Berthod holds a doctorate in economics from the Freie Universität Berlin, as well as master's degrees in organization theory and management from the Université Paris-Dauphine, France, and the Växjö Universitet, Sweden. Since 2011 Berthod has been an honors fellow of the Dahlem Research School in the Freie Universität Berlin. His research reports on crises and societal conflicts; he currently is investigating food-borne disease outbreaks in global food production networks.

Nicole Woolsey Biggart is director of the Energy Efficiency Center at the University of California at Davis. She holds the Chevron Chair in Energy Efficiency and is a professor at the UC Davis Graduate School of Management. Interested in the impact of social networks, Biggart has researched the auto industry, the U.S. commercial building industry, and network direct selling. She received her PhD in sociology at UC Berkeley.

Lenora Bohren is a senior research scientist, director of the National Center for Vehicle Emissions Control and Safety (NCVECS), and director of research for the Institute of the Built Environment (IBE) at Colorado State University. She holds a doctorate and for more than twenty years has worked on environmental issues with organizations such as the USEPA and the USDA. Bohren helped develop methodologies to assess the socioeconomic characteristics of tropical soils management and has conducted air pollution studies throughout the United States and Mexico. Working with the Natural Resource Ecology Laboratory (NREL) at CSU, she assessed farmers' and ranchers' attitudes toward adaptation to climate change and global warming; and with the Institute for the Built Environment (ISE) at CSU, she assessed indoor air quality in the Pine Ridge Indian Reservation school system.

Bertien Broekhans holds a PhD in science and technology studies and is an assistant professor at Delft University of Technology since 2007. Her research, teaching, and consulting deal with issues of expertise, governance, and the management of uncertainties. Her research projects examine questions about the roles of experts and expertise in decision making processes. Her teaching areas include courses on decision making.

William Browning launched Rocky Mountain Institute's Green Development Services in 1991, and was a founding member of the U.S. Green Building Council's Board of Directors. In 2006, Browning founded a new research and consulting firm, Terrapin Bright Green, LLC with partners Bob Fox, Rick Cook, and Chris Garvin. Terrapin crafts high-performance environmental strategies for corporations, governments, and large-scale real estate developments. Browning received a bachelor's degree in environmental design from the University of Colorado, and a master of science degree in real estate development from the Massachusetts Institute of Technology.

Zinta S. Byrne is an associate professor of industrial and organizational psychology at Colorado State University with interests in employee engagement and organizational transformation and culture. She coedited *Purpose and Meaning in the Workplace*, forthcoming in 2013. In addition to her academic responsibilities, Byrne consults with several organizations on their change efforts.

Michael Conger is a Ph.D. candidate in management and entrepreneurship at the University of Colorado. His research interests lie in the areas of social entrepreneurship, entrepreneurial identity, institutional entrepreneurship, social movements, the role of business in society, and ethics. Conger recently received the 2011 Founders' Award from the Society for Business Ethics. He has presented his research at the Darden Entrepreneurship and Innovation Research Conference, BCERC, NYU/Stern Social Entrepreneurship Conference, SEE Conference on Sustainability, Ethics and Entrepreneurship, Society for Business Ethics, and the Academy of Management.

Jennifer E. Cross is an associate professor in the Department of Sociology at Colorado State University. She received her PhD in sociology from the University of California at Davis. Her areas of expertise include social change and innovation, where she studies conservation behaviors and decision-making, inter-organizational networks and knowledge networks, and behavior change tools for increasing sustainability in organizations.

David Deal is a sustainability consultant at the firm evolve environment::architecture. His degrees in environmental science and management inform his expertise in environmental law and regulatory issues, organization and sustainable business practices, and environmental management systems. Deal taught sustainability and built environment issues at Slippery Rock University in Pennsylvania.

Beth M. Duckles is an assistant professor of sociology at Bucknell University. Her current research project is an organizational ethnography of the USGBC's development of the LEED 2012/version 4 rating system. Research interests include environmental sociology, nonprofit organizations, and mixed methods. She received her doctorate in sociology at the University of Arizona.

Brian Dunbar is director of the Institute for the Built Environment (IBE) and professor emeritus at Colorado State University. Dunbar has guided project work for the National Park Service, cities, school districts, and universities, and created the Sustainable Building graduate program at CSU. A USGBC LEED faculty member and LEED fellow, he coauthored *147 Tips on Teaching Sustainability* and has been honored by the AIA, USGBC, and the governor of Colorado. Dunbar earned his bachelor's and master's degrees in architecture from the University of Michigan.

Robert G. Eccles is a professor of management practice at the Harvard Business School. His current research interests fall into three domains. The first is new models of collaboration for making cities more sustainable broadly defined in economic, environmental, and social terms. The second is innovations in processes, products, and business models that are needed for a company to create a sustainable strategy, one that creates value for shareholders over the long term by contributing to a sustainable society. Third, he is studying how professional service firms can survive and prosper over the long term by building their capabilities through a disciplined and integrated management of the client and talent markets.

Amy C. Edmondson is the Novartis Professor of Leadership and Management at Harvard Business School. She received her PhD in organizational behavior from Harvard University, and studies interpersonal dynamics that affect learning and innovation in organizations. Her current research focuses on interdisciplinary collaboration in the pursuit of innovation in the built environment. Before her academic career, Edmondson was director of research at Pecos River Learning Centers, and chief engineer for architect/inventor Buckminster Fuller.

Bill Franzen is president of SAGE 2 and Associates, LLC, a consulting firm focused on advancing sustainable and green-built educational environments. Franzen is a retired executive director of operations from Poudre School District in Fort Collins, Colorado. The Poudre School District is a recognized national leader in energy conservation, environmental stewardship, and green building programs.

Ronald Fry is professor and chair, Department of Organizational Behavior at Case Western Reserve University (CWRU). He is coeditor of the *Journal of Corporate Citizenship* and directs the CWRU Fowler Center for Sustainable Value's World Inquiry Project. A co-creator of the Appreciative Inquiry change theory and method, he is widely published in the domain of organization change and development, group dynamics, and executive education.

Jock Herron holds a doctorate from the Harvard Graduate School of Design where he co-leads the Smart Cities and Wellness project sponsored by the Humana Corporation and the Responsive Environments and Artifacts Lab at the GSD. He has a longstanding interest in global health, urbanization, technology and design, writing his doctoral dissertation on "Global Health and Design: Learning from Tanzania and Abbott Payson Usher" and receiving a Derek Bok Teaching Award as a teaching fellow for the Harvard College course The Design of American Cities. A former partner at a global investment firm, Herron has a MDesS with distinction and a MPA from Harvard and decades ago an MBA from the University of Chicago and a BA from Hampshire College.

Stephen Hockley is a sustainability consultant with the firm evolve environment::architecture. He earned degrees in psychology and his master's in business administration with a focus on sustainability, which inform his expertise in organizational sustainability and change management. Hockley has extensive project management experience with a wide range of organizations including multinational corporations, regional nonprofits, municipalities, and universities.

Kathryn B. Janda is an interdisciplinary, problem-based scholar and senior researcher at the Environmental Change Institute at Oxford University. Her research encompasses three principal areas: (1) social, economic, and environmental implications of ecological design; (2) social dimensions of energy use; and (3) the relationship between environmental practice and organizational decision making. She received undergraduate degrees in electrical engineering and English literature from Brown University and her MS and PhD from the Energy and Resources Group at the University of California, Berkeley. She has worked in the Energy Analysis Program at Lawrence Berkeley National Laboratory; served as an American Association for the Advancement of Science Environmental Policy Fellow at the U.S. Environmental Protection Agency; and taught Environmental Studies at Oberlin College.

Nitin R. Joglekar is an associate professor of operation and technology management at Boston University School of Management. His research interests span a variety of innovation and technology commercialization challenges: ratcheting design complexity, startup and valuation of energy supply chains, and the implementation of sustainable hospitality services. Prior to his academic career, Joglekar worked in the energy and IT industries and has founded a software startup.

Gavin Killip joined the Environmental Change Institute at Oxford University in 2004 after working for ten years on energy efficiency and building-integrated renewable energy projects in the voluntary and public sectors. In his own 1908 home he has reduced energy consumption and CO_2 emissions by 60 percent through major refurbishment. In 2011 he completed his PhD at Oxford University, examining the implications of long-term climate change targets on the practices of small and medium-sized enterprises in the UK construction industry. He has a BA in linguistics from York University and an MSc degree in advanced environmental and energy studies from the University of East London, in association with the Centre for Alternative Technology.

Alison G. Kwok is a professor of architecture at the University of Oregon. She is a licensed architect, and received her PhD and MArch from the University of California, Berkeley. Her current research includes identifying adaptation and mitigation strategies for climate change, and developing building performance case studies.

Larissa Larsen is an associate professor in the Urban and Regional Planning Program at the University of Michigan. She received her PhD from the University of Illinois at Urbana-Champaign. Her research focuses on environmental inequities in the built environment and advancing issues of urban sustainability and social justice.

Michelle A. Meyer is a PhD candidate in the Department of Sociology at Colorado State University. Her research and teaching interests encompass society-environment interactions to include disaster resilience and mitigation, climate change displacement, environmental sociology, community sustainability, research methods and statistics, and social stratification.

Christine Mondor is an eternal optimist about the advancement of sustainability through the design of places, processes, and products. She is a strategic principal of evolve environment::architecture and a registered architect. Her research and design work focuses on the relationship between sustainability, culture and the built environment and her projects have received numerous regional and national awards. She is adjunct associate professor in practice at Carnegie Mellon University in the School of Architecture.

Monica Ponce de Leon is the dean of the University of Michigan's Taubman College of Architecture and Urban Planning, and the design director of the architecture firm MPdL Studio. As a founding partner of the former architecture firm Office dA she has received over sixty international awards including the Cooper-Hewitt National Design Award for Architecture in 2007.

Nicholas B. Rajkovich is a PhD candidate in the Urban and Regional Planning Program at the University of Michigan. He received an MArch from the University of Oregon. His current research evaluates the effectiveness of energy efficiency and weatherization programs to help communities adapt to climate change.

Stuart Reeve is energy manager for the Poudre School District (PSD) in Fort Collins, Colorado. For the past thirty-eight years, he has held several positions in facility services and business services. In 1993 under his leadership, PSD created an energy management plan and an energy efficiency team, which has evolved into an award-winning program, recognized both by the U.S. Department of Energy and the U.S. Environmental Protection Agency. PSD passed a $120 million bond election in 2010 and is now embracing major infrastructure replacements in all existing schools. Reeve has been an active member of the Association of Energy Engineers since 1996 and a Certified Energy Manager. He is a founding member and past chairperson for the Colorado Association of School District Energy Managers established in 1998.

Johnny Rodgers is a master of science graduate from Simon Fraser University's School of Interactive Arts & Technology, and a member of the Human Centered Systems for Sustainable Living research group. His research focuses on supporting sustainable resource use in the home through information visualization, interaction design, and ubiquitous computing.

Garima Sharma is a doctoral candidate in organizational behavior at the Weatherhead School of Management, Case Western Reserve University. Sharma's research interests focus on the underlying processes and outcomes of corporate social responsibility and sustainability. Her current research unpacks how organizations integrate multiple and often competing profit and social/environmental logics. She received her bachelor's degree in engineering (equivalent to the BS) and master's in business administration (MBA) from India, after which she worked as an internal organizational development consultant for a software consulting firm in India.

Geoffrey Thün is an associate professor at the Taubman College of Architecture and Urban Planning at the University of Michigan, and a partner in the research-based practice RVTR, whose recognition includes the 2009 Canadian Prix de Rome in Architecture, a 2010 R&D Award from *Architecture* Magazine and a 2011 RAIC Award of Excellence for Innovation in Architectural Practice. His research engages the application of systems theory and ecological methods to design questions across a range of scales, from that of the megaregion to that of the building envelope.

Ellen van Bueren holds a PhD in Technology, Policy and Management and is an assistant professor at Delft University of Technology. Her research and teaching focus on decision making for complex problems that require multiple actors to collaborate. She started her career in sustainable urban planning, which is still one of her favorite research domains. She is editor of *Bestuurskunde* (the Dutch journal of public administration) and regional associate editor of the journal *Smart and Sustainable Built Environment*.

Kathy Velikov is an assistant professor at the Taubman College of Architecture and Urban Planning at the University of Michigan, a registered architect and a partner in the award-winning research-based practice RVTR. Her research ranges from urban systems to high-performance buildings to prototype development of kinetic envelopes that mediate energy, atmosphere, and social space. She holds an MA from the University of Toronto and a BArch and BES from the University of Waterloo.

Rohit Verma is a professor of service operations management at the School of Hotel Administration, Cornell University. His research interests include new product/service design, quality management and process improvement, and operations/marketing interrelated issues.

Robert Woodbury is a professor at Simon Fraser University. He is currently the director, Art and Design Practice of the Graphics, Animation, and New Media Network. His research is in computational design, visual analytics, and human-centered systems for sustainable living. He holds a PhD and MSc from Carnegie Mellon and a BArch from Carleton.

Jeffrey G. York is an assistant professor of management and entrepreneurship at the University of Colorado Boulder. He received his PhD from the Darden School of Business at the University of Virginia in 2010. York's teaching and research are focused on environmental entrepreneurship and the simultaneous creation of ecological and economic goods. He teaches classes in business planning, entrepreneurial thinking, and environmental ventures. His research has appeared in the *Best Paper Proceedings of the Academy of Management*, *Journal of Business Venturing*, *Journal of Business Ethics*, and *Journal of Management*.

Jie J. Zhang is an assistant professor of operations management and information systems at the University of Vermont School of Business Administration. Her research is in the area of service operation management with a focus on sustainability-related issues, including the measurement of environmental sustainability, performance analysis of sustainable practices, and coordination of sustainability investment through contracts.

Index

Note: References to tables are denoted by t, and references to figures by f.

Abbott, Andrew, 42
Accelerating Change (Egan), 46
Accountability, 21
  flat teams and, 230
  generativity and, 240
  green schools and, 230
  Green Teams and, 230
  high-performance, 230
  quality assurance and, 271
  titanium buildings and, 83
Actor-network theory (ANT), 40–42, 47–48
Adam Joseph Lewis Center, 57–60, 73n4
Adaptive Living Interface System (ALIS)
  ambient information and, 187–189
  direct information and, 184–186
  feedback reception and, 184–186
  folk units and, 186
  as information ecosystem, 188
  integration of, 180–181, 187–189
  leveraging and, 186–187
  lighting and, 180–181, 189
  mobile devices and, 187–188
  Neighborhood Network views and, 184–186, 188
  North House project and, 178, 180–190
  one-step optimization and, 180
  Resource Dashboard and, 181, 185f
  Sleep mode and, 181
  social interaction and, 186–187
  spatial/lifestyle integration and, 187–189
  "Spinner" interface and, 188
Aerosols, 313
Air quality, 5, 65, 271
  bioeffluents and, 313
  biophilic design and, 312–319, 322–325
  cigarettes and, 313
  Clean Air Act and, ix
  dismissal of, ix
  efficiency and, 77
  formaldehyde and, 313
  green schools and, 227
  hydrocarbons and, ix
  indoor, 312–315
  motor vehicles and, ix–xii
  plants and, 312–315
  radon and, 313
  Sick Building Syndrome and, 313
  smog and, ix
  volatile organic compounds (VOCs) and, 153–156, 159, 162, 165
Air Quality Solutions, 316
Aliveness, 241, 243, 245–247, 249, 252, 254
Ambec, Stefan, 108
Amelung, Wolfgang, 316
American Institute of Architects (AIA), 5–6, 12
  conventional building practices and, 334

green schools and, 222
Institute Honor Award in Architecture and, 334
LEED (Leadership in Energy and Environmental Design) and, 57, 62, 63t
titanium buildings and, 79, 86–87, 91, 97n7
Top Ten Green Project award and, 334
American Society of Heating, Refrigerating and Air-Conditioning Engineers (ASHRAE), xi, 69
American Society of Landscape Architects (ASLA), 91
Anderson, Ray, 269–270
ANSI/ASHRAE/USGB/IES Standard 189, 69
Appleton, Jay, 346
Appreciative inquiry (AI), 239–240, 243–244, 253–254
Architects, 2, 23
Adam Joseph Lewis Center and, 58–59
American Institute of Architects (AIA), 5–6, 12, 57, 62, 63t, 79, 86–87, 91, 97n7, 222, 334
biophilic design and, 309, 343, 345, 347, 349n1
contractors and, 14–15
conventional building practices and, 334
generativity and, 247
green building supply industry and, 140
green schools and, 222–223, 227–228, 231–233, 236
LEED (Leadership in Energy and Environmental Design) and, 57–62, 65–66, 69–70
niche innovations and, 142–144, 159, 162
organizational sustainability and, 197
renovation and, 36, 41–43
sustainable buildings and, 173–174, 180–181, 191–192, 264, 267, 273
system of professions approach and, 41–42
temporary organization and, 3
titanium buildings and, 79–81, 83t, 84, 86–93, 96, 97nn6,8
Waldschlößchenbridge and, 290, 295, 300
World Heritage Committee and, 290
Artifacts, 1, 14, 146, 173–174, 191, 252, 264, 285
Arts and Crafts movement, 336
Association for the Advancement of Sustainability in Higher Education (AASHE), 204
Atkins, Douglas, 63–66
Autodesk, 91
Axelrod, Robert, 118

B&Q, 44
Bank of America Tower, 347
Barhydt, Lauren, 19, 171–195
Barnard, Chester, xiv
Barrett, Frank J., 244
Bartczak, Clayton, 22, 307–329, 344
Bartram, Lyn, 19, 171–195
Baudelaire, Charles, 302
Baumgartner, Frank R., 67
Beamish, Thomas, xii
Benzene, 313–314
Berthod, Olivier, 21–22, 285–306
Beston, Henry, 341
Biederman, Irving, 345
Biggart, Nicole Woolsey, ix–xiv
Bioeffluents, 313
Biofiltration, 316
BioHome, 315
Biophilic design
architects and, 309
basis of, 343–346
benzene and, 313–314
biofiltration and, 316
BioHome and, 315
built environment and, 307–315, 326–327, 343–344, 348
Central Park and, 345
connection with nature and, 309–311, 314, 316, 341

constructing community of, 341–349
consumption and, 313, 316, 321
cultural issues and, 345–348
definition of, 307
ecological issues and, 310, 341, 349
economic issues and, 317–323, 326, 345
efficiency and, 314, 342–343
electricity and, 36, 49, 316, 325, 342–343
emotion and, 310
energy and, 316, 319, 321, 342–343
engineers and, 36, 39, 41–43, 51, 348
environmental issues and, 312–314, 321, 343, 349
expertise and, 312
firms and, 341
forest bathing and, 311
fractals and, 345
Fromm and, 309
Hawthorne effect and, 342
health issues and, 307–315, 319, 322–324, 341, 343, 349
historical perspective on, 307
housing and, 312
indoor air quality and, 312–314
institutions and, 324, 341, 343
Japan and, 311
LEED (Leadership in Energy and Environmental Design) and, 322, 347
lighting and, 309–311, 325, 342–343
living walls and, 307–327
management and, 321, 342
markets and, 316–319, 322
materials and, 313, 345–346
narratives and, 318
natural analogs and, 344–347
nature in the space and, 344–345
nature of the space and, 346–347
niche innovations and, 316–317
operating costs and, 342
phyto-remediation and, 316
pollution and, 312–314, 349
in practice, 347–349
productivity and, 23, 310–311, 314
psychology and, 22, 307–323, 344–346
regulation and, 314, 343
renovation and, 342–343
resources and, 310, 318, 349
Rocky Mountain Institute (RMI) and, 341–343, 349n1
Savannah Hypothesis and, 346
Sick Building Syndrome and, 313
social issues and, 309, 321, 327, 342
stress reduction and, 310–311, 344
student performance and, 310–311
sustainability and, 341, 343
technology and, 307–309, 316, 320, 326–327, 342
water and, 311, 314–316, 324–325, 344–348
Wilson and, 309, 341
worker productivity and, 314
Biophilic Design (Kellert, Heerwagen, and Mador), 311
Bohren, Lenora, 22, 307–329
Booher, David E., 61–62
Boyce, Mary, 9
Boyer, Kenneth K., 108
Brand equity, 109, 111
Breathing Wall Ecosystem, 316
Bricks, 287, 338, 344
British Defense Ministry, 86–87
Broekhans, Bertien, 18–19, 145–167
Brounen, Dirk, 7
Browning, William, 22–23, 341–350
Brundtland Commission Report, 199, 288
Bryant Park, 347
Buffalo Organization for Social and Technological Innovation (BOSTI), 342
Buildability, 47–49
Building applied photovoltaics (BAPVs), 178
Building codes, xii
ANSI/ASHRAE/USGBC/IES Standard 189 and, 69
Dutch, 151t
energy codes and, 17, 57, 68, 343

LEED and, 60, 67, 69–71 (*see also* LEED [Leadership in Energy and Environmental Design])
local, 321
organizational sustainability and, 200
punctuated equilibrium of, 67
Building Dashboard, 79
Building energy, 50, 62, 79, 171–172, 174, 178, 181, 184
Building information modeling (BIM), 3
LEED (Leadership in Energy and Environmental Design) and, 57, 62, 73n1
titanium buildings and, 78, 84–88, 90, 92–94, 96, 97nn8,9
Building integrated photovoltaics (BIPVs), 178
Building materials
bricks, 287, 338, 344
concrete, 152, 154, 162–163, 275, 287, 302, 343, 345–346
glass, 6, 202, 334, 347
metal, 81–122, 151, 348
steel, 334
wood, 7, 275, 287, 336–337, 345
Building Owners and Managers Association (BOMA), xi, 10, 91
Building Research & Information journal, 39
Building Research Establishment's Environmental Assessment Method (BREEAM), 5, 80, 97n4
Buildings
ceilings and, 342, 347
as cultural artifacts, 1 (*see also* Cultural issues)
energy consumption of, 3 (*see also* Energy)
floors and, 7, 43, 49, 150, 151t, 156, 176, 343, 347
furniture and, 263, 267, 276, 313, 336–337
HVAC and, 172, 222, 231, 267, 270
lighting and, 8 (*see also* Lighting)
plumbing and, 43, 97n6, 222, 227, 231
Sick Building Syndrome and, 313
smart, 5, 336
windows and, 7, 43, 174, 178, 272, 310, 343–344
Built environment, 1, 3, 6
biophilic design and, 307–315, 326–327, 343–344, 348
changing social concepts of, 14–15
environmental issues and, 4 (*see also* Environmental issues)
generativity and, 254
green building supply industry and, 129
multinational corporations and, 103–106, 116, 118, 120–121
social integration and, ix–xiv (*see also* Social issues)
sustainability and, 171 (*see also* Sustainability)
Bullitt Foundation, 91
Butcher, Andrew C., 35–36
Buttel, Frederick, 286
Byrne, Zinta S., 20, 219–238

Cable, Daniel M., 117–118
CAFE (Corporate Average Fuel Economy), xi
Calatrava, Santiago, 345
Capital E Analysis Group, 8–9
Carbon
biophilic design and, 313–314
deforestation and, 337
dioxide, 4, 35, 46–48, 151, 313–314, 333
green schools and, 224–225, 235
low and zero (LZC) technologies and, 35, 43, 49
monoxide, ix
neutrality, 199
niche innovations and, 147
organizational sustainability and, 199, 203f
refurbishment, 36–37, 43–52, 72
renovation and, 36–39, 42–52
savings, 35

sustainable buildings and, 172
titanium buildings and, 77
Case Western Reserve University, 245
Center for Sustainable Landscapes (CSL), 202–203
Central Park, 345, 347
Chartwell School, 58, 63–67, 71
Chesapeake Bay Foundation, 85, 91
Churchill, Winston, 1, 6, 171
Cigarettes, 313
Clarity
contractual, 86
green building supply industry and, 140
individual thriving and, 241, 246–247, 249, 254
organizational sustainability and, 208
Clean Air Act, ix
Climate change, 4, 35, 39, 42, 50, 147, 220–221, 235, 333
Cole, Ray, 173
Collaboration, 17, 20
collective action and, 128, 134–135, 138, 140
flat teams and, 219–237
generativity and, 239
green building supply industry and, 134–135
integrated project delivery (IPD) and, 62, 78, 86–88, 90, 97nn6,7,9
LEED (Leadership in Energy and Environmental Design) and, 57–58, 61–72, 73nn1,5
multi-agent, 116–117
multinational corporations and, 104, 109, 116
niche innovations and, 153–156, 159, 162, 165
organizational sustainability and, 211–216, 221, 224, 227–231, 234–235
renovation and, 46
self-interest and, 65
sustainable buildings and, 174, 186, 192
titanium buildings and, 77–94, 97nn6,7
Collaborative project management (CPM), 86–88, 92–94
Collaborative rationality, 58, 61–63, 67, 71
Columbia Forest Products, 337
Comprehensive Assessment System for Building Environmental Efficiency (CASBEE), 5, 97n4
Concrete, 152, 154, 162–163, 275, 287, 302, 343, 345–346
Conger, Michael, 18, 127–144
Conservation
carbon refurbishment and, 36–37, 43–52, 72
energy and, 20, 103, 172, 178, 184, 221, 224, 226, 236
goals for, 187
insulation and, 6–7, 35–36, 43, 47, 49, 120, 151t, 163, 176, 267, 269–270, 313
organizational process and, 38
water and, 9, 111, 135, 174, 202, 223, 269
World Heritage Sites and, 22, 286–287, 290–292, 298–303, 304nn2,3,4
zero net energy and, 59, 66, 73nn3,4, 172 (*see also* Energy)
Construction Industry Research and Information Association (CIRIA), 48
Consumer behavior cost factor (CBCF), 105, 107, 110–115
Consumption, 3, 6
Adaptive Living Interface System (ALIS) and, 178, 180–190
biophilic design and, 313, 316, 321
consumer behavior cost factor (CBCF) and, 105, 107, 110–115
consumerism and, 264
conventional building practices and, 333–338
electricity and, 4 (*see also* Electricity)
empowering the inhabitant and, 171–172, 180–181, 184–188

green building supply industry and, 137
green schools and, 223, 236
multinational corporations and, 103–107, 110–116, 120–121
renovation and, 35, 40
sustainability and, 223 (*see also* Sustainability)
titanium buildings and, 79
water and, 344–348 (*see also* Water)
Contractors, 3, 14–15
generativity and, 239, 247
green building supply industry and, 128
green schools and, 222–223, 232
LEED (Leadership in Energy and Environmental Design) and, 57, 59, 62, 68
niche innovations and, 153, 155–159
organizational sustainability and, 212
performance-based contracting and, 96n1
renovation and, 36, 45, 48
titanium buildings and, 79–83, 89, 96n1, 97n8
Conventional building practices
architects and, 334
challenging, 333–338
consumption and, 333–338
cultural issues and, 337
design and, 333–338
economic issues and, 334, 338
education and, 247
efficiency and, 334, 336, 338
electricity and, 333–335
engineers and, 336–338
environmental issues and, 333–338
Forest Stewardship Council (FSC) and, 337–338
formaldehyde and, 337
health issues and, 337
institutions and, 335–337
LEED (Leadership in Energy and Environmental Design) and, 334–335
life-cycle assessment (LCA) and, 337
markets and, 333–338
materials and, 22, 334–338
polyvinyl acetate (PVAc) and, 337
regulation and, 333, 335
resources and, 333
stakeholders and, 334
sustainability and, 333, 336–337
technology and, 333–335, 338
United States Green Building Council (USGBC) and, 334–335
waste and, 333–334, 336, 338
water and, 333–337
CookFox Architects, 347–348
Cool roofs, xii
Cross, Jennifer E., 20, 219–238
Cultural issues
artifacts and, 1, 14, 146, 173–174, 191, 252, 264, 285
biophilic design and, 345–348
buildings and, xii
connection with nature and, 309–311, 314, 316, 341
conventional building practices and, 337
corrective actions and, 11
empowering the inhabitant and, 172, 186
ethics and, 11, 16, 129
generativity and, 254
gravitational assist and, 206–207, 210
green schools and, 227, 235
identity and, 4, 13, 264
intellectual loyalties and, 2
internal forces and, 206
niche innovations and, 145–147, 152, 160, 161t
operational change and, 19–21
organizational change and, 19–21, 72
organizational sustainability and, 197, 202, 206–212
realignment and, 3–6
Waldschlößchenbridge and, 286–302, 289–291
World Heritage Sites and, 22, 286–287, 290–292, 298–303, 304nn2,3,4

Dashboards, 79, 89, 181, 185f
Daylight, 9
  biophilic design and, 311, 342–344, 348
  green building supply industry and, 135
  green schools and, 219, 226, 232
  LEED (Leadership in Energy and Environmental Design) and, 62–66
  organizational sustainability and, 203f
  sustainable buildings and, 178, 271, 277
  titanium buildings and, 78
Deal, David, 19–20, 197–217
Decision making
  city administration and, 22
  collaborative, 90 (*see also* Collaboration)
  diversity and, 17
  efficiency and, 287
  ethics and, 11, 16, 129
  flat team approach and, 220
  green building supply industry and, 131
  innovation and, 22, 37, 57, 61, 71, 90, 94, 112, 131, 162–163, 189, 199, 208, 212–214, 219, 267, 287, 293, 309, 320, 323, 326
  LEED (Leadership in Energy and Environmental Design) and, 57, 61, 71, 90
  legitimacy and, 288
  linear, 17, 57
  living walls and, 309, 317, 320, 323, 326
  niche innovation and, 162–163
  quality and, 267
  real-time, 94
  renovation and, 37
  sustainability and, 112, 189, 199, 208, 212–214, 267
  technical rationality and, 21, 287
  titanium buildings and, 90, 94
  Waldschlößchenbridge and, 287, 293
Deep green, 21, 207, 268, 273–274, 278t, 279
Deforestation, 337
Delmas, Magali A., 104
Design, 7
  Adaptive Living Interface System (ALIS) and, 178, 180–190
  biophilic, 307–327, 341–349
  building information modeling (BIM) and, 78, 84–88, 90, 92–94, 96, 97nn8,9
  competitive, 149–152, 223
  conventional building practices and, 333–338
  craft training and, 2
  custom commercial, xi–xii
  expertise and, 16 (*see also* Expertise)
  future issues and, 22–23
  generativity and, 240
  green building supply industry and, 128, 135, 138, 141
  green schools and, 219–236
  Hannover Principles and, 58, 73n2
  incentive, 117–120, 122
  innovation and, 4, 17, 322 (*see also* Innovation)
  integrated, 3, 152, 156, 162, 212–213, 222, 232
  LEED and, 57 (*see also* LEED [Leadership in Energy and Environmental Design])
  life-cycle assessment and, 73n1, 77, 79, 81–85, 89–91, 94, 120, 128, 155, 161t, 162, 203f, 221, 223, 337
  limited housing, xi
  multinational corporations and, 103, 108, 110, 117–118, 121–122
  natural analogs and, 345–346
  niche innovations and, 146–165
  North House project and, 174–190
  principal-agent relationships and, 155–156, 162
  quality and, 335 (*see also* Quality)
  regime change and, 146–149
  renovation and, 36–40, 43, 46, 48
  request for qualification (RFQ) and, 58, 64–65
  role of projects and, 146–149

sequential, 38
social issues and, 9–15 (*see also* Social issues)
"state of the art" approach and, 35–36, 84–85
Susanka and, xii–xiii
sustainability and, 171 (*see also* Sustainability)
synergies and, 6
titanium buildings and, 77–98
traditions of, 3
waste and, 5 (*see also* Waste)
Waldschlößchenbridge and, 293
Design for Ecological Democracy (Hester), 310
Deutsche Bank, 8
Dew, Nicholas, 131
DiMaggio, Paul J., 287–288
Dion, Douglas, 118
Dissidence, 41, 48
Distributed Responsive System of Skins (DReSS), 176–180
Diversity
decision making and, 17
expertise and, 231
generativity and, 246
geographical, 50
green schools and, 20, 220–221, 224, 228, 232, 235–236
internal, 20, 220–221, 224, 228, 235–236
LEED (Leadership in Energy and Environmental Design) and, 57, 61, 67
organizational sustainability and, 203f
renovation and, 50
species, 337
Dreiseitl, Herbert, 348
Dresden. *See* Waldschlößchenbridge
Duckles, Beth M., 21–22, 207, 263–284
Dunbar, Brian, 22, 307–329, 344
Dutton, Jane E., 243–244

Earth Day, 77
Eccles, Robert G., 17, 77–100
Ecological issues
biophilic design and, 310, 341, 349
green building supply industry and, 127–132, 138
niche innovations and, 153–154, 161t
organizational sustainability and, 199, 202
recycling and, 84, 154–155, 209, 214, 225, 229, 235, 263, 265, 267, 270, 274, 276
sustainable buildings and, 180, 188, 191, 273–274, 276, 279
titanium buildings and, 96
Eco-Mart, 343
Economic issues, 1, 3
biophilic design and, 317–323, 326, 345
building sector and, 2, 4, 7, 15, 35, 128, 132–142, 172, 216n1
consumer behavior cost factor (CBCF) and, 105, 107, 110–115
conventional building practices and, 334, 338
corporate motivations to adopt sustainability and, 107–110
efficiency and, 38 (*see also* Efficiency)
elevated attention to, 10–11
first cost savings and, 7–8, 68, 242, 318–319
for-profit firms and, 17, 127, 265–267, 274, 276, 278
generativity and, 241–242
green building supply industry and, 127–134, 138–141
green schools and, 220, 226–227
gross domestic product (GDP) and, 2, 17
high-performance buildings and, 8
imperative to build and, 285
LEED (Leadership in Energy and Environmental Design) and, 7–9, 68, 81, 90
life-cycle assessment (LCA) and, 73n1, 77, 79, 81–85, 89–91, 94, 120, 128, 155, 161t, 162, 203f, 221, 224, 234, 337

living walls and, 317–319
multinational corporations and, 103–104, 108–109, 111, 116–117, 120
net present value (NPV) and, 8, 90, 267
niche innovations and, 147–148, 153
operating costs and, 8, 68, 90, 105, 107–116, 242, 342
organizational sustainability and, 199, 206, 208
payoffs and, 117–118, 120
Prisoner's Dilemma and, 105, 117–118
productivity and, 342–343 (*see also* Productivity)
profit-driven green and, 21, 207, 268–271, 278t
progress and, 7–9
property value and, 9, 242
renovation and, 38, 42, 45, 51, 342–343
sustainability and, 269 (*see also* Sustainability)
titanium buildings and, 90
transactional burdens and, 120
uncertainty and, 129–134
Waldschlößchenbridge and, 294, 302, 304n1
water costs and, 8, 242
Edmondson, Amy C., 17, 77–100
Efficiency, xiii, 11, 22, 303
Adaptive Living Interface System (ALIS) and, 178, 180–190
biophilic design and, 314, 342–343
CAFE standards and, xi
conventional building practices and, 334, 336, 338
decision making and, 287
Distributed Responsive System of Skins (DReSS) and, 176–180
electricity and, 58–59 (*see also* Electricity)
Energy Star program and, xii, 91, 221
generativity and, 239–241, 254
green building supply industry and, 130
green schools and, 221, 225–226, 229, 231
housing and, 35
influence of ratings and, 79–81
LEED (Leadership in Energy and Environmental Design) and, 5–7, 58–60, 65, 67, 69, 73nn3,4, 270, 334
life-cycle assessment and, 73n1, 77, 79, 81–85, 89–91, 94, 120, 128, 155, 161t, 162, 203f, 221, 223, 337
living walls and, 314
motor vehicles and, 77
multinational corporations and, 105–107, 110–112, 116, 120–121
niche innovations and, 150–153, 161t, 163
professional service firms (PSFs) and, 14
renovation and, 35, 38–43, 46
resource, 18, 69, 105–107, 110–112, 116, 120
solar panels and, 7
sustainability and, 171–172, 176, 178, 189, 201–208, 213, 270
titanium buildings and, 77–91, 94–95, 97n4, 98n12
Waldschlößchenbridge and, 293
waste and, 4–8, 12, 20–21, 46, 77, 79, 84–88
water and, 5, 60, 203f
Egan, John, 46
EHDD Architecture, 65
Eichholtz, Piet, 7
Electricity, 4, 6, 8
biophilic design and, 316, 325, 342–343
conventional building practices and, 333–335
green schools and, 222–223, 225, 227, 231, 236–237
LEED (Leadership in Energy and Environmental Design) and, 58–59
multinational corporations and, 106–107, 111

niche innovations and, 146
renovation and, 36, 49
sustainable buildings and, 178, 181, 269
titanium buildings and, 95
Electric vehicles, x
Empire State Building, 90
Energy
Adaptive Living Interface System (ALIS) and, 178, 180–190
biophilic design and, 316, 319, 321, 342–343
building, x, 4, 6, 50, 62, 79, 171–172, 174, 178, 181, 184
coal and, x
codes for, 17, 57, 68, 343
conservation of, 20, 103, 172, 178, 184, 221, 224, 225, 236
consumption and, 3–4 (*see also* Consumption)
cost of, 8
crisis of 1970s and, 313
Distributed Responsive System of Skins (DReSS) and, 176–180
efficiency and, 11, 22 (*see also* Efficiency)
electricity and, 4, 6, 8 (*see also* Electricity)
embodied, 336
fossil fuels and, x, 321
generativity and, 239–250, 254
green building supply industry and, 127–128
green schools and, 221–229, 233, 235–236
housing and, 35
insulation and, 6–7, 35–36, 43, 47, 49, 120, 151t, 163, 176, 267, 269–270, 313
LEED (Leadership in Energy and Environmental Design) and, 5, 57–62, 65–68, 71, 73nn3,4, 263, 270, 272, 285, 334–335
lighting and, 8
living walls and, 316, 319, 321
loss of, 7
multinational corporations and, 103–105, 108, 110–111, 113–114, 117, 121
National Renewable Energy Laboratory and, 59
niche innovations and, 147, 150–152, 163, 165n3
North House project and, 174–176, 178, 180–181, 184, 186–189, 191
petroleum and, x
plywood and, 336
policy for, 35
reducing consumption of, 3, 6 (*see also* Sustainability)
renewable, xiii, 43, 59, 73n4, 127, 151t, 192, 203f
renovation and, 35, 38–44, 48–51
residential retrofits and, 17
savings of, 8–9
solar, 6–7 (*see also* Solar energy)
sustainable buildings and, 59, 66, 73nn3,4, 171–181, 184–192, 263, 270
thermal, 7, 36, 42, 120, 176, 178, 234, 274–275
titanium buildings and, 17, 77–79, 88–92, 95
U.S. consumption rates and, 3, 333
U.S. Department of Energy (DOE) and, x–xi, 8, 12, 19, 73n4, 79, 335
vehicle, ix–xii
vitality and, 242, 245–247, 249–250, 252, 254
waste and, 4–8, 12, 20–21, 46, 77, 79, 84–88 (*see also* Waste)
wood and, 336–337
zero net, 59, 66, 73nn3,4, 172
Energy Performance Certificates, 42
Energy Star program, xii, 91, 221
Engineered Wood Association, 337
Engineers, 2, 4
biophilic design and, 314, 348
conventional building practices and, 336–338
generativity and, 251
green schools and, 222, 223, 228, 233, 236

imperative to build and, 300
LEED (Leadership in Energy and Environmental Design) and, 59–62, 65, 69
niche innovations and, 149, 153, 155, 159
renovation and, 36, 39, 41–43, 51
sustainable buildings and, 173–174, 191–192, 273
titanium buildings and, 78–93, 96
Enrollment, 41
EnterNOC, 103
Entrepreneurs, 18
biophilic design and, 320
collective action and, 134–135
green building supply industry and, 127–142
institutional, 12, 134–135
Environmental issues
benzene and, 313–314
bioeffluents and, 313
biophilic design and, 343, 349
cigarettes and, 313
climate change and, 4, 35, 39, 42, 50, 147, 220–221, 235, 333
consumption and, 3–4 (*see also* Consumption)
conventional building practices and, 333–338
deforestation and, 337
Earth Day and, 77
formaldehyde and, 313, 315, 337
generativity and, 240–245, 251–254
GPR score and, 150, 151t, 153–155, 158–160, 162, 164
green building supply industry and, 127–130, 134–135, 140–141
greenhouse gases and, 171, 321, 333
green schools and, 220–221, 224–226, 228–237
LEED and, 5 (*see also* LEED [Leadership in Energy and Environmental Design])
living walls and, 312–314, 321
multinational corporations and, 103–121
niche innovations and, 145–159, 162–165
pollution and, 312–314, 349 (*see also* Pollution)
radon and, 313
recycling and, 84, 154–155, 209, 214, 225, 229, 235, 263, 265, 267, 270, 274, 276
renovation and, 47
resource efficiency and, 18, 69, 105–107, 110–112, 116, 120
responsibility and, 219, 220, 223, 226, 231 (*see also* Responsibility)
sustainability and, 1 (*see also* Sustainability)
titanium buildings and, 77–80, 84, 89, 91–92, 95–96, 97n4
toxic materials and, 84, 108, 135, 225, 277
volatile organic compounds (VOCs) and, 6–7, 206, 263, 313
waste and, 7 (*see also* Waste)
World Heritage Sites and, 22, 286–287, 290–292, 298–303, 304nn2,3,4
Environmental Leader, 9
Equity
brand, 109, 111
social, 5–6, 66, 199, 201, 208
Ethics, 11, 16, 129
Evolutionary organizations, 210–211
Expertise, 18–19
diversity and, 231
emerging professions and, 16–17
green schools and, 223, 228, 230–236
Green Teams and, 232–234
jurisdiction and, 14–16, 41–43
LEED (Leadership in Energy and Environmental Design) and, 216n2
legitimacy and, 127, 129, 132–138, 140
living walls and, 312
multinational corporations and, 109–110, 116
niche innovations and, 146, 149–150, 153–155, 159–160

organizational sustainability and, 200, 211–212, 216n2
renovation and, 36–37
role of, 13
skills and, 227 (*see also* Skills)
sustainable buildings and, 192
system of professions approach and, 16, 36–37, 40–42, 47–52
titanium buildings and, 86, 91
vocational training and, 49–51
World Heritage Committee and, 300

Farmer, Graham, 70
Federation of Master Builders, 38
Firms
biophilic design and, 341
collective action and, 128, 134–135, 138, 140
economic focus of, 127
economic uncertainty and, 131–133
entrepreneurial theory and, 131–134
evolution of new industries and, 130–135
expertise and, 16 (*see also* Expertise)
for-profit, 17, 127, 265–267, 274, 276, 278
generativity and, 250
green building supply industry and, 127–142
green schools and, 223
housing innovation and, 37–38
incumbent, 130–140
institutional uncertainty and, 133–134
LEED (Leadership in Energy and Environmental Design) and, 58, 60, 65, 128
legitimacy and, 127, 129, 132–138, 140
markets and, 17 (*see also* Markets)
multinational corporations and, 103–122
niche innovations and, 150, 152, 158–162, 165
professional service (PSFs), 14
public goods and, 129
renovation and, 37–38, 44–47, 50
sustainable buildings and, 265, 275–278
titanium buildings and, 78, 87, 96
First cost savings, 7–8, 68, 242, 318–319
Fisk, William, 314
Flat teams
accountability and, 229–230
decision making and, 220
diversity and, 220–221, 232
economic issues and, 226–227
educational concerns and, 227
expertise and, 233–234
green schools and, 219–237
Green Team and, 221–237
inclusive vision development and, 224–228
increased building literacy and, 231–234
increased participation and, 228–231
interactional patterns and, 220, 235–236
motivation and, 231
Operations Department and, 221–223, 224–225, 227–237
professionals and, 227–228
research groups and, 223
responsibility and, 220
shared participation and, 220
unified goals and, 225
Floors, 7, 43, 49, 150, 151t, 156, 176, 343, 347
Forest bathing, 311
Forest Stewardship Council (FSC), 337–338
Formaldehyde, 313, 315, 337
For-profit firms, 17, 127, 265–267, 274, 276, 278
Fossil fuels, x, 321
Fowler Center for Sustainable Value, 245
Fractals, 345
Fragmentation, 2–3, 13
imperative to build and, 286
integrated project delivery (IPD) and, 62, 78, 86–88, 90, 97nn6,7,9

LEED (Leadership in Energy and Environmental Design) and, 62, 63t, 70
niche innovations and, 155
renovation and, 36–37, 43, 46
titanium buildings and, 79, 83, 87
Franzen, Bill, 20, 219–238
Fromm, Eric, 309
Fry, Ronald, 20–21, 239–259
Furniture, 263, 267, 276, 313, 336–337

General benefits, 9
General Services Administration, 69
Generativity, 20–21
accountability and, 240
aliveness and, 241, 243, 245–247, 249, 252, 254
appreciative inquiry (AI) and, 239–240, 243–244, 253–254
architects and, 247
benefits of, 239–254
built environment and, 254
capacity-creating dynamics and, 243
civic responsibility and, 244
clarity and, 241, 246–247, 249, 254
collaboration and, 239
concept of, 243–245
contractors and, 239, 247
cultural issues and, 254
design and, 240
diversity and, 246
economic issues and, 241–242
efficiency and, 239–241, 254
energy and, 239–250, 254
engineers and, 251
environmental issues and, 240–245, 251–254
firms and, 250
health issues and, 241–243, 249, 252
high-performance buildings and, 253
implications of, 252–254
individual thriving and, 245–254
innovation and, 245, 249
learning and, 241–242, 245–254
LEED (Leadership in Energy and Environmental Design) and, 241, 247–252
lighting and, 241
narratives and, 251, 253
operating costs and, 242
organizational capacity building and, 240, 244–245, 249–252
pollution and, 241
positive meaning and, 251–252
positive organizing and, 239, 243, 250–251
practice perspective and, 241–242
productivity and, 239, 242, 244, 251–254
property value and, 242
psychology and, 243, 248
regulation and, 242
resources and, 240–244, 249–250, 254
responsibility and, 240–241, 244–245, 251–254
skills and, 239, 245, 248–249
social issues and, 241–243, 253
stakeholders and, 239–244, 248, 251–254
sustainability and, 241–247, 251–255
team context and, 240–250, 253–254
United States Green Building Council (USGBC) and, 241
vitality and, 242, 245–247, 249–250, 252, 254
waste and, 242
water and, 242
Genetron Systems, 316
Gentry, Thomas A., 37
Genzyme Corporation, 91
Geothermal energy, 7, 120, 233–234, 274
Gestalt theory, 251
Gieryn, Thomas F., 264–265, 267
Gilman, Robert, 6
Glass, 6, 202, 334, 347. *See also* Windows

Global Reporting Initiative (GRI), 199–200, 204
Glynn, Mary Ann, 243–244
Godish, Thad, 315
Goodwin, Georgia, 310
Governance structures, 13–14, 130, 198, 203, 211, 214
GPR score, 150, 151t, 153–155, 158–160, 162, 164
Gravitational assist
  business knowledge kinds and, 210
  competitive forces and, 206–207
  cultural issues and, 206–207, 210
  motivating forces and, 204–210
  organizational sustainability and, 19, 197–198, 200, 203–216
  regulation and, 206–210
  sharing knowledge to create, 210–214
  vectors of action and, 203–204
Greater Pittsburgh Community Food Bank, 201–202, 207–209, 212
Green, Stephen, 199
Green building
  actor-network theory (ANT) and, 40–42, 47–48
  Adam Joseph Lewis Center and, 57–60, 73n4
  Adaptive Living Interface System (ALIS) and, 178, 180–190
  ANSI/ASHRAE/USGBC/IES Standard 189 and, 69
  biophilic design and, 307–327, 341–349
  buildability and, 47–49
  buildings as dynamic entities and, 77
  built environment and, 15–16, 19, 35–36, 42, 51–52, 80 (*see also* Built environment)
  Chartwell School and, 58, 63–67, 71
  climate change and, 4, 35, 39, 42, 50, 147, 220–221, 235, 333
  collaboration and, 17, 20 (*see also* Collaboration)
  concept of, 1–2, 6, 240–241
  cultural issues and, 11 (*see also* Cultural issues)
  deep green and, 21, 207, 268, 273–274, 278t, 279
  deeply held beliefs over, 15
  ethics and, 11, 16, 129
  expertise and, 13–14 (*see also* Expertise)
  first cost savings and, 7–8, 68, 242, 318–319
  fiscal waste and, 20
  general benefits and, 9
  generativity and, 239–255
  governance structures and, 13–14, 130, 198, 203, 211, 214
  green innovators and, 21, 207, 268, 274–280, 278
  health issues and, 5, 8–9, 22, 42, 60, 63 (*see also* Health issues)
  hidden green and, 21, 207, 268, 276–279
  high-performance, 5–6, 8, 17, 20–21, 57, 63, 69, 71, 172, 174, 176, 178, 221, 224–231, 235, 253, 272
  increased building literacy and, 231–234
  increasing buildability and, 48–49
  individual thriving and, 245–254
  influence of ratings and, 79–81
  inspectors and, 15, 36
  insulation and, 6–7, 35–36, 43, 47, 49, 120, 151t, 163, 176, 267, 269–270, 313
  integrated project delivery (IPD) and, 62, 78, 86–88, 90, 97nn6,7,9
  LEED and, 57–58 (*see also* LEED [Leadership in Energy and Environmental Design])
  life-cycle assessment (LCA) and, 73n1, 77, 79, 81–85, 89–91, 94, 120, 128, 155, 161t, 162, 203f, 221, 224, 337
  living laboratories and, 202–203
  multinational corporations and, 103–106, 116, 118, 120–121
  negative publicity over, 58–60
  niche innovations and, 147, 162
  North House project and, 174–190, 335

operating costs and, 8, 68, 90, 105, 107, 109, 112–116, 242, 342
organizational theory and, 12
practical green and, 21, 207, 268, 271–273, 278t
practice perspective and, 241–242
professionals and, 13–14
profit-driven green and, 21, 207, 268–271, 278t
renovation and, 35–37, 42, 51–52
request for qualification (RFQ) and, 58, 64–65
social issues and, 3–10 (*see also* Social issues)
social-technical system for, 47–51
"state of the art" approach and, 35–36, 84–85
sustainability and, 11, 171 (*see also* Sustainability)
titanium buildings and, 77–98
Green Building Information Gateway, 71, 79
Green building supply industry
architects and, 140
built environment and, 129
collaboration and, 134–135
collective action and, 128, 134–135, 138, 140
consumption and, 137
contractors and, 128
decision making and, 131
ecological issues and, 127–132, 138
economic uncertainty and, 131–133
efficiency and, 130
emergence of, 129, 136–140
energy and, 127–128
entrepreneurs and, 127–142
environmental issues and, 127–13l, 134–135, 140–141
evolving new industries and, 130–135
for-profit firms and, 127
health issues and, 132, 135
incumbent firms and, 130–140
innovation and, 131, 137
institutions and, 127–131, 133–141
LEED (Leadership in Energy and Environmental Design) and, 128–129, 132–134, 136–140
legitimacy and, 127, 129–130, 132–138, 140
management and, 130
markets and, 127–133, 136–142
materials and, 128–129, 135, 137, 139, 141
niche innovations and, 129
policy and, 132, 141
politics and, 134
psychology and, 145
public good and, 127–134, 137–141
regulation and, 140–141, 146
social movements and, 127–138, 141
solar energy and, 138
spillovers and, 127–134
sustainability and, 127, 135, 140–142
technology and, 131, 135, 137, 141
United States Green Building Council (USGBC) and, 130–141
waste and, 135
water and, 135
Greenhouse gases, 171, 321, 333
Greenhouses, 202, 208
Greening the Building and the Bottom Line (Rocky Mountain Institute), 341–343
Green innovators, 21, 207, 268, 274–280, 278
Green schools
accountability and, 230
architects and, 222, 227, 228, 232–233, 236
bond cycles and, 221
case methodology for, 221–222
consumption and, 223, 236–237
contractors and, 222–224, 233
cultural issues and, 227–228, 235
design and, 128, 135, 138, 141, 219–237
diversity and, 20, 220–221, 224, 228, 232, 235–236
economic issues and, 220, 226–227
educational concerns and, 227

efficiency and, 221, 226–227, 229, 231
electricity and, 222–223, 225, 227, 231, 236–237
energy and, 221–229, 233, 235–236
engineers and, 222, 223, 228, 233, 236
environmental issues and, 219–221, 224–226, 228–237
expertise and, 223, 228, 230–236
firms and, 223
flat team approach and, 219–237
Green Team and, 220–237
health issues and, 219, 224
high-performance buildings and, 20, 221–231, 235
housing and, 236
inclusive vision development and, 224–228
increased building literacy and, 231–234
interactional processes and, 234–236
LEED (Leadership in Energy and Environmental Design) and, 68, 220–221, 227, 233
life-cycle assessment (LCA) and, 221, 224, 234
lighting and, 8–9, 227, 233
living walls and, 324
management and, 20
materials and, 225
motivation and, 231
Operations Department and, 220–222, 224, 227–237
politics and, 220–221, 224–226
pollution and, 225
productivity and, 8–9, 227
professionals and, 227–228, 230
research groups and, 223
resources and, 219, 221–222, 223, 224–225, 229, 232
responsibility and, 220, 223, 226, 231
skills and, 228
stakeholders and, 21
sustainability and, 221, 223–225, 231, 235
technology and, 234
unified goals and, 225
United States Green Building Council (USGBC) and, 219, 236
utility costs and, 221–222, 223
waste and, 222, 223, 226
water and, 221, 223, 225, 232
Green Teams
accountability and, 230
diversity and, 222–223, 232
economic issues and, 226–227
educational concerns and, 227
expertise and, 233–234
green schools and, 220–237
inclusive vision development and, 224–228
increased building literacy and, 231–234
increased participation and, 228–231
interactional processes and, 234–236
motivation and, 231
Operations Department and, 220–222, 224, 227–237
organizational sustainability and, 20, 214
professionals and, 227–228
research groups and, 223
unified goals and, 225
Gross domestic product (GDP), 2, 17
Guindon, Carlos, 315
Guy, Simon, 70

Hahn, Christopher, 198, 213–214
Hannover Principles, 58, 73n2
Hannoversch Munden, 348
Hassler, Uta, 38
Hawken, Paul, 269
Hawthorne effect, 342
Hayek, F. A., 131
Hayter, Sheila J., 61
Health issues, 5, 8–9, 22
benzene and, 313–314
bioeffluants and, 313
biophilic design and, 307–315, 319, 322–324, 341, 343, 349
cigarettes and, 313

conventional building practices and, 337
formaldehyde and, 313, 315, 337
generativity and, 241–243, 249, 252
green building supply industry and, 132, 135
green schools and, 219, 224
indoor air quality and, 312–314
LEED (Leadership in Energy and Environmental Design) and, 60, 63, 67–68
multinational corporations and, 113
natural light and, 309–311, 325, 342–343
niche innovations and, 150, 165n3
organizational sustainability and, 203f
psychological, 4, 19, 22, 37, 39, 117, 145, 172, 178–180, 192, 243, 248, 307–312, 317–320, 322, 344, 346
radon and, 313
renovation and, 42
safety and, 67, 201, 203f, 248
Sick Building Syndrome and, 313
sustainable buildings and, 178, 180, 271–272, 277
toxic materials and, 84, 108, 135, 225, 277
volatile organic compounds (VOCs) and, 6–7, 206, 263, 313
Heerwagen, Judith, 311, 346
Henn, Rebecca L., 1–32, 241
Herman Miller SQA, 343–344
Herron, Jock, 17, 77–100
Heschong, Lisa, 345
Heschong Mahone Group, 8–9, 345
Hester, Randolph, 310
Hidden green, 21, 207, 268, 276–279
High-performance buildings
concept of, 5–6, 21
economic issues and, 8
generativity and, 253
green schools and, 20, 221–231, 235
innovation and, 17
LEED (Leadership in Energy and Environmental Design) and, 57, 63, 69, 71
North House project and, 174–190
sustainability and, 172, 174, 176, 178, 272
titanium buildings and, 17
Hilton (hotel chain), 110
Hirsch, Paul M., 286
Hockley, Stephen, 19–20, 197–217
Hoffman, Andrew J., 1–32, 286
Honeywell and Johnson Controls, 89
Horman, Michael, 71
Hospitality industry
brand equity and, 109, 111
consumer behavior cost factor (CBCF) and, 105, 107, 110–115
data availability and, 104
EPA Portfolio Manager and, 120–121
guest type and, 112–113
incentive design and, 117–120, 122
industry commitment and, 104
information sharing and, 118
innovation and, 104, 108, 110–111, 114
LEED (Leadership in Energy and Environmental Design) and, 121
location and, 111–112
multi-agent collaboration and, 116–117
multinational corporations and, 18, 106–122
operating cost factor (OCF) and, 105, 107, 110–115
ownership structure and, 110–111
paired-case comparisons and, 114–116
policy implications for, 116–121
regional ambient temperature and, 113–114
revenue per available room and, 106
site-specific alignment and, 118
supply chains and, 106–109, 120, 122
sustainability and, 106–122
transactional burdens and, 120
"House as a system" approach, 35, 43
House of Commons Chamber, 1

Housing, xii
biophilic design and, 312
buildability and, 47–49
carbon refurbishment and, 36–37, 43–52
cohousing and, 15
energy efficiency and, 35
green schools and, 236
imperative to build and, 285
innovation and, 37–40, 147
North House project and, 174–190, 335
redefining green home and, 174
renovation and, 35–40, 43–51
sustainable buildings and, 171–172, 186
Howlett, Owen, 9
HSBC, 199
Hudson River, 347
Hunter, Claudia, 9
HVAC (heating, venting, and air conditions), 172, 220–222, 231, 267, 270
Hybrid vehicles, x
Hydrocarbons, ix

Identity, 4, 13, 264
Illuminating Engineering Society of North America (IESNA), 69
Imperative to build
Adam Joseph Lewis Center and, 58–60
challenging, 285–304
competing views over, 289–304
competition of potential futures and, 293–294
defending of, 286–288, 290–293, 298–302
determination of, 286, 290–293, 302
Dresden case study and, 289–304
engineers and, 300
fragmentation and, 286
housing and, 285
LEED as Lunatic Environmentalists Enthusiastically Demolishing, 285
life-cycle assessment and, 73n1, 77, 79, 81–85, 89–91, 94, 120, 128, 155, 161t, 162, 203f, 221, 224, 337
management and, 286, 290, 299, 304nn1,2
narratives and, 287–295, 299, 302
policy and, 297, 303
rapidity of environmental changes and, 285–286
reality and, 287
resources and, 288, 297, 303
shedding ambiguity on, 287–289
stabilization of, 286, 290–293, 295–298, 302
stakeholders and, 22, 288, 293, 295
uncertainty and, 286
Waldschlößchenbridge and, 286–302
World Heritage Sites and, 22, 286–287, 290–292, 298–303, 304nn2,3,4
Incentive design, 117–120, 122
Individual thriving
aliveness and, 241, 243, 245–247, 249, 252, 254
generativity and, 245–254
learning and, 241–242, 245–254
organizational capacity building and, 240, 244–245, 249–252
positive meaning and, 251–252
positive organizing and, 250–251
vitality and, 242, 245–247, 249–250, 252, 254
Indoor air pollution, 312–315
Indoor Environmental Department, 314
Innes, Judith E., 61–62
Innovation, x
Buffalo Organization for Social and Technological Innovation (BOSTI) and, 342
decision making and, 22, 37, 57, 61, 71, 90, 94, 112, 131, 162–163, 189, 199, 208, 212–214, 220, 267, 287, 293, 309, 320, 323, 326
developing regional networks for, 50–51
diffusion of, 13
flat team approach and, 220–237
generativity and, 245, 249

green building supply industry and, 131, 137
green innovators and, 21, 207, 268, 274–280
green schools and, 220–237
high-performance buildings and, 17
hospitality industry and, 104, 108, 110–111, 114
housing and, 37–40, 147
informational access and, 333
inhibition of, 37
knowing one's audience and, 21
LEED (Leadership in Energy and Environmental Design) and, 5, 57, 60, 78, 322
living walls and, 322
low-carbon refurbishment and, 50–51
mainstreaming, 10, 16, 18–19, 44–50, 140, 145–165, 221, 277
niche, 9–10, 18–19, 22, 129–165, 316
organizational theory and, 12
professionals and, 39–43, 52
renovation as, 35–52
sustainability and, 171, 176, 206, 270, 276 (*see also* Sustainability)
titanium buildings and, 78, 84, 90, 95–96
UK construction industry and, 46–47
Innovation in Design (ID), 322
Inspectors, 15, 36
Institutions, 16–17, 187, 266, 279
as barrier to innovation, 37
biophilic design and, 324, 341, 343
building sector and, 2, 4, 7, 15, 35, 128, 132–142, 172, 216n1
collective action and, 128, 134–135, 138, 140
conventional building practices and, 335–337
entrepreneurs and, 12, 134–135
green building supply industry and, 127–141
imperative to build and, 21, 285, 287, 302–303
LEED (Leadership in Energy and Environmental Design) and, 57, 60–62, 71
markets and, 8 (*see also* Markets)
multinational corporations and, 108, 112, 120
multiple institutional logics and, 14, 23, 129
niche innovations and, 145–148, 160
organizational sustainability and, 199, 212, 215
realignment of, 3, 6, 10–13, 133–134, 138
renovation and, 36–37, 42, 51
titanium buildings and, 79, 81, 83, 86–87, 91, 97nn5,6
uncertainty and, 133–134
Insulation, 6–7, 35–36, 43, 47, 49, 120, 151t, 163, 176, 267, 269–270, 313
Insurance, 8, 242
Integrated project delivery (IPD), 62, 78, 86–88, 90, 97nn6,7,9
Integrative Design Guide to Green Building (7group and Reed), 212
Interface Carpeting, 270

Janda, Kathryn B. 16–17, 35–55
Joglekar, Nitin R., 18, 103–125
Johnson Wax Company, 347
Jones, Bryan D., 67
Jurisdiction
fragmentation and, 13 (*see also* Fragmentation)
LEED (Leadership in Energy and Environmental Design) and, 69
professionals and, 14–16, 41–43
renovation and, 41–43
system of professions approach and, 16, 36–37, 40–42, 47–52

Kador Group, 9
Kats, Gregory E., 9–10
Kellert, Stephen R., 311
Kelly, Michael J., 39–40
Kibert, Charles J., 241–242
Killip, Gavin, 16–17, 35–55

Koebel, C. Theodore, 37
Kohler, Niklaus, 38
Kok, Nils, 7
Korkmaz, Sinem, 71
Kuo, Frances, 312, 349
Kwok, Alison G., 17, 57–76

Lanoie, Paul, 108
Larsen, Larissa, 17, 57–76
Laszlo, Christopher, 210
Laszlo, Ervin, 210
Laszlo, Kathia, 210–211
Lawrence Berkeley National Laboratory, 314
LEED (Leadership in Energy and Environmental Design), xii, 21
  Adam Joseph Lewis Center and, 57–60, 73n4
  ANSI/ASHRAE/USGBC/IES Standard 189 and, 69
  architects and, 57–62, 65–66, 69–70
  authentic dialogue and, 58, 61
  biophilic design and, 322, 347
  building information modeling (BIM) and, 57, 62, 73n1
  bureaucracy of, 68–69
  certification by, 5, 7, 9, 17, 61, 66–71, 81, 97n8, 128, 133, 136–137, 140, 200–202, 216, 220, 241, 248–250, 263–264, 267–268, 322, 334–335
  Chartwell School and, 58, 63–67, 71
  contractors and, 57, 59, 62, 68
  conventional building practices and, 334–335
  decision making and, 57, 61, 71, 90
  development of rating system of, 60–63
  diversity and, 57, 61, 67
  early adopters of, 212, 264, 268, 275–279
  economic issues and, 7–9, 68, 81, 90
  efficiency and, 5–7, 58–60, 65, 67, 69, 73nn3,4, 270, 334
  electricity and, 58–59
  energy and, 57–62, 65–68, 71, 73nn3,4, 263, 270, 272, 285, 334–335
  engineers and, 59–62, 65, 69
  Existing Buildings, Operations, and Maintenance (LEED EBOM) and, 103, 201, 203, 208–209, 213, 215
  expertise and, 216n2
  firms and, 58, 60, 65, 128
  fragmentation and, 62, 63t, 70
  future research and, 70–72
  generativity and, 241, 247–252
  Gold rating and, 5, 61, 66, 81, 221, 334
  green building supply industry and, 128–129, 132–134, 136–140
  green schools and, 68, 220–221, 227, 233
  health issues and, 60, 63, 67–68
  high-performance buildings and, 57, 63, 69, 71
  hospitality industry and, 121
  imperative to build and, 285, 302–303
  influence of ratings and, 79–81
  innovation and, 5, 57, 60, 78, 322
  institutions and, 57, 60–62, 71
  jurisdiction and, 69
  lighting and, 62–66, 78, 263, 271
  limitations of, 57
  living walls and, 322
  as Lunatic Environmentalists Enthusiastically Demolishing, 285
  management and, 72
  markets and, 57, 60, 67, 96n2, 97n8
  materials and, 5, 60, 70, 201, 250, 263, 267, 270, 275
  multinational corporations and, 103, 121
  New Construction (LEED NC) and, 201, 208, 268, 272, 280n3
  next steps for, 69–70
  operating costs and, 68, 90
  organizational narratives and, 21–22, 207, 263–279
  Pilot Project Program and, 60–61

Platinum rating and, 5, 7, 61, 78, 81, 84–87, 90, 202, 247, 276, 347
politics and, 67, 220
pollution and, 66
productivity and, 63–64, 68
professionals and, 61, 69–72, 81, 92–94, 97n3, 136–137, 140, 200, 268, 272, 276–277
promotion of collaboration by, 57–58, 61–72, 73nn1,5
as public policy, 67–68
request for qualification (RFQ) and, 58, 64–65
resistance to, 67–69
resources and, 60, 69–70, 216n1
responsibility and, 62, 69
Silver rating and, 5, 61, 81, 220
skills and, 72, 94
social issues and, 60, 66, 70, 77–81
solar energy and, 58–59, 263, 267
stakeholders and, 57, 60, 65, 67, 70, 73n1, 264–267, 271–272, 276–279
supply chains and, 78, 80, 83t, 86, 88
sustainability and, 60–61, 65–66, 69–70, 73n2, 174, 197, 200–204, 208–216, 263–279, 280n3, 285
technology and, 57, 62, 68, 73n3
titanium buildings and, 78–94, 97nn3,4,8
training and, 93–94
waste and, 263
water and, 263, 269, 274
Legitimacy
decision making and, 288
green building supply industry and, 127, 129, 132–138, 140
Levitt, Raymond E., 38
Lewis, Marianne W., 108
Life-cycle assessment (LCA)
building information model (BIM) and, 73n1
conventional building practices and, 337
green schools and, 221, 224, 234
multinational corporations and, 120, 128
niche innovations and, 155, 161t, 162
organizational sustainability and, 203f
titanium buildings and, 79, 81–85, 89–91, 94
Lighting
Adaptive Living Interface System (ALIS) and, 180–181, 189
automatic shut-off, 241
biophilic design and, 309–311, 325, 342–343
bounced, 348
daylight, 178 (*see also* Daylight)
energy usage of, 8
fluorescent, 7
generativity and, 241
green schools and, 8–9, 226, 232
IESNA and, 69
LED, 156, 189, 263
LEED (Leadership in Energy and Environmental Design) and, 62–66, 78, 263, 271
motion-sensor corridors and, 7
natural, 9, 58, 62–66, 78, 135, 178, 203f, 219, 227, 233, 271, 277, 311, 342–344, 348
niche innovations and, 151, 154, 156
productivity and, 8–9, 341–344
recessed, 49
reducing artificial, 178
reflective, 95
skylights and, 343–344
sustainable buildings and, 178, 180–181, 189
workplace productivity and, 9
Light pollution, 66
Lindsey, Gail, 61
Living Building Challenge, 202–203, 208–209
Living laboratories, 202–203
Living PlanIT Urban Operating System, 95
Living walls
benzene and, 313–314
biofiltration and, 316
concerns about, 324–325

connection with nature and, 309–311, 314, 316
conventional building practices and, 334
decision making and, 309, 317, 320, 323, 326
economic issues and, 317–323, 326
efficiency and, 314
emotion and, 310
energy and, 316, 319, 321
environmental issues and, 312–314, 321
examples of, 308f, 315–317
expertise and, 312
green schools and, 324
important factors of, 322
indoor air quality and, 312–314
innovation and, 322
interest of upper management in, 321
LEED (Leadership in Energy and Environmental Design) and, 322
light and, 309–311, 325
novelty of, 320–321
objective attributes of, 323–324
phyto-remediation and, 316
praise for, 324
professionals and, 317
psychology and, 307–314, 317–323
resources and, 310, 318
Sick Building Syndrome and, 313
student performance and, 310–311
suggestions for future projects of, 325
three definitive conclusions for, 326–327
water and, 311, 314–316, 324–325
worker productivity and, 310–311, 314
Lohr, Virginia, 310, 314–315
Loisos, George, 66–67
Loorbach, Derk, 211
Lounsbury, Michael, 286
Lovins, Amory, 269
Low and zero carbon (LZC) technologies, 35, 43, 49
Low-carbon refurbishment
increasing buildability and, 47–49
innovation and, 46–47, 50–51
integration capacity and, 49–50
regional networks and, 50–51
socio-technical system of professions and, 47–51
United Kingdom and, 44–51
Lucid Design Group, 79

Macallen Building, 334–335
Mador, Martin L., 311
Malcolm Baldrige National Quality Award, 91
Management, 2, 23
biophilic design and, 321, 342
collaboration and, 86–88 (*see also* Collaboration)
Distributed Responsive System of Skins (DReSS) and, 176–180
green building supply industry and, 130
green schools and, 20
imperative to build and, 286, 290, 299, 304nn1,2
LEED (Leadership in Energy and Environmental Design) and, 72
multinational corporations and, 103–111, 114–115, 118, 120–121
niche innovations and, 18–19, 145–152, 159, 163
organizational sustainability and, 202, 210–212, 214
professionals and, 14 (*see also* Professionals)
renovation and, 38, 43, 46, 48
social issues and, 13
sustainable buildings and, 186 (*see also* Sustainability)
titanium buildings and, 78–79, 84, 86–90, 94–96, 97nn5,7
Mankoff, Jennifer, 187
Manufacturing
biophilic design and, 342–343
conventional building practices and, 334, 337
declining base of, 14

efficiency and, 80 (*see also* Efficiency)
green schools and, 222, 231, 233, 245
motor vehicles and, xi–xii
multinational corporations and, 108
niche innovations and, 153
organizational sustainability and, 198
renovation and, 38, 49
sustainable buildings and, 176, 270
titanium buildings and, 80, 90
Markets
biophilic design and, 316–319, 322
building sector and, 2, 4, 7, 15, 35, 128, 132–142, 172, 216n1
conventional building practices and, 333–338
green building supply industry and, 127–133, 136–142
LEED (Leadership in Energy and Environmental Design) and, 57, 60, 67, 96n2, 97n8
multinational corporations and, 104, 108–110, 116, 118, 121
NAFTA and, 206
niche innovations and, 152, 157–160, 161t
organizational sustainability and, 197, 200, 206, 210, 214, 216n1
profit-driven green and, 21, 207, 268–271, 278t
renovation and, 38, 45, 47, 49, 51
strategies for, 17–19
structures of, 17–19
supply chains and, 6 (*see also* Supply chains)
sustainable buildings and, 266–270, 275, 277
transition of, 5, 8, 10, 12, 17–19
uncertainty and, 129–134
Marriot (hotel chain), 110
Materials
biophilic design and, 313, 345–346
consumption and, 4 (*see also* Consumption)
conventional building practices and, 22, 334–338
disposal of, 22
formaldehyde and, 313, 315, 337
generativity and, 241, 248, 250
green building and, 3 (*see also* Green building)
green building supply industry and, 128–129, 135, 137, 139, 141
green schools and, 225
LEED (Leadership in Energy and Environmental Design) and, 5, 60, 70, 201, 250, 263, 267, 270, 275
multinational corporations and, 106
natural analogs and, 345–346
niche innovations and, 154–155, 161t, 165n3
organizational sustainability and, 201, 203f, 204, 207, 209
plywood and, 336–337
polyvinyl acetate (PVAc) and, 337
recycling and, 84, 154–155, 209, 214, 225, 229, 235, 263, 265, 267, 270, 274, 276
regionally sourced, 7
renovation and, 48–50
soy-based, 337
sustainable buildings and, 173–178, 189, 191, 263, 267, 270, 275
titanium buildings and, 17, 78, 84
toxic, 84, 108, 135, 225, 277
Matthiessen, Lisa Fay, 7–8
Mayne, Thom, 69
McDonough, William, 269
McGraw Hill Construction, 8, 91
Meadows, Donella, 206
Meyer, Michelle A., 20–21, 219–238
Mills, C. Wright, xiv
Mobile devices, 187–188
Mondor, Christine, 19–20, 197–217
Moore, Steven A., 70
Morris, Peter, 8
Motor Vehicle Air Pollution Control Act, ix
Motor vehicles
air pollution and, ix–xii
CAFE standards and, xi

efficiency and, 77
electric, x, 146
urban flow and, 95
Waldschlößchenbridge and, 298
Multinational corporations
brand equity and, 109, 111
built environment and, 103–106, 116, 118, 120–121
collaboration and, 104, 109, 116
consumer behavior cost factor (CBCF) and, 105, 107, 110–115
consumption and, 103–107, 110–116, 120–121
contractual bridges and, 108
data availability and, 104
design and, 103, 108, 110, 117–118, 121–122
economic issues and, 103–104, 108–109, 111, 116–117, 120
efficiency and, 105–107, 110–112, 116, 120–121
electricity and, 106–107, 111
energy and, 103–105, 108, 110–111, 113–114, 117, 121
environmental issues and, 103–121
expertise and, 109–110, 116
Global Reporting Initiative (GRI) and, 199–200, 204
green building and, 103–106, 116, 118, 120–121
health issues and, 113
hospitality industry and, 18, 106–122
HSBC and, 199
incentive design and, 117–120, 122
industry commitment and, 104
information sharing and, 118
institutions and, 108, 112, 120
LEED (Leadership in Energy and Environmental Design) and, 103, 121
life-cycle assessment (LCA) and, 120, 128
location and, 111–112
management and, 103–111, 114–115, 118, 120–121
markets and, 104, 108–110, 116, 118, 121
operating costs and, 105, 107–116
ownership structure and, 110–111
policy and, 104–105, 112, 116–121
pollution and, 108
regulation and, 108
resources and, 103–116, 120–121
social issues and, 103, 117, 266
solar energy and, 105, 114, 116–117, 120
stakeholders and, 108, 116, 120
supply chains and, 104, 106–109, 120, 122
sustainability and, 103–122
technology and, 108, 120–121
transactional burdens and, 120
United States Green Building Council (USGBC) and, 103
waste and, 108, 112
water and, 106–107, 111, 114, 116–118
Munich Re, xiii

Narratives, 16
analysis methodology for, 267–268
biophilic design and, 318
deep green and, 21, 207, 268, 273–274, 278t, 279
generativity and, 251, 253
green innovators and, 21, 207, 268, 274–280
hidden green and, 21, 207, 268, 276–279
imperative to build and, 287–295, 299, 302
LEED (Leadership in Energy and Environmental Design) and, 21–22
market, 266
organizational, 21, 207, 263–279
practical green and, 21, 207, 268, 271–273, 278t
process of becoming and, 265
profit-driven green and, 21, 207, 268–271, 278t
stakeholder theory and, 265
sustainable buildings and, 21–22, 207

National Aeronautics and Space Administration (NASA), 315
National Association of Home Builders (NAHB), 12
National Environmental Education and Training Foundation (NEETF), 184
National Renewable Energy Laboratory, 59
National Socialists, 301
Nattrass, Brian, 210–211
Natural analogs, 344–347
Nature in the space, 344–345
Nature of the space, 346–347
Nedlaw Living Walls, 316
Netherlands
  design competition and, 149–152
  GPR score and, 150, 151t, 153–155, 158–160, 162, 164
  niche innovations and, 146–161
  organizational sustainability and, 211
  principal-agent relationships and, 155–156, 162
  regime influence and, 146–149, 152–163
  Regional Environmental Service and, 162
  town hall project in, 146–164
Net present value (NPV), 8, 90, 267
New Urbanism, xiii
Niche innovations, 9–10, 18–19, 22
  architects and, 142–144, 159, 162
  barriers to, 145
  biophilic design and, 316–317
  case methodology for, 148–149
  collaboration and, 153–156, 159, 162, 165
  concept of niche and, 145–147
  contractors and, 153, 155–159
  cultural issues and, 145–147, 152, 160, 161t
  decision making and, 162–163
  design and, 146–165
  ecological issues and, 153–154, 161t
  economic issues and, 147–148, 153
  efficiency and, 150–153, 161t, 163
  electricity and, 146
  energy and, 147, 150–152, 163, 165n3
  engineers and, 149, 153, 155, 159
  environmental issues and, 145–159, 162–165
  expertise and, 146, 149–150, 153–155, 159–160
  firms and, 150, 152, 158–162, 165
  fragmentation and, 155
  GPR score and, 150, 151t, 153–155, 158–160, 162, 164
  green building supply industry and, 129
  guiding principles and, 153–154
  health issues and, 150, 165n3
  industry structure and, 155–156
  institutions and, 145–148, 160
  knowledge and, 159–160
  landscape of, 145–148, 153
  life-cycle assessment (LCA) and, 155, 161t, 162
  lighting and, 151, 154, 156
  long-term visions and, 148
  mainstreaming, 145–165
  management and, 18–19, 145–152, 159, 163
  markets and, 152, 157–160, 161t
  materials and, 154–155, 161t, 165n3
  Netherlands and, 146–161
  policy and, 146, 150–152, 158–160, 161t
  politics and, 147, 150, 158–159
  principal-agent relationships and, 155–156, 162
  quality and, 145, 158
  reflexive capacity and, 148
  regime influence and, 146–149, 152–163
  regulation and, 150, 152, 158, 160, 161t
  resources and, 151, 153–154, 161t
  role of projects and, 146–149
  socio-technical regimes and, 145–149, 152–165
  solar energy and, 151, 154
  stakeholders and, 149

sustainability and, 145–165
town hall project and, 146–164
waste and, 151, 155, 165n3
water and, 151t, 158, 165n3
Nitrogen oxides, ix
North American Free Trade Agreement (NAFTA), 206
North House project
Adaptive Living Interface System (ALIS) and, 178, 180–190
ambient information and, 187–189
communication technologies and, 176
conventional building practices and, 335
Distributed Responsive System of Skins (DReSS) and, 176–180
energy and, 174–176, 178, 180–181, 184, 186–189, 191
feedback technologies and, 174, 176, 178, 181, 184–190
mobile devices and, 187–188
photovoltaics and, 178
smaller living spaces of, 176
solar energy and, 176, 178, 187
spatial/lifestyle integration and, 187–189
Nösperger, Stanislas, 52
Not So Big House, The (Susanka), xii–xiii

Oberlin College, 57–61, 66–67, 70–71
Operating cost factor (OCF), 105, 107, 110–115
Operating costs
biophilic design and, 342
generativity and, 242
green building and, 8, 68, 90, 105, 107, 109, 112–116, 242, 342
LEED (Leadership in Energy and Environmental Design) and, 68, 90
multinational corporations and, 105, 107–116
titanium buildings and, 90
Organizational sustainability
architects and, 197
business knowledge kinds and, 210
capacity building and, 249–252
collaboration and, 211–216, 221, 224, 227–231, 234–235
competitive forces and, 206, 207
contractors and, 212
cultural issues and, 19–21, 72, 197, 202, 206–212
decision making and, 199, 208, 212–214
definition for, 198–200
diversity and, 203f
ecological issues and, 199, 202
economic issues and, 199, 206, 208
efficiency and, 201–208, 213
energy and, 197, 202–204, 209–210, 213
environmental issues and, 197–206, 211–214
evolutionary corporations and, 210–211
expertise and, 200, 211–212, 216n2
framework selection and, 215
Global Reporting Initiative (GRI) and, 199–200, 204
gravitational assist and, 19, 197–198, 200, 203–216
Green Team and, 20, 214
health issues and, 203f
institutions and, 199, 212, 215
leadership and, 216
LEED (Leadership in Energy and Environmental Design) and, 197, 200–204, 208–216
life-cycle assessment (LCA) and, 203f
living laboratories and, 202–203
management and, 202, 210–212, 214
markets and, 197, 200, 206, 210, 214, 216n1
materials and, 201, 203f, 204, 207, 209
motivating forces and, 204–210
Netherlands and, 211
organizational capacity building and, 240, 244–245, 249–252
people, process, place, and, 20, 198, 200–204, 208

policy and, 199, 203f, 206, 209
positive meaning and, 251–252
positive organizing and, 250–251
principles for action and, 214–216
productivity and, 203f
quality and, 200–201, 206
regulation and, 199, 203, 206–210
resources and, 199–200, 204, 207, 211, 215, 216n2
sharing knowledge and, 210–214
smaller organizations and, 200–203
social issues and, 197–201, 212
stakeholders and, 212–213, 224
supply chains and, 201
systemic approach to, 19, 197–198
technology and, 204
United States Green Building Council (USGBC) and, 200
vectors of action and, 203–204
waste and, 202, 203f, 208–210
water and, 202, 203f, 207
Organizational theory, 2, 12
Organization-as-brain approach, 211
Organization-as-machine approach, 210–211
Orr, David W., 58–61, 173–174
Outermost House, The (Beston), 341
Ownership structure, 110–111

Paint, 6–7, 206, 263, 313, 319
Paley Park, 348
Passive House, xii
Paumgartten, Paul V., 241
Pearson-Mims, Caroline H., 310, 314–315
People, process, place, 20, 198, 200–204, 208
Petrichenko, Ksenia, 35–36
Petroleum, x
Phipps Conservatory and Botanical Gardens, 202–203, 207–213
Photovoltaics, 58–59, 73n4, 178, 270
Phyto-remediation, 316
Piacentini, Richard, 198, 202, 209, 213
Pittsburgh Opera, 201–202, 207–209, 212–214
PKF Hospitality Research, 106
Plants. *See also* Biophilic design
air quality and, 312–315
BioHome and, 315
carbon dioxide and, 314–315
physical benefits of, 314–315
psychological benefits of, 22, 307–322, 344–346
stress reduction and, 310–311
student performance and, 310–311
worker productivity and, 310–311, 314
Plumbing, 43, 97n6, 222, 227, 231
Plywood, 336–337
Policy, 23
challenging conventional building design and, 333–338
Clean Air Act and, ix
energy, 35
green building supply industry and, 132, 141
hospitality industry and, 116–121
imperative to build and, 297, 303
incentive design and, 117–120, 122
LEED and, 57, 69–70 (*see also* LEED [Leadership in Energy and Environmental Design])
Motor Vehicle Air Pollution Control Act and, ix
multi-agent collaboration and, 116–117
multinational corporations and, 104–105, 112, 116–121
NAFTA and, 206
niche innovations and, 146, 150–152, 158–160, 161t
organizational sustainability and, 199, 203f, 206, 209
regulation and, 158 (*see also* Regulation)
renovation and, 35, 38, 40, 46–52
sustainability and, 11 (*see also* Sustainability)
transactional burdens and, 120
USGBC and, 17, 57–61, 67–72 (*see also* United States Green Building Council [USGBC])

Politics
empowering the inhabitant and, 171, 187
green building supply industry and, 134
green schools and, 220–221, 224–226
LEED (Leadership in Energy and Environmental Design) and, 67, 220
National Socialists and, 301
niche innovations and, 147, 150, 158–159
renovation and, 50
social issues and, 2–3, 10–11, 21 (*see also* Social issues)
special interests and, 286
Waldschlößchenbridge and, 289–291, 296–297, 300–301
Pollution, 5–6, 12
air, ix–xii, 77, 312–315
biophilic design and, 312–314, 349
generativity and, 241
green schools and, 224
LEED (Leadership in Energy and Environmental Design) and, 66
light, 66
multinational corporations and, 108
Sick Building Syndrome and, 313
titanium buildings and, 77
Polyvinyl acetate (PVAc), 337
Ponce de Leon, Monica, 22, 333–339
Positive organization scholarship (POS), 239, 243
Powell, Walter, 287–288
Practical green, 21, 207, 268, 271–273, 278t
Principal-agent analysis, xiii, 155–157, 162
Prisoner's Dilemma, 105, 117–118
Productivity
biophilic design and, 23, 310–311, 314, 341–344, 349
efficiency and, 20 (*see also* Efficiency)
generativity and, 239, 242, 244, 251–254
green schools and, 8–9, 226
Hawthorne effect and, 342
Heschong Mahone Group and, 8–9
LEED (Leadership in Energy and Environmental Design) and, 63–64, 68
lighting and, 8–9, 341–344
organizational sustainability and, 203f
productivity and, 310–311, 314, 341–344, 349
renovation and, 47
sustainable buildings and, 8–9, 271–272, 277
titanium buildings and, 78–79, 83–84, 96n2
well-being and, 4
Professionals, 2, 12
accredited (APs), 61, 72, 81, 94, 97n3, 136–137, 140, 268
competencies and, 43–44
expertise and, 13 (*see also* Expertise)
green schools and, 227, 230
influence of, 14–16
jurisdiction and, 14–16, 41–43
LEED (Leadership in Energy and Environmental Design) and, 61, 69–72, 92–94, 136–137, 139, 200, 268, 272, 276–277
living walls and, 317
professionals and, 136–137, 139
quality standards and, 140
regulation and, 141
renovation and, 39–43, 52
role of, 13
skills and, 227 (*see also* Skills)
social-technical system for, 47–51
sustainability and, 266 (*see also* Sustainability)
system of professions approach and, 16, 36–37, 40–42, 47–52
titanium buildings and, 81, 87, 90–94, 96, 97n5
World Heritage Center and, 290
Professional service firms (PSFs), 14
Profit-driven green, 21, 207, 268–271, 278t
Property value, 9, 242

Psychology
biophilic design and, 22, 307–323, 344–346
connection with nature and, 309–311, 314, 316, 341
forest bathing and, 311
fractals and, 345
generativity and, 243, 248
green building supply industry and, 145
Hawthorne effect and, 342
health issues and, 19, 22, 37, 39, 117, 145, 172, 178–180, 192, 243, 248, 307–314, 317–320, 322, 344–346
institutional barriers and, 145
living walls and, 307–312, 317–323
natural analogs and, 344–347
natural light and, 309–311, 325, 342–343
nature in the space and, 344–345
nature of the space and, 346–347
personal behavior and, 39
positive meaning and, 251–252
Prisoner's Dilemma and, 105, 117–118
renovation and, 37, 39
Savannah Hypothesis and, 346
social barriers and, 37
stress and, 147, 164, 227, 310–311, 344, 349
sustainable buildings and, 19, 172, 178–180, 192
Public goods
firms and, 129
green building supply industry and, 127–134, 137–141
spillover and, 18, 21, 127–134, 141
Pyle, Michael, 312

Quality, 335
accountability and, 271
air, ix–xii, 5, 65, 77, 227, 271, 312–319, 322–325
award-based, 91
buildability and, 48
communication and, 19
competitive priorities and, 108
construction, xiii
conversational space and, 248
decision making and, 267
feedback and, 186
indoor, 60
of life, 60
Malcolm Baldridge National Quality Award and, 91
niche innovation and, 145, 158
organizational sustainability and, 200–201, 206
phase change in, 95
of place, 42
practical green rationale and, 271
professional standards for, 140
real-time auditing of, 95
risk mitigation and, 82
skilled installation and, 36
workplace, 4, 200–201, 206, 255, 342
workspace, 200
Quigley, John M., 7

Radon, 313
Rajkovich, Nicholas B., 17, 57–76
Real estate investment trusts (REITs), 118
Recycling
green schools and, 225, 229, 235
niche innovations and, 154–155
organizational sustainability and, 209, 214
sustainable buildings and, 263, 265, 267, 270, 274, 276
titanium buildings and, 84
Reeve, Stuart, 20, 219–238
Regulation, xi, 265
ANSI/ASHRAE/USGBC/IES Standard 189 and, 69
biophilic design and, 314, 343
BOMA and, xi
BREEAM and, 5, 80, 97n4
building codes and, 17 (*see also* Building codes)
CASBEE and, 5, 97n4

conventional building practices and, 333, 335
energy codes and, 17, 57, 68, 343
general benefits and, 9
generativity and, 242
gravitational assist and, 206–210
green building supply industry and, 140–141, 146
imperative to build and, 292t, 295, 303–304
increased, 22
inspectors and, 15, 36
LEED and, 5 (*see also* LEED [Leadership in Energy and Environmental Design])
multinational corporations and, 108
niche innovations and, 150, 152, 158, 160, 161t
organizational sustainability and, 199, 203, 206–210
professionals and, 141
renovation and, 37, 41, 46, 50
titanium buildings and, 92
USGBC and, 5 (*see also* United States Green Building Council [USGBC])
voluntary, 17
Renewable energy, xiii, 43, 59, 73n4, 127, 151t, 203f
Renovation
actor-network theory (ANT) and, 40–42, 47–48
architects and, 36, 41–43
biophilic design and, 342–343
buildability and, 47–49
carbon refurbishment and, 36–37, 43–52
collaboration and, 46
consumption and, 35, 40
contractors and, 36, 45, 48
decision making and, 37
design and, 36–40, 43, 46, 48
dissidence and, 41, 48
diversity and, 50
economic issues and, 38, 42, 45, 51, 342–343
efficiency and, 35, 38–43, 46
Empire State Building and, 90
energy and, 35, 38–44, 48–51
entrepreneurs and, 320
environmental issues and, 47
expertise and, 36–37
firms and, 37–38, 44–47, 50
fragmentation and, 36–37, 43, 46
green building and, 35–37, 42, 51–52
health issues and, 42
"house as a system" approach and, 35, 43
housing and, 35–40, 43–51
increased integration capacity and, 49–50
increasing buildability and, 48–49
as innovation, 35–52
inspectors and, 36
institutions and, 36–37, 42, 51
jurisdiction and, 41–43
LZC technologies and, 35, 43, 49
management and, 38, 43, 46, 48
materials and, 48–50
organizational process and, 38
policy and, 35, 38, 40, 46–52
politics and, 50
productivity and, 47
as professionalism, 39–43, 52
psychology and, 37, 39
regulation and, 37, 41, 46, 50
responsibility and, 42
retrofits and, 17, 35, 38–39, 47, 50, 73n4, 79, 82, 90, 94, 172, 324, 342–343, 347
skills and, 36–38, 43–51
social issues and, 36–37, 40–42, 47–52
socio-technical systems (STSs) and, 37, 47–51
solar energy and, 35, 38, 42
stakeholders and, 49, 51
"state of the art" approach and, 35–36
supply chains and, 38, 41, 48–49
sustainability and, 35–36, 38, 46–47
system of professions approach and, 16, 36–37, 40–42, 47–52

technology and, 37, 43, 46–51
United Kingdom and, 35–39, 42, 44–51
waste and, 46
Request for qualification (RFQ), 58, 64–65
Resource Dashboard, 181, 185f
Resources
biophilic design and, 310, 318, 349
conventional building practices and, 333
dependencies and, 21
efficiency and, 18 (*see also* Efficiency)
energy and, 20 (*see also* Energy)
generativity and, 240–244, 249–250, 254
green schools and, 219, 221–222, 223, 224–225, 229, 232
imperative to build and, 288, 297, 303
LEED (Leadership in Energy and Environmental Design) and, 5, 60, 69–70, 216n1
multinational corporations and, 103–116, 120–121
niche innovations and, 151, 153–154, 161t
North House project and, 174–190
organizational sustainability and, 199–200, 204, 207, 211, 215, 216n2
recycling and, 84, 154–155, 209, 214, 225, 229, 235, 263, 265, 267, 270, 274, 276
research on, 10–11
sustainable buildings and, 171–174, 178, 180–181, 186, 189–191, 273–274
titanium buildings and, 79, 84, 88, 95
Responsibility
civic, 244
corporate, 12, 266
flat teams and, 220
generativity and, 240–241, 244–245, 251–254
green schools and, 220, 223, 226, 231
LEED (Leadership in Energy and Environmental Design) and, 62, 69
renovation and, 42
social, 6, 12, 42, 62, 69, 199, 220, 223, 226, 231, 240–241, 244–245, 251–252, 254, 266
Rethinking Construction (Egan), 46
Riley, David, 71
Rockefeller Foundation, 8
Rocky Mountain Institute (RMI), 341–343, 349n1
Rodgers, Johnny, 19, 171–195
Rothermel, Joyce, 198
Rypkema, Donovan, 285–306

Saarinen, Eero, 345
Safety, 67, 201, 203f, 248
Savannah Hypothesis, 346
Schendler, Auden, 67–69
SC Johnson Administration Building, 347
Scotland, 47, 50
Scott, W. Richard, 13–14
Selznick, Philip, xiv
Shane, Scott, 117–118
Sharma, Garima, 20–21, 239–259
Shell, Scott, 65, 66
Sick Building Syndrome, 313
SITES rating systems, 202–203
Skills, 3, 14
continuous learning and, 92
generativity and, 239, 245, 248–249
green schools and, 228
installation and, 36
LEED (Leadership in Energy and Environmental Design) and, 72, 94
management, 94
renovation and, 36–38, 43–51
titanium buildings and, 94
Skylights, 343–344
Small and medium-sized enterprises (SMEs), 45, 47, 49–51, 357
Smart buildings, 5, 336
Smith, Adrian, 152–153
Smog, ix

Social issues, 17–19, 23
actor-network theory (ANT) and, 40–42, 47–48
biophilic design and, 309, 321, 327, 342
buildings as dynamic entities and, 77
building sector and, 2, 4, 7, 15, 35, 128, 132–142, 172, 216n1
collective action and, 128, 134–135, 138, 140
concepts of man-made environments and, 14–15 (*see also* Built environment)
connection with material objects and, 1
connection with nature and, 309–311, 314, 316, 341
cultural loyalties and, 2 (*see also* Cultural issues)
empowering the inhabitant and, 171–172, 176, 186–187
environmental issues and, 9–22 (*see also* Environmental issues)
equity and, 5–6, 66, 199, 201, 208
ethics and, 11, 16, 129
forcing mechanisms and, 86
future research in, 10–15
generativity and, 241–243, 253
governance structures and, 13–14, 130, 198, 203, 211, 214
green building supply industry and, 127–138, 141
health issues and, 341, 343, 349 (*see also* Health issues)
identity and, 4, 13, 264
imperative to build and, 285–304
influence of ratings and, 79–81
LEED (Leadership in Energy and Environmental Design) and, 60, 66, 70, 77–81
leveraging and, 186–187
missing component of, 9–10
multinational corporations and, 103, 117, 266
multiple institutional logics and, 14, 23, 129
organizational sustainability and, 197–201, 212
people's connection to technology and, 40–41
Prisoner's Dilemma and, 105, 117–118
professionals and, 13
realignment and, 3–6
reality and, 287
recycling and, 84, 154–155, 209, 214, 225, 229, 235, 263, 265, 267, 270, 274, 276
renovation and, 36–37, 40–42, 47, 52
responsibility and, 6, 12, 42, 62, 69, 199, 220, 223, 226, 231, 240–241, 244–245, 251–252, 254, 266
shifting ideals and, 4–10
societal objectives and, 13
stress and, 147, 164, 226, 310–311, 344
sustainability and, 171 (*see also* Sustainability)
technology and, 10–11 (*see also* Technology)
titanium buildings and, 77–81, 86, 89–90, 95–96
Waldschlößchenbridge and, 297–299
World Heritage Sites and, 22, 286–287, 290–292, 298–303, 304nn2,3,4
Social media, 19, 253
Social movements, 12–13, 23
challenging imperative to build and, 290, 297–299
green building supply industry and, 127–134, 138, 141
sustainable buildings and, 187, 265–266
Social networking, 186–187
Socio-technical systems (STSs), 37
Solar Decathlon, 19, 335
Solar energy
Distributed Responsive System of Skins (DReSS) and, 176–180
efficiency and, 7

green building supply industry and, 138
LEED (Leadership in Energy and Environmental Design) and, 58–59, 263, 267
multinational corporations and, 105, 114, 116–117, 120
niche innovations and, 151, 154
North House project and, 176, 178, 187
passive, 6, 38, 151, 176
photovoltaics and, 58–59, 73n4, 178, 270
renovation and, 35, 38, 42
solar income and, 58–59, 73nn2,4
sustainability and, 176, 187, 206, 263, 267
U.S. Department of Energy (DOE) and, 19, 335
Spillovers, 18, 127–134, 141
Spreitzer, Gretchen M., 243, 245
Stakeholders
building information model (BIM) and, 73
collaboration and, 57 (*see also* Collaboration)
conventional building practices and, 334
diversity and, 67, 86, 212–214, 225
formation processes and, 116
generativity and, 239–244, 248, 251–254
green schools and, 21
imperative to build and, 22, 288, 293, 295
integrated design and, 212
LEED (Leadership in Energy and Environmental Design) and, 57, 60, 65, 67, 70, 73n1, 264–267, 271–272, 276–279
multinational corporations and, 108, 116, 120
multiple, 20, 239–242, 267
networks and, 51
niche innovations and, 149
nontraditional, 60
organizational sustainability and, 212–213, 225
performance-based contracting and, 96n1
performance goals and, 57
renovation and, 49, 51
silo-based, 97n7
theory of, 265
titanium buildings and, 7n7, 82–84, 86, 96n1
turnover issues and, 84
"State of the art" approach, 35–36, 84–85
Steel, 334
Stress, 147, 164, 226, 310–311, 344, 349
Subcontractors, 36, 80, 82, 83t
Suddaby, Roy, 14
Sullivan, William, 2, 312, 349
Sunlight, 58, 178
Supply chains, 6, 18
hospitality industry and, 106–109, 120, 122
LEED (Leadership in Energy and Environmental Design) and, 78, 80, 83t, 86, 88
multinational corporations and, 104, 106–109, 120, 122
organizational sustainability and, 201
renovation and, 38, 41, 48–49
Susanka, Sarah, xii–xiii
Sustainability
biophilic design and, 314, 326–327, 341, 343
Brundtland Commission Report and, 199, 288
concerns over, 1–5, 9–20
conventional building practices and, 333, 336–337
corporate motivations to adopt, 107–110
data availability and, 104
decision making and, 112, 189, 199, 208, 212–214, 267
Earth Day and, 77

efficiency and, 171–172, 176, 178, 189, 201–208, 213, 270
Forest Stewardship Council (FSC) and, 337–338
generativity and, 241–247, 251–255
Global Reporting Initiative (GRI) and, 199–200, 204
GPR score and, 150, 151t, 153–155, 158–160, 162, 164
green building supply industry and, 127, 135, 140–142
green schools and, 221, 223–225, 231, 235
Hannover Principles and, 58, 73n2
high-performance buildings and, 172, 174, 176, 178, 272
hospitality industry and, 106–122
imperative to build and, 285–286, 288, 293, 302–303
industry commitment and, 104
LEED (Leadership in Energy and Environmental Design) and, 60–61, 65–66, 69–70, 73n2, 174, 197, 200–204, 208–216, 263–279
life-cycle assessment (LCA) and, 73n1, 77, 79, 81–85, 89–91, 94, 120, 128, 155, 161t, 162, 203f, 221, 224, 234, 337
lighting and, 181
multinational corporations and, 103–122
niche innovations and, 145–165
organizational, 197–216
recycling and, 84, 154–155, 209, 214, 225, 229, 235, 263, 265, 267, 270, 274, 276
renovation and, 35–36, 38, 46–47
resource efficiency and, 18, 69, 105–107, 110–112, 116, 120
solar energy and, 176, 187, 206, 263, 267
as sustainable development, 5, 35, 38, 47, 73n2, 111, 160, 199, 211, 288, 303
titanium buildings and, 77–98
triple bottom line of, 6, 199, 203, 210
U.S. Environmental Protection Agency (EPA) and, 4, 12, 108, 120, 174, 313
waste and, 263 (*see also* Waste)
World Business Council for Sustainable Development (WBCSD) and, 35, 42, 146–147, 155
World Heritage Sites and, 285–286, 288, 293, 302–303
Sustainability Tracking and Assessment Rating System (STARS), 204
Sustainability Victoria, 9
Sustainable buildings
active design research prospects and, 191–192
Adaptive Living Interface System (ALIS) and, 178, 180–190
ambient information and, 187–189
architects and, 173–174, 180–181, 191–192, 264, 267, 273
aware home and, 173–174
biophilic design and, 314, 326–327
collaboration and, 174, 186, 192
computational intelligence and, 174
deep green and, 21, 207, 268, 273–274, 278t, 279
Distributed Responsive System of Skins (DReSS) and, 176–180
ecological issues and, 180, 188, 191, 273–274, 276, 279
electricity and, 178, 181, 269
empowering the inhabitant and, 171–192
energy usage and, 59, 66, 73nn3,4, 171–181, 184–192, 263, 270
engineers and, 173–174, 191–192, 273
expertise and, 192
firms and, 265, 275–278
green innovators and, 21, 207, 268, 274–280
health issues and, 178, 180, 271–272, 277
hidden green and, 21, 207, 268, 276–279
housing and, 171

inhabitant intelligence and, 173–174, 176
LEED (Leadership in Energy and Environmental Design) and, 263–264, 267–268, 272–274, 277, 280n3
management and, 186, 272
markets and, 266–270, 275, 277
materials and, 173–178, 189, 191, 263, 267, 270, 275
mobile devices and, 187–188
narratives and, 207
North House project and, 174–190, 335
organizational narratives for, 263–279
policy and, 173
practical green and, 21, 207, 268, 271–273, 278t
productivity and, 8–9, 271–272, 277
profit-driven green and, 21, 207, 268–271, 278t
psychology and, 19, 172, 178–180, 192
reading space of, 264–265
redefining green home and, 174
residential, 171–172, 186
resources and, 171–174, 178, 180–181, 186, 189–191, 273–274
responsive interfaces and, 172, 176, 180–182, 186, 188
smaller living spaces of, 176
social movements and, 187, 265–266
spatial/lifestyle integration and, 187–189
systems intelligence and, 171, 191
technology and, 172–176, 180, 186–188, 191, 263, 274
titanium buildings and, 77–98
water and, 174, 181, 263, 269, 274
zero net energy and, 59, 66, 73nn3,4, 172
Sustainable cities, 77, 94–96
Swiss Re, xiii
System of professions, 16, 36–37, 40–42, 47–52

Taylor, John E., 38
Technical rationality, 21, 287
Technology, 1
actor-network theory (ANT) and, 40–41
Adaptive Living Interface System (ALIS) and, 178, 180–190
biophilic, 307–327, 341–349
communication, 176, 214
conventional building practices and, 333–335, 338
dashboards and, 79, 89, 181, 185f
Distributed Responsive System of Skins (DReSS) and, 176–180
empowering the inhabitant and, 172–177, 180, 187–188, 191
feedback, 174, 176, 178, 181, 184–190
green building supply industry and, 131, 135, 137, 141
green schools and, 233–234
improved, 5–7
innovation and, 21 (*see also* Innovation)
LEED (Leadership in Energy and Environmental Design) and, 57, 62, 68, 73n3
low and zero carbon (LZC), 35, 43, 49
multinational corporations and, 108, 120–121
new, 5–7, 16, 43, 46, 49, 57, 84, 95, 141, 320, 326
niche, 22, 145–147, 152, 154–157, 160–163
North House project and, 174–190
organizational sustainability and, 204
people's connection to, 40–41
renovation and, 35–37, 40–41, 43, 46, 49
responsive interfaces and, 172, 176, 180–182, 186, 188
social change and, 10–11
socio-technical systems (STSs) and, 37

sustainable buildings and, 172–176, 180, 186–188, 191, 263, 274
titanium buildings and, 79, 84, 89, 95
Terrapin Bright Green, 347–348
Tesco, 44
Thermal energy, 7, 36, 42, 120, 176, 178, 234, 274–275
Thün, Geoffrey, 19, 171–195
Tiefensee, Wolfgang, 300
Titanium buildings
accountability and, 83
architects and, 79–81, 83t, 84, 86–93, 96, 97nn6,8
building information modeling (BIM) and, 78, 84–88, 90, 92–94, 96, 97nn8,9
buildings as dynamic entities and, 77
collaboration and, 77–94, 97nn6,7
consumption and, 79
continuous learning and, 92
contractors and, 79–83, 89, 96n1, 97n8
decision making and, 90, 94
design and, 77–96
ecological issues and, 96
economic issues and, 90
efficiency and, 77–91, 94–95, 97n4, 98n12
electricity and, 95
energy and, 17, 77–79, 88–92, 95
engineers and, 78–93, 96
environmental issues and, 77–80, 84, 89, 91–92, 95–96, 97n4
expertise and, 86, 91
firms and, 78, 87, 96
fragmentation and, 79, 83, 87
green building and, 80
high-performance buildings and, 17
implementation issues and, 90–94
influence of ratings and, 79–81
innovation and, 78, 84, 90, 95–96
institutions and, 79, 81, 83, 86–87, 91, 97nn5,6
integrated project delivery (IPD) and, 62, 78, 86–88, 90, 97nn6,7,9
LEED (Leadership in Energy and Environmental Design) and, 78–94, 97nn3,4,8
life-cycle assessment (LCA) and, 79, 81–85, 89–91, 94
management and, 78–79, 84, 86–90, 94–96, 97nn5,7
materials and, 78, 84
operating costs and, 90
pilot project for, 91–92
Platinum rating and, 84–85
pollution and, 77
productivity and, 78–79, 83–84, 96n2
professionals and, 81, 87, 90–94, 96, 97n5
rating categories and, 77–78, 81–84
regulation and, 92
resources and, 79, 84, 88, 95
scaling strategy and, 92
skills and, 94
social issues and, 77–81, 86, 89–90, 95–96
stakeholders and, 82–84, 86, 96n1
sustainable cities and, 77, 94–96
technology and, 79, 84, 89, 95
training and, 93–94
United Kingdom and, 96n2
United States Green Building Council (USGBC) and, 78–81, 91
updated required capabilities and, 91
waste and, 77, 79, 84–88
water and, 88–89, 95, 98n12
Todd, Joel Ann, 61
Toffel, Michael, 104
Town halls, 348
design competition for, 149–152
Netherlands project and, 146–164
regime influence and, 146–149, 152–163
Toxic materials, 84, 108, 135, 225, 277
Triple bottom line, 6, 199, 203, 210
True, James L., 67
Turner Green Buildings, 8–9

Udall, Randy, 67–69

UK Department for Communities and Local Government, 39
Ulrich, Roger, 310, 344
UNESCO, 22, 286–287, 290, 292t, 304n4, 398–401
United Kingdom, 17
  British Defense Ministry and, 86–87
  innovation in construction industry and, 46–47
  low-carbon refurbishment and, 44–51
  renovation and, 35–39, 42, 44–51
  titanium buildings and, 96n2
United States
  biophilic design and, 313–314, 323, 342
  building codes and, 67
  conventional building practices and, 333–334, 337
  energy codes and, 17, 57
  energy savings and, 8
  green building supply industry and, 128, 135–142, 136
  green schools and, 20, 219–220
  growth of LEED certifications and, 5
  hospitality industry and, 105–122
  notions of home and, 15
  population growth in, 4
  professional societies and, 69
  request for qualification (RFQ) procedure and, 65
  sustainable buildings and, 172
  titanium buildings and, 78–81, 86, 96n2, 97n4
University of Guelph-Humber, 308f, 316
University of Illinois, 349
University of Michigan, 348–349
University of Southern California, 345
University of Toronto, 308f, 316
Urban Green Council, 8
Urbanization, 4, 111–112, 302, 304n1
Ürge-Vorsatz, Diana, 35–36
U.S. Department of Energy (DOE), x–xi, 8, 12, 19, 73n4, 79, 335
U.S. Department of Transportation, xi
U.S. Environmental Protection Agency (EPA), 4, 12, 108, 120, 174, 313
U.S. Green Building Council (USGBC)
  ANSI/ASHRAE/USGBC/IES Standard 189 and, 69
  conventional building practices and, 334–335
  conveying greenness and, 263–264
  corporate nature of, 6
  generativity and, 241
  Green Building Information Gateway and, 71, 79
  green building supply industry and, 130–141
  green schools and, 219, 236
  hidden green and, 276
  as institutional entrepreneur, 12, 17
  LEED and, 241 (*see also* LEED [Leadership in Energy and Environmental Design])
  mission of, 60
  multinational corporations and, 103
  organizational sustainability and, 200
  promotion of collaboration by, 57–61, 67–72
  titanium buildings and, 78–81, 91
  waste and, 5

Van Bueren, Ellen, 18–19, 145–167
Vectors of action, 203–204
Velamuri, Ramakrishna, 131
Velikov, Kathy, 19, 171–195
Venkataraman, Sankaran, 131
Ventresca, Marc, 286
Verma, Rohit, 18, 103–125
Vitality, 242, 245–247, 249–250, 252, 254
Vocational training, 49–51
Volatile organic compounds (VOCs), 6–7, 206, 263, 313

Waldschlößchenbridge
  architects and, 290, 295, 300
  competing views over, 289–304
  competition of potential futures and, 293–294

decision making and, 287, 293
design and, 293
Dresden case study and, 289–304
economic issues and, 294, 302, 304n1
efficiency and, 293
Elbe Valley and, 289, 299–301
Higher Administrative Court and, 301
imperative to build and, 286–302
motor vehicle traffic and, 298
politics over, 289–291, 296–297, 300–301
social issues and, 297–299
Wales, 50
Walls Fargo Bank, xiii
Walmart, 343
Walraven, Brenna, 10
Waste, 8, 12
conventional building practices and, 333–334, 336, 338
corporate motivations to adopt sustainability and, 108
disposal of, 7, 242
environmental issues and, 7
fiscal, 20–21
generativity and, 242
green building supply industry and, 135
green schools and, 221–222, 223, 226
LEED (Leadership in Energy and Environmental Design) and, 263
management of, 151, 202, 286
multinational corporations and, 108, 112
niche innovations and, 151, 155, 165n3
nonindustrial, 4
organizational sustainability and, 202, 203f, 208–210
reduction of, 85t, 210
renovation and, 46
titanium buildings and, 77, 79, 84–88
United States Green Building Council (USGBC) and, 5
Water, 4
ALIS system and, 181
biophilic design and, 311, 314–316, 324–325, 344–348
conservation of, 9, 111, 135, 174, 202, 223, 269
conventional building practices and, 333–337
costs of, 8, 242
efficiency and, 5, 60, 203f
generativity and, 242
green building supply industry and, 135
green schools and, 222–223, 225, 232
LEED (Leadership in Energy and Environmental Design) and, 263, 269, 274
multinational corporations and, 106–107, 111, 114, 116–118
niche innovations and, 151t, 158, 165n3
organizational sustainability and, 202, 203f, 207
rainwater collection and, 7
retention capacity and, 151t
sustainable buildings and, 174, 181, 263, 269, 274
titanium buildings and, 88–89, 95, 98n12
waterless urinals and, 232, 263
Western Electric, 342
Westinghouse Air Brake, 201
Whole Systems Integrated Process Guide (ANSI), 212
William McDonough and Partners, 58, 344
Williams, Roy, 66
Wilson, Alex, 9
Wilson, Edward O., 309, 341
Windows, 7, 43, 174, 178, 272, 310, 343–344
Wolverton, B. C., 312–313
Wood, 7, 275, 287, 336–337, 345
Woodbury, Robert, 19, 171–195
World Business Council for Sustainable Development (WBCSD), 35, 42, 146–147, 155

World Heritage Committee, 290, 292, 300–301, 304n2, 304n3
World Heritage Sites
 sustainability and, 285–286, 288, 293, 302–303
 Waldschlößchenbridge and, 22, 286–287, 290–292, 298–303, 304nn2,3,4
World Inquiry project, 245
Wyndham (hotel chain), 110

York, Jeffrey G., 18, 127–144

Zero Energy Homes, 172
Zero net energy use, 59, 66, 73nn3,4, 172
Zhang, Jie J., 18, 103–125

Urban and Industrial Environments

Series editor: Robert Gottlieb, Henry R. Luce Professor of Urban and Environmental Policy, Occidental College

Maureen Smith, *The U.S. Paper Industry and Sustainable Production: An Argument for Restructuring*

Keith Pezzoli, *Human Settlements and Planning for Ecological Sustainability: The Case of Mexico City*

Sarah Hammond Creighton, *Greening the Ivory Tower: Improving the Environmental Track Record of Universities, Colleges, and Other Institutions*

Jan Mazurek, *Making Microchips: Policy, Globalization, and Economic Restructuring in the Semiconductor Industry*

William A. Shutkin, *The Land That Could Be: Environmentalism and Democracy in the Twenty-First Century*

Richard Hofrichter, ed., *Reclaiming the Environmental Debate: The Politics of Health in a Toxic Culture*

Robert Gottlieb, *Environmentalism Unbound: Exploring New Pathways for Change*

Kenneth Geiser, *Materials Matter: Toward a Sustainable Materials Policy*

Thomas D. Beamish, *Silent Spill: The Organization of an Industrial Crisis*

Matthew Gandy, *Concrete and Clay: Reworking Nature in New York City*

David Naguib Pellow, *Garbage Wars: The Struggle for Environmental Justice in Chicago*

Julian Agyeman, Robert D. Bullard, and Bob Evans, eds., *Just Sustainabilities: Development in an Unequal World*

Barbara L. Allen, *Uneasy Alchemy: Citizens and Experts in Louisiana's Chemical Corridor Disputes*

Dara O'Rourke, *Community-Driven Regulation: Balancing Development and the Environment in Vietnam*

Brian K. Obach, *Labor and the Environmental Movement: The Quest for Common Ground*

Peggy F. Barlett and Geoffrey W. Chase, eds., *Sustainability on Campus: Stories and Strategies for Change*

Steve Lerner, *Diamond: A Struggle for Environmental Justice in Louisiana's Chemical Corridor*

Jason Corburn, *Street Science: Community Knowledge and Environmental Health Justice*

Peggy F. Barlett, ed., *Urban Place: Reconnecting with the Natural World*

David Naguib Pellow and Robert J. Brulle, eds., *Power, Justice, and the Environment: A Critical Appraisal of the Environmental Justice Movement*

Eran Ben-Joseph, *The Code of the City: Standards and the Hidden Language of Place Making*

Nancy J. Myers and Carolyn Raffensperger, eds., *Precautionary Tools for Reshaping Environmental Policy*

Kelly Sims Gallagher, *China Shifts Gears: Automakers, Oil, Pollution, and Development*

Kerry H. Whiteside, *Precautionary Politics: Principle and Practice in Confronting Environmental Risk*

Ronald Sandler and Phaedra C. Pezzullo, eds., *Environmental Justice and Environmentalism: The Social Justice Challenge to the Environmental Movement*

Julie Sze, *Noxious New York: The Racial Politics of Urban Health and Environmental Justice*

Robert D. Bullard, ed., *Growing Smarter: Achieving Livable Communities, Environmental Justice, and Regional Equity*

Ann Rappaport and Sarah Hammond Creighton,

Degrees That Matter: Climate Change and the University

Michael Egan, *Barry Commoner and the Science of Survival: The Remaking of American Environmentalism*

David J. Hess, *Alternative Pathways in Science and Industry: Activism, Innovation, and the Environment in an Era of Globalization*

Peter F. Cannavò, *The Working Landscape: Founding, Preservation, and the Politics of Place*

Paul Stanton Kibel, ed., *Rivertown: Rethinking Urban Rivers*

Kevin P. Gallagher and Lyuba Zarsky, *The Enclave Economy: Foreign Investment and Sustainable Development in Mexico's Silicon Valley*

David N. Pellow, *Resisting Global Toxics: Transnational Movements for Environmental Justice*

Robert Gottlieb, *Reinventing Los Angeles: Nature and Community in the Global City*

David V. Carruthers, ed., *Environmental Justice in Latin America: Problems, Promise, and Practice*

Tom Angotti, *New York for Sale: Community Planning Confronts Global Real Estate*

Paloma Pavel, ed., *Breakthrough Communities: Sustainability and Justice in the Next American Metropolis*

Anastasia Loukaitou-Sideris and Renia Ehrenfeucht, *Sidewalks: Conflict and Negotiation over Public Space*

David J. Hess, *Localist Movements in a Global Economy: Sustainability, Justice, and Urban Development in the United States*

Julian Agyeman and Yelena Ogneva-Himmelberger, eds., *Environmental Justice and Sustainability in the Former Soviet Union*

Jason Corburn, *Toward the Healthy City: People, Places, and the Politics of Urban Planning*

JoAnn Carmin and Julian Agyeman, eds., *Environmental Inequalities Beyond Borders: Local Perspectives on Global Injustices*

Louise Mozingo, *Pastoral Capitalism: A History of Suburban Corporate Landscapes*

Gwen Ottinger and Benjamin Cohen, eds., *Technoscience and Environmental Justice: Expert Cultures in a Grassroots Movement*

Samantha MacBride, *Recycling Reconsidered: The Present Failure and Future Promise of Environmental Action in the United States*

Andrew Karvonen, *Politics of Urban Runoff: Nature, Technology, and the Sustainable City*

Daniel Schneider, *Hybrid Nature: Sewage Treatment and the Contradictions of the Industrial Ecosystem*

Catherine Tumber, *Small, Gritty, and Green: The Promise of America's Smaller Industrial Cities in a Low-Carbon World*

Sam Bass Warner and Andrew H. Whittemore, *American Urban Form: A Representative History*

John Pucher and Ralph Buehler, eds., *City Cycling*

Stephanie Foote and Elizabeth Mazzolini, eds., *Histories of the Dustheap: Waste, Material Cultures, Social Justice*

David J. Hess, *Good Green Jobs in a Global Economy: Making and Keeping New Industries in the United States*

Joseph F. C. DiMento and Clifford Ellis, *Changing Lanes: Visions and Histories of Urban Freeways*

Joanna Robinson, *Contested Water: The Struggle Against Water Privatization in the United States and Canada*

William B. Meyer, *The Environmental Advantages of Cities: Countering Commonsense Antiurbanism*

Rebecca L. Henn and Andrew J. Hoffman, eds., *Constructing Green: The Social Structures of Sustainability*